A Structural Ricardian Valuation of Climate Change Impacts on Agriculture in Pakistan

Schriften zur Internationalen Entwicklungs- und Umweltforschung

Herausgegeben vom

Zentrum für internationale
Entwicklungs- und
Umweltforschung

der Justus-Liebig-Universität Gießen

Band 31

PL ACADEMIC RESEARCH

Mirza Nomman Ahmed

A Structural Ricardian Valuation of Climate Change Impacts on Agriculture in Pakistan

PL ACADEMIC RESEARCH

Bibliographic Information published by the Deutsche Nationalbibliothek
The Deutsche Nationalbibliothek lists this publication in the Deutsche Nationalbibliografie; detailed bibliographic data is available in the internet at http://dnb.d-nb.de.

Zugl.: Gießen, Univ., Diss., 2013

Library of Congress Cataloging-in-Publication Data
Ahmed, Mirza Nomman, 1982-
A structural Ricardian valuation of climate change impacts on agriculture in Pakistan / Mirza Nomman Ahmed. – [1st edition].
pages cm – (Schriften zur internationalen Entwicklungs- und Umweltforschung, ISSN 1615-312X ; Band 31)
ISBN 978-3-631-65014-1
1. Agriculture–Economic aspects–Pakistan. 2. Agriculture–Environmental aspects–Pakistan. 3. Climatic changes–Economic aspects–Pakistan. 4. Crops and climate–Pakistan. I. Title. II. Series: Schriften zur internationalen Entwicklungs- und Umweltforschung ; Bd. 31.
HD2075.5.A64 2013
338.1'3095491--dc23

2013041474

D 26
ISSN 1615-312X
ISBN 978-3-631-65014-1 (Print)
E-ISBN 978-3-653-03910-8 (E-Book)
DOI 10.3726/978-3-653-03910-8

PL Academic Research is an Imprint of Peter Lang GmbH.
Peter Lang – Frankfurt am Main · Bern · Bruxelles · New York · Oxford · Warszawa · Wien

This book is part of an editor´s series of PL Academic Research and was peer reviewed prior to publication.

www.peterlang.com

This thesis is dedicated to my parents.

Sadiqa Naeem & Mirza Naeem

Preface

In his work the author addresses a highly interesting topic from the perspective of agricultural economics and environmental policy. The contribution discusses the impacts of climate change on agriculture in developing countries, particularly in Pakistan. Pakistan's agriculture is considered to be notably vulnerable to changes in temperature, weather extremes and altered seasonal distribution of precipitation. In terms of content, above all emphasis is laid on addressing the question of 'how farmers adapt to changing climatic conditions'. Concerning this only sparse literature is available, especially for agriculture in developing countries. The economic valuation of changes in climate is complicated by the existence of significant temporal and spatial differences and that in many cases appropriate data for an empirically sound analysis is often unavailable. This is where Mr. AHMED's work commences from. On the basis of a comprehensive dataset which has been made available to him by the Pakistan Bureau of Statistics especially for the purpose of the research, the author works out which adaptation strategies farmers in different Agro-Ecological-Zones/ Districts/ Provinces select and which role i.e. annual seasons, water availability, soil quality, infrastructure, available technology, education, household size and the frequent flood catastrophes thereby play. With the help of sophisticated econometric models and a choice modeling approach he identifies the most important explanatory factors for adaptation behavior, derives the respective income effects and simulates the impacts of a future change in temperature and precipitation. In modeling the impacts of climate change on agriculture a prominent role is played by agronomic models, computable general equilibrium models and the Ricardian Approach. The author directs particular attention to the latter and points out the importance that the adaptation strategies of farmers and Local Traditional Knowledge (LTK) have and how the Standard-Ricardian Approach and the Structural-Ricardian Approach are distinguished. While the former rests upon land values as the dependent variable, the latter under consideration of a choice modeling approach tries to explain net crop revenues. For South America and Africa initial and extensive estimations on the basis of the Structural-Ricardian Approach are already available, whereas for South Asia no study exists. This is where the author by using the example of Pakistan breaks new ground.

This work presents a wealth of highly interesting findings regarding the impacts of climate change on agriculture of a developing country, which clearly extends beyond the literature so far. Methodologically as well new ground is broken by simultaneously applying the Standard-Ricardian Model and the Structural-Model on the basis of a uniquely comprehensive dataset to a South Asian developing country. This book distinctly contributes to advances in knowledge from a scientific perspective and can be recommended as reading matter to everyone interested in the topic.

Giessen, September 2013 Prof. Dr. Dr. h.c. P. M. SCHMITZ

Acknowledgements

This book is the outcome of my research engagement at the Justus Liebig University Giessen. I started my PhD studies in 2008 at the department of Agricultural and Development Policy under the supervision of venerated Prof. Dr. Dr. h.c. P. M. Schmitz. I would herewith like to take the opportunity to thank those who guided me through this challenging episode of my life and supported me whenever the 'downs' dominated the 'ups', and beyond. Foremost, I would like to express my gratitude to the lord of the heavens and the earth for giving me the strength, health, family and wisdom to achieve this milestone. From the bottom of my heart I thank my beloved parents Sadiqa and Mirza Naeem for making me the person that I am and facilitating this great success in every respect. Thank you for your prayers and giving me encouragement and hope whenever I needed it most. Words can't describe how thankful I am to my beloved wife Usma and daughters Umaiza and Aliza for being endlessly tolerant and giving me the extra time in office that I needed. Thanks to my lovely sister Nadia for her support and advice. Thanks to my niece Serena and my brother (in-law) Shafaat for believing in me. I would also like to thank my parents-in-law Hafeez and Razia Bhatti for their kind support. My academic achievements would not have been possible without the valuable guidance and support of my mentor, teacher and supervisor, honorable Prof. Schmitz. He has paved the way to this success by granting me valuable advice right from the initial period of my academic studies. I thank him for his trust and belief in me. Surely, it is his understanding and support that has made this academic career so seamless and neat, my gratitude to him for trusting in my abilities and granting me a direct job entry at his department. By making me work on challenging tasks he has contributed to uncovering my hidden skills, nurturing my abilities and strengthening my self-awareness. I am also grateful to Prof. Schmitz for granting me the opportunity to coordinate the Climate Change Network for Central Asia. I would also like to acknowledge the support of my dear friends. Foremost, I am thankful to Tasleem Ahmed for having a sympathetic ear for everything that I had on my mind and heart. He always supported me and was always around whenever I needed him, thanks for sharing my grief as well as my joy. For his valuable input and discussions I am grateful to him and also would like to thank him for believing in me and pushing me to realize my true potential. I am also thankful to my friends Arslan, Puran, Ali, Zeeshan, Shahbaz, Jam, Sayeed and Ashok for their support, especially to Puran for helping me prepare for the final PhD defense. Without my dear colleagues this task would not have been possible to achieve, it is their support that has immensely contributed to this accomplishment. Foremost credit goes to respected Dr. Joachim Hesse for his advice, support and fruitful discussions that were instrumental for finalizing this piece of work. I am truly grateful to Viktoriya, Hendrik, Doniyor, Shavkat, Manuchehr, Juliane, Ira, Natasha, Nadia, Margot, Mrs. Bender, Nino, Sabine, Dr. Höher and Eli for making this working environment so wonderful.

Giessen, September 2013 Mirza Nomman AHMED

Table of Contents

List of Figures and Tables

Tables

1 Introduction and Problem Statement

Today there is a broad consensus on the anthropogenic involvement in climate change. The connection between human activity related green house gas emissions and their impacts on temperature and precipitation regimes has been subject of numerous studies. There seems to be enough evidence that changing climatic patterns will impact on economic well-being (DESCHENES AND GREENSTONE, 2007). Harmful effects of climate change are expected worldwide. However, especially to developing countries it poses a far more serious threat as many of their environmental and developmental problems are at risk of being exacerbated (UNFCCC, 2007; CLINE, 2007; MENDELSOHN AND WILLIAMS, 2004). Moreover, developing countries heavily rely on climate sensitive sectors to generate income, most importantly the agricultural sector. Labor in developing countries is highly abundant and relatively inexpensive, thus the economy mainly relies on labor intensive technologies, leaving less room for advanced adaptation options (MENDELSOHN et al., 2001). From a physiographic perspective developing countries located in tropical regions usually have a large share of soils that are unsuitable for agricultural purposes; this additionally increases their vulnerability to potential damage from environmental changes (MENDELSOHN AND DINAR, 1999).

Pakistan, situated in the South Asian region between 24-37°N of latitude and 61-76°E of longitude, with agriculture as its mainstay and responsible for almost 70% of the livelihoods of the population, directly or indirectly linked to the sector and home to a population of approximately 170 million with 32% living below the poverty line, is one such developing country with a high vulnerability towards present and future climate change (ESP, 2007). CRUZ et al. (2007) report evidence on serious increases in the frequency and intensity of extreme weather related events such as extended drought periods, tropical cyclones, flash floods and severe dust storms in the Asian region. In this context, both the frequency and the intensity of climate related extreme events in Pakistan have increased in the recent past. From 1998-2002 the province of Balochistan was hit by severe drought conditions, affecting 84% of the population directly, killing 76% of the province's livestock and causing mass migration due to widespread hunger and disease. Of late in 2010, the country was struck and devastated in large parts by epic floods. Millions were left homeless and important harvest was destroyed, exacerbating issues of food security. In 2008 out of 309 million tonnes (mt) of carbon dioxide (CO2) equivalent total Greenhouse Gas (GHG) Emissions, 39% were contributed by the agricultural sector (TFCC, 2010). Despite the fact that the country's share in global GHG emissions is marginal, there is enough reason for concern in future when having a look at the annual percentage growth rate of per capita carbon dioxide emissions from 1990 to 2007 (Figure 1.1).

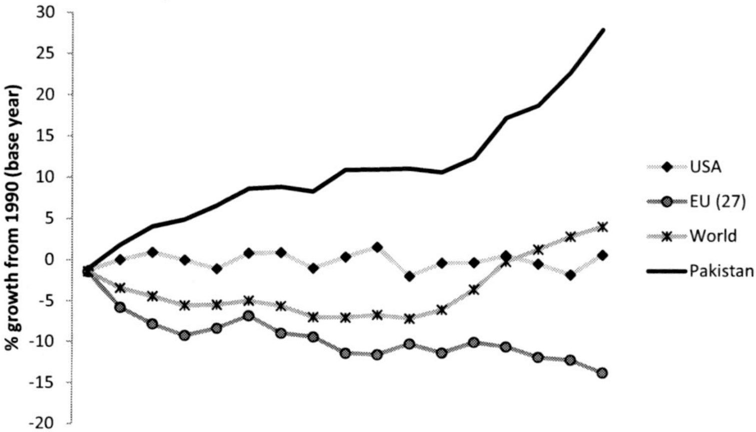

Figure 1.1 Per Capita CO_2 Emissions in selected regions of the world, 1990-2007
Source: own illustration after CAIT 8.0 (CAIT, 2011)

A detailed study conducted by the (Pakistan) Global Change Impact Studies Centre (GCISC) for the period from 1970 to 2000 found significant increases in both, daily minimum and daily maximum Temperatures in 80 percent of the sampled stations (SHEIKH et al., 2009). According to the final report of the country's Task Force on Climate Change (TFCC) published in 2010, mean annual temperature in Pakistan over the period of the last century, in accordance with the global trend, has increased by 0.6 degrees Celsius. Alarmingly, the last fifty years clearly indicate an increased rate of decadal warming, averaging 0.24 degrees Celsius/ decade. Furthermore, data from the Climate Research Unit (CRU) in the United Kingdom relative to the national scale indicate a higher increase in mean annual temperature for Northern Pakistan. In addition, based on data from 1951 to 2000 the TFCC report highlights the general warming trend in mean and maximum temperatures for the summer season (April and May), this throughout the country. For the same time period, the Monsoon Season that spans from July to September has generally shown a decreasing trend in temperatures.

The LEAD Climate Change Action Plan of Pakistan declares the country to be highly vulnerable to climate change. According to the vulnerability index

Pakistan is ranked 12th globally, economic losses of approximately 4.5 billion dollars are anticipated, grassland productivity and consequently crop and livestock yields are expected to suffer severely from climatic change manifested in significantly higher temperatures and decreased surface water availability and changing precipitation patterns (LP, 2008).

In spite of these concerns and forecasts, not many studies have been undertaken in Pakistan on the economic losses and social welfare impacts that are expected to result from climate damage to agriculture, more or less owing to the fact that in the majority of the developing world subsistence farming still remains a pivotal part of agriculture, complicating the economic analysis of environmental change (CHAMBWERA AND STAGE, 2010). Despite an internationally extensive interest in the measurement of the economic impacts of climate change, the empirical research on Asia remains scarce. By using two different climate response functions, one derived from a cross sectional Ricardian study of India (MENDELSOHN et al., 2001) and the other estimated from agricultural-economic simulation results (ADAMS et al., 1999), the Yale University study entitled "Climate Change Impacts on Southeast Asian Agriculture" based on a GIM (Global impact Model) tries to compute the economic impacts of climate change and extrapolates the results to all countries in the region (Bangladesh, Bhutan, Cambodia, China, India, Indonesia, Laos, Malaysia, Nepal, Pakistan, Philippines, Taiwan, Thailand, Vietnam). In summary the study finds that the agricultural impacts of a change in climate are dependent on four factors: the response function, the size of the agricultural sector, the initial temperature and precipitation, and the climate scenario. The study provides some initial evidence on the impact of climate change in this region, predicting losses in the range of two to sixteen billion dollars (mild & extreme scenario) for Pakistan's agriculture until 2100, while at the same time stressing the need for further empirical research in this region, particularly mentioning the need to include more countries than just India, which was used as a proxy for all Southeast Asian Countries. More research is clearly needed to refine the estimates of impacts in this vulnerable region. Potential adaptation measures have to be indentified for both, farmers and governments (MENDELSOHN, 2005). The agriculture sector is facing several issues that can be further aggravated by climate change. Amongst others these problems include: low relative productivity of most crops, slow productivity growth, expansion of cultivable land is no longer possible, fertilizer use is leveling off (less returns to intensification), and productivity growth through technological progress appears to have lost momentum since the green revolution. Increasing temperatures, changes in average precipitation, water stress, monsoon variability, constrained irrigation water availability and increased frequency of extreme events are of particular concern for the agriculture sector and define the magnitude of its vulnerability. Although climate change studies for Pakistan have been carried out, they so far have concentrated on warming science itself, not addressing the economic impacts of climatic change on the agricultural sector. Crop simulation approaches have assessed future impacts on the basis of current yields, not including adaptations made by farmers. Therefore, these estimates from an economic perspective appear to be rather crude, as they assume that farmer's will continue with the same practices, regardless of climate change. Developing country studies are in short supply, especially studies that address the costs of climate change and analyze likely adaptation measures (MARGULIS et al., 2008).

Although models have been constructed to assess the economic impact of global warming on agriculture, a clear consensus on the methodology and impacts has not yet evolved. According to MENDELSOHN (2000) the literature so far (not much has changed) suggests that over the next hundred years global food supplies in aggregate will not be harmed. In an extensive study for the FAO MENDELSOHN (2000) also states that "*..warming is not expected to affect aggregate production in most developing countries*". However, at the same time results suggest that productivity declines will be inevitable, especially in regions where temperatures are expected to increase and precipitation is predicted to decline. Using an extensive literature survey WASHINGTON et al. (2006) found that Asia alongside Africa, the Middle East and South America has been a neglected region as far as climate research is concerned. They concluded, that this shortcoming is detrimental to climate risk assessment, planning of adaptation and decision making in developing countries. For Pakistan a deficit in climate change research has been identified. To this effect government authorities and research institutions have appealed to the wider research community to fill this gap. On the occasion of a conference organized by the World Meteorological Organization (WMO) the director of the Pakistan Meteorological Department (PMD) clearly stressed the need for intensifying research to study the adverse impacts of climate change on different socio-economic sectors such as water resources and agricultural production in Pakistan (ALAM, 2009). Moreover, he has recommended to enhance capacity building in the use, development and modification of mathematical models for use in climate change related studies. Based on research on climate change in a multiplicity of sectors, the Global Change Impact Studies Centre (Pakistan) has called for developing clear cut government policies to counter the adverse impacts from climate change (ALI et al., 2009). Changes in yield have to be studied further as no clear evidence exists in which direction aggregate effects will move. Furthermore, shifting rainfall patterns have to be studied. Detailed impact assessment studies are called for, as without research in this direction adaptation strategies will be hard to formulate (WORLD BANK, 2011). In general, the shortage of studies for Pakistan is also reasoned in the fact that it is highly diverse with respect to climate, socio-economy, and environmental characteristics. Not only on the country level but also on the inter and intra provincial scale.

1.1 Objectives

With reference to the above mentioned problems and gaps in the research on the economic impacts of climate change on agriculture for developing countries, and in particular for Pakistan, this study's mandate is to address these shortcomings by developing two different econometric models to assess the economic impacts of climate change on the country's agricultural sector. By using the Ricardian valuation approach developed by MENDELSOHN et al. (1994) as a basis, first a model is constructed for Pakistan, accounting for a multiplicity of shortcomings of the initial model. A unique model is constructed which is adapted to the country's socio-economic and environmental circumstances. The first model's mandate is to unveil the economic impacts for farmers on the aggregate district scale. For this farm level data is aggregated. The analysis focuses on estimating the impacts and their spatial spread, covering districts, provinces and Agro-Ecological Zones. The standard model thus captures implicit adaptations made by farmers and unravels climate sensitivities. Using a micro-level farm dataset a second two-staged structural model is build to understand the true nature of adaptations for the crop sector of the country. The structural model provides additional insights as it models farm level data using a revealed-preference (choice) approach. After estimating the farm-specific choices given certain climatic and socio-economic features, the model also estimates the expected impacts on farm incomes, conditional on the choice made by the farmer. The study specifically mandates to uncover adaptations that farmers are most likely to make with a change in climate and create options for developing relevant policies to facilitate these beneficial adaptations. Global Circulation Model predictions computed by the IPCC are used to simulate future impacts on farmer choices and on farm incomes. The choice model is constructed in order to explore the results from different modeling perspectives, this in particular to the background of a variety of different approaches that are used for modeling climate impacts on the sector with their respective strengths and weaknesses. This study examines the intensity of the impacts to farmers, tries to pinpoint the different zones and regions in the country that will suffer most and elaborate on farmer adaptations. By considering two different approaches, adaptations outside agriculture shall also be analyzed (*note: land values capture adaptations outside the agriculture sector*). Departing from an intelligent farmer's scenario, where adaptation has ever since taken place, and in conjunction with the research question of how farmers will be affected by Climatic Change and how they will respond, the specific objectives of the study are to identify the impacts of climate change on agricultural production, specifically incomes, to understand the spatial patterns of the impacts, to analyze the seasonality of the impacts (beneficial vs. detrimental impacts), to understand important features that besides climate play a key role in determining agricultural performance, understanding the observed preferences of farmers conditional on climate and other factors, to model their possible future adaptation behavior, and last but not least, to simulate the impacts of climate change on the agricultural sector using GCM scenarios. All these objectives are directed towards the goal of providing a useful starting point for policy interventions on the farm or district level. Given the anticipated changes in climate policymakers and analysts can use these projections to identify policy measures that can make it easier for farmers to switch to new production patterns. Ultimately, the identification of the production patterns that famers are likely to switch to in the wake of global warming shall serve as valuable input for designing a policy guide for planned adaptation by understanding autonomous adaptations.

1.2 Conceptual Approach

Upfront a detailed account on the importance of the Agricultural sector is provided, including geophysical, climatic, hydrological, economic and structural dimensions. This thorough review is followed up by deliberations on the nexus between climate and agriculture, with the aim to clarify interdependencies and present plant-physiological processes that are related to climate and how they can be altered. Ensuing, a historical review on the country's past climate aims at uncovering the vulnerability of the region to specifically climate related extreme events. Subsequent deliberations discuss the different available techniques to estimate the drawn interdependency between agriculture and climate. In fact, the idea is to depict approaches to analyzing the vulnerability of the sector using quantitative research methods. After the review of empirical approaches to the estimation of the economic impacts of climate change, the selected model is presented in detail using a thorough review of existing model related literature. This is followed up by a conceptual description of the standard model. Particular relevance is given to adaptation with its different manifestations in an intersecting literature review. Following this review, and given the weakness of the first model not to reveal specific adaptations, a conceptual description of the advanced approach termed "Structural Ricardian Model" is presented. After presenting the theory and the revised models for the application to Pakistan (model specification), the data and study area are described. Further deliberations present the empirical or econometric strategy, model estimations alongside specification and robustness tests and discussion of results. A final chapter runs climate change simulations to assess the future sensitivity of the sector to global warming. Ultimately, the study concludes by summarizing the key findings, presenting policy implications, addressing shortcomings of the study and highlighting relevant fields where further research attention is greatly required.

2 The Agriculture Sector of Pakistan

Pakistan is a predominantly agricultural country with a diverse climate that can be distinguished into ten broad agro-climatic zones. The topography ranges from high mountain peaks in the north to wide stretching plains in the south. A significant portion of the population depends on the agricultural sector with its backward and forward linkages. To this effect, the country's agricultural policy plays a pivotal role for guaranteeing the functioning of markets and the facilitation of trade, all directed towards the ultimate goal of providing food security for the approximately 170 million inhabitants. Moreover, the country is home to the largest irrigation network that at the same time is also the largest glacial melt fed system worldwide. Key crops are wheat, cotton, rice, sugarcane, maize and vegetables. This chapter is devoted to revealing the importance of the sector by starting off with the explanation of the country's geography and soils and climatic diversity, which is followed by an overview on the water resources. Furthermore, land use patterns, major crops, land reforms and tenurial systems are discussed alongside deliberations on the importance of the sector for the economy, which also includes an overview on agricultural policies.

2.1 Geography and Soils

Pakistan is located in Southern Asia between 24-37°N of latitude and 61-76°E of longitude. The 36th largest country of the world, spreads over a total geographical area of 796.000 km^2 and is bordered by the four independent states India (East), Iran (Southwest), Afghanistan (West) and China (Northeast). In the south the country has a 1046 km stretching coastline on the Arabian Sea (CIA WORLD FACTBOOK, 2011).

The country has various diverse landscapes, ranging from the extremely arid deserts and rocky plateaus of the provinces Balochistan (West) and Sindh (South) and the fertile irrigated areas of the Indus Basin to the high snow-covered mountain peaks (e.g. Godwin-Austen K2 – 8611m) and seven large glaciers of the Himalayan belt. Pakistan geologically expands over the Indian and the Eurasian tectonic plates. The provinces Balochistan and NWFP lie on the Eurasian Plate and the provinces Punjab and Sindh are located on the northwestern corner of the Indian plate (AQUASTAT, 2011). Physiographically the country can be divided into three different regions or two different provinces, respectively (Figure 2.1).

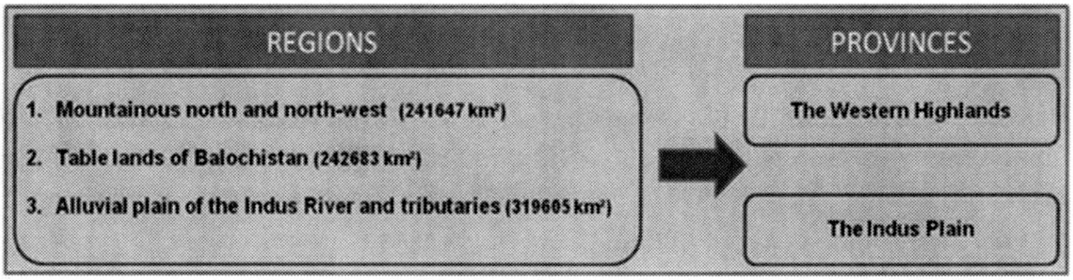

Figure 2.1 Physiographic Division of Pakistan
Source: own illustration after FRAMJI et al. 1982 and AKRAM, 2004

Both physiographic provinces of Pakistan are very distinct. Stretching from the Makran coast in the south to the Pamir Plateau in the north, the Western Highlands cover significant parts of the provinces North-West-Frontier-Province (NWFP) and Balochistan and also some parts of the Punjab province. The Indus Plain is the country's most important region in terms of agricultural production. More than 20 million hectares of land are located in this physiographic province and large quantities of water are available through the Indus River and its tributaries flowing across the agriculturally most important administrative provinces Sindh and Punjab. The majority of the total area consists of plateau and mountain, with a comparatively smaller share of plains and deserts (Figure 2.2).

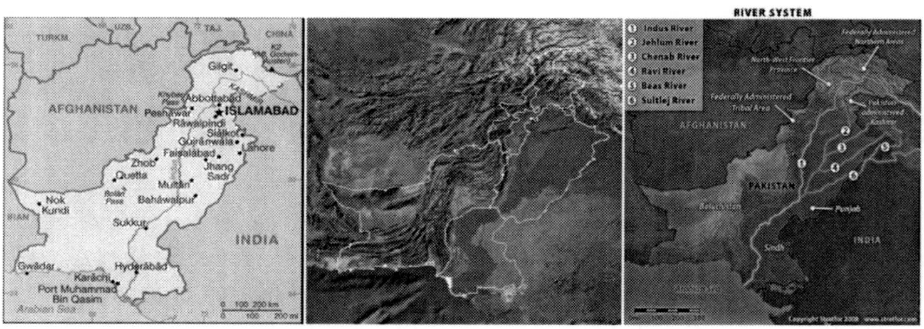

Figure 2.2 Pakistan – Boundaries and Geographical Features
Sources: CIA WORLD FACTBOOK (2011); ESRI (1999); STRATFOR (2008)

The country's diverse landscape and its long latitudinal stretch from the southern coastline with the Arabian Sea to the Himalayan mountain ranges are important factors that have contributed to the soil development and diversity in the country. In this regard, specific factors important for soil development in general can be summarized as follows: time, topography, parent material, flora, fauna, climate, human settlement, water availability and their interactions (PLASTER, 2008). The following review is based on the findings of the study by the Soil Survey of Pakistan (SSP, 1985). As far as soil types are concerned the country can be divided into nine ecologic or geomorphic zones (Table 2.1):

Table 2.1 Soil Related Ecological-Geomorphologic Division of Pakistan

1	Northern Mountainous Region
2	Western Mountainous Region
3	Potwar Upland
4	Sandy Deserts
5	Piedmont Plains
6	Old River Terraces
7	Subrecent River Plains
8	Recent River Plains
9	Indus Delta

Source: AHMAD (1985) and MIAN AND SYAL (1986)

The northern mountainous region over most parts is characterized by higher precipitation levels, lower temperatures and a distinct flora, setting it apart from the rest of the country's terrain. As far as the geological pattern is concerned, the area is split into high mountain ranges and narrow valleys. The valleys normally throughout the year do not freeze, whereas in the high mountains perennial snows are common. The natural vegetation distribution is changing from the foothills of the terrain to the slopes, with open scrub in the lowest parts followed by a coniferous trees zone that lastly transitions into areas of alpine pastures. Considerable soil leaching is observed. The soil types are decalcified with low base saturation. Soil salinity or sodicity is absent. Loamy soils with a high amount of rock fragments are found in the mountain slope zones, whereas the valley soil material is deeper and mostly of alluvial nature. The loamy soils, although high in rock and gravel content are rich in organic matter, whereas the valley soils are generally calcic and cambic.

The western mountainous region is the largest geographic division of the country. The region is characterized by moderately high mountains and wide valleys. Rocky outcrop is most present and therefore the soil parent material contains calcite. The area is generally arid with some semi-arid parts. The central and northern parts of this zone are characterized by harsh winters and relatively mild summers. As a result of several biotic disturbances the vegetation comprises only low shrubs and grasses. This constrained biological activity is also reflected in the soil pattern. The mountain slopes are affected by heavy erosion, the respective soils are therefore thin and patchy. Erosion originating from the mountain slopes gives the intervening valleys their characteristic soil pattern, dominated by a mixture of stones and gravel and out-washed alluvial fill. With distance from the mountain ranges, the soil pattern changes to silty and clayey. The overall soil profile is weakly developed and entirely base saturated. The organic content is low and salinization processes are active. Desert soils (Aridisols) are most prevalent in this particular ecological division. The Potwar upland Zone's surface is characterized by low hilly ranges and flat surface. Loess deposits, sandstone and bedrock based surface material, silty and loamy alluvium form the soil material. Eroding waters have shaped the Potwar zone's landscape. The northern part of this zone is located at the foothill of the outer Himalayan range, with higher amounts of precipitation. Temperatures favor plant growth throughout the year, thus extensive dry-farmed arable cropping is practiced. Soil development differs with the amount of precipitation. In areas closer to the Himalayan Ranges, where the area is wetter, soil development is more pronounced. The flat table lands are characterized by well-developed deep soil profiles. The slopes closer to the mountain ranges show a weaker soil development. The major diagnostic soil features are cambic and calcic horizons. The dominant soil types comprise Inceptisols, Entisols and Aridisols. The Thal, Chagai-Kharan and Thar-Cholistan deserts form the Sandy desert zone, characterized by a ridge valley pattern, which was formed by seasonal winds. A further characteristic is the vegetation made up of scrub and perennial grasses that are mainly used for extensive grazing. The soil material is highly permeable with a fine textural gradation from heavy sandy loam in the subsoil to lighter fine sands in the lower stratum. The area is mainly dominated by aridisols and entisols. Alluvial sediments are the main feature of the vast Piedmont plains zone, which in the upper reaches is dominated by a mixture of sandy loam and gravel. In the middle ranges of the zone, sandy deposits are the most common picture, whereas further down silty sediments dominate. A large part of the Piedmont plains zone is uncultivated, specifically owing to the limited water availability. Seasonal cropping is practiced subject to water availability. Low shrubs and grasses represent the main vegetation in this zone. Water scarcity and low vegetation have led to a weak soil development with occasional presence of high salinity levels. From a soil genetic perspective cambic and calcic horizons are a dominant feature of this zone. Generally aridisols typify this zone as a dominant soil type. A flat surface characterizes the old River Terraces zone, which is formed by river alluvia from the Himalayas. The northern parts are rich in loamy sediments, whereas the southern part is silty. Diversified irrigated cropping is the dominant land use type in this zone, owing to the morphologic features of the soil. Inceptisols, Aridisols, Alfisols and Vertisols are the characteristic soil types of this particular zone. Young soils with a moderate depth are most prevalent in the Subrecent River Plains zone, whereas Recent River plains zone is dominated by Entisols. Dominant soils of the Indus Delta zone include silt loams with a relatively weak structure and saline soils. Re-irrigation in this area is complicated by the rapid rate of re-salinizaton. Aridisols and Entisols are common in this zone. The upcoming chapters are devoted to the analysis of Pakistan's climate and water resources.

2.2 Climate and Water Resources of Pakistan

Pakistan's geographical location gives it a unique climatological character, with great temporal and spatial variation. With summer temperatures averaging 32°C and in parts winter temperatures below the freezing point the country can be classified into 5 different climatic zones. The range is from arid in most parts to humid in the northern region. As far as water resources are concerned, the country heavily depends on glacial melt to feed the Indus River System. In the wake of climate change water scarcity is of growing concern.

2.2.1 Climate

Pakistan is located in the sub-tropics and partially in the temperate region between latitudes 24° and 37°N and longitudes 61° and 76°E. It has a long latitudinal stretch from the Arabian Sea in the south to the Himalayan Mountains in the north. As alteration in latitude is associated with variations in atmospheric composition, this large latitudinal stretch amidst a diverse geography with different topographical characteristics is responsible for a great variability in climatic parameters on both temporal and spatial scales (BARRY AND CHORLEY, 1987). There are many approaches to classifying a regions climate such as empirical, genetic, functional technical, to name a few. A recent study conducted by KHAN et al. (2010) has classified the climate of Pakistan by focusing on the distribution of temperature and precipitation and other relevant climatic variables such as i.e. evapotranspiration. In following, based on the study by KHAN et al. (2010), contributions from the Pakistan Meteorological Department concerning the country's seasons (PMD, 1985) and the study by ZAHID AND RASUL (2011) on the thermal classification of Pakistan, the country's important climatic features shall be framed.

Although other variables are likewise important in the classification of climate, first and foremost priority is given to precipitation and temperature for delimiting climatic boundaries. Thus, using long term observations of monthly and annual means of temperature and precipitation with reference to the KÖPPEN (1936) climate classification, the country can be divided into five different broader zones comprising tropical semi-arid climate, tropical arid climate, cold semi-arid climate, snow forest climate and extreme cold climate. In vast parts of the country in the south and north arid climates are found. Arid climates are generally characterized by significantly low levels of precipitation, with annual total rainfall well below 250 mm. Depending on humidity levels, evapotranspiration, sunshine and the temporal extent or stretch of the seasons several climate subtypes and micro-regions can be distinguished. Thus, for the arid climates, 7 subtypes and 23 micro-regions have been defined (KHAN et al., 2010). In the arid zones the maximum temperature measured in June can exceed 32°C. Average humidity ranges between 35% and 60%, with higher humidity levels recorded for the districts Dera Ismail Khan (D.I. Khan) of the NWFP, Hyderabad of the Sindh province and Skardu and Gilgit of the Pakistan administered Kashmir region. Generally lower humidity levels for the arid climate zone are found for the Balochistan province's districts Khuzdar and Nushki. Mean annual summer temperatures for the arid climates can range between 20°C and 25°C, whereas winter temperatures can vary between 0°C in the northern mountainous region of Kashmir and 15°C in the district of Chitral in the NWFP. Dry summers are encountered in most of the Balochistan province's districts such as amongst others Kalat and Loralai. Precipitation in these regions generally occurs in the winter months. Dry winters are common in most parts of the Sindh province. The semi-arid climates are characterized by a total annual rainfall between 250 mm and 500 mm, comprising parts of the upper Punjab province, lower NWFP and Balochistan. A further division into 4 climate subtypes and 10 micro-regions has been made. Semi-arid climates generally have higher humidity levels between 50% and 65%, such as for instance in the Punjab districts of Faisalabad and Sargodha. Another characteristic feature of the semi- arid zones is that the precipitation or rainfall patterns are much more uniform as compared to arid climates. However, local seasonal maxima exist, that are fluctuating between summer (Punjab and Balochistan) and winter (Northern Areas).

As far as the temperature regimes in the semi –arid climates are concerned, they more or less follow the same path as in the arid zones. Areas with an annual rainfall between 500 mm and 1000 mm, such as parts of the upper Punjab, Kashmir and upper NWFP are referred to as sub-humid, thus these regions are parts of the sub-humid climate zone. The sub-humid climate zone can be further divided into 3 climate subtypes and 6 micro-regions. Generally humidity levels are high with 60-65%. Dry areas do not exist; rainfall is received throughout the climatic zone with highest total annual rainfall recorded for the Punjab province's districts Sialkot and Jhelum and the northern district of Jammu. A considerable part of the rainfall is received in the summer season, winter maxima are for instance found for the NWFP's districts Malakand, Lower Dir and Kohat. Humid climates are found in the extreme north of the Punjab province, NWFP's Malakand Division and both parts of Kashmir. The total annual rainfall ranges between 1000 mm and 2000 mm. Relative humidity varies between 65% and 70%. Generally the northern parts of the country are located in the humid zone including the capital city of Islamabad and further northern situated districts Upper Dir and Balakot. The summer temperatures range between 20°C and 25°C, with occasional maxima exceeding 30°C in June. In the winter season temperatures vary between 0°C and 15°C, with January being the coldest month. Undifferentiated highland climates are encountered in the mountainous north of the country at altitudes above the 2500 m level. These climates are generally characterized by perpetual snow and ice with temperatures around or below 0°C throughout the year. As the name suggests, the climate is rather uniform at these altitudes.

Figure 2.3 Climatic Classification of Pakistan
Source: modified after KHAN et al. (2010)

Further classified, the country possesses four clearly defined seasons. The hot (summer) season stretches from April to June and is followed by the Monsoon season from July to September. The sequencing autumn months are referred to as the monsoon retreat period, which is marked from October to November. The cold or winter season in Pakistan commences in the middle of December and lasts until the end of March. March and April can be further classified into the spring season, whereas the demarcation is not as clear as in the other cases. The hot season is characterized by very high temperatures, which in the months of May and June, where the territory has already been continuously exposed to overhead sun for several months starting from April, can easily exceed the 50°C mark. Mean daily maximum temperatures generally fluctuate between 41°C and 46°C. In contrast to the inner-country, the coastal areas are blessed by the sea breeze that generally keeps maximum temperatures in the hot summer season down to 35°C. However, occasionally if due to climatic disturbances the sea breeze is cut off, the temperatures can easily rise above 45°C. Generally the months from April to June are relatively dry. In the plains the total rainfall in this season amounts to 76 mm maximum and 27 mm minimum, depending on the location. As far as precipitation is concerned, the districts in close proximity to the mountains receive a total seasonal rainfall between 102 mm and 127 mm. Not for many parts of the country the months from April to June are important, however for the districts of Peshawar and Kashmir the rain received in these months represents a considerable share of the total annually received amount. These rainfalls in Peshawar and Kashmir originate from the western disturbances that travel over the northern latitudes and are accompanied by thunderstorms. The hot-summer season is followed up by the wettest season of the year, the monsoon period. In this context, the word monsoon stands for seasonal changes in the wind direction (Figure 2.4).

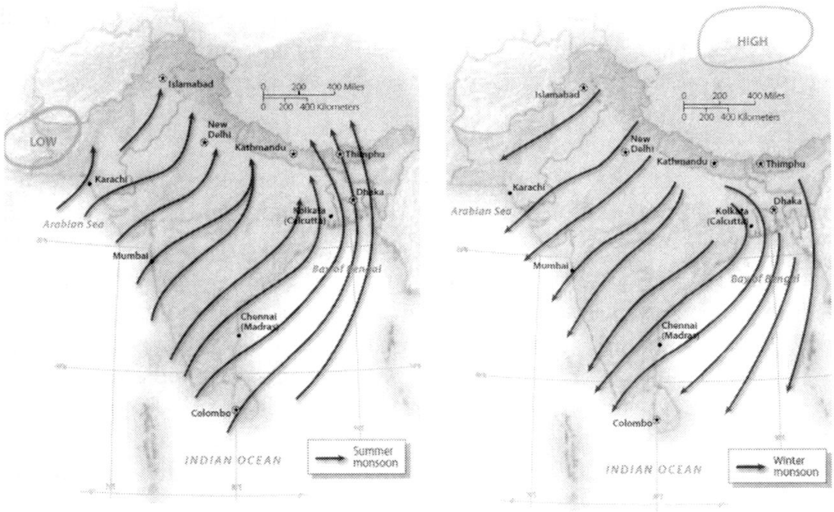

Figure 2.4 The South-East-Asian Monsoons
Source: ROWNTREE et al. (2011)

The hot summer months preceding the Monsoon leave the country dry and hot, with average temperatures over land exceeding the 40°C mark. To this effect, a thermal low named ‘Indus-low’ establishes over the Sindh, southern-central Punjab and central-Balochistan region. This development is understood to be responsible for the southwest monsoon over Pakistan and other parts of the subcontinent. In the middle of May the monsoon appears over Sri Lanka and progressively moves northwards. By the beginning of July it reaches the west of the Punjab province in Pakistan. The Balochistan and Sindh provinces are only brushed by monsoon incursions on an occasional basis, regular moist currents do not occur. Approximately 80% of the annual rainfall is received during the monsoon period. The rainfall during the monsoon season occurs in spells, which at times can be of torrential nature. These heavy summer rains are common in the northern Punjab only, reasoned in the fact that Pakistan is situated at the western edge of the monsoon system. Steady throughout the month of July, the monsoon current retreats slowly at the end of August and at times can be active even in September. In years with the monsoon current lasting till the end of September, heavy floods have been recorded in the Indus Basin. The torrential rains are mainly caused by a series of tropical depressions originating from the northeastern Arabian Sea. The normal rainfall for this season fluctuates around 150 mm. The spatial variability of rainfall in Pakistan is very high which becomes evident when comparing 1500 mm in the northern mountainous areas and less than 100 mm in the south (SHEIKH et al., 2011). For Pakistan the summer monsoon is of great relevance, the winter monsoon however does not affect the country; winters in the country are generally dry with some exceptions.

After the monsoon season, the monsoon retreat season sets in by the middle of September. In the so-called transitory period until the middle November temperatures in the country start rising again, peaking at approximately 37°C. By the end of November temperatures generally drop by approximately 6°C. Clear skies, light winds and the absence of rainfall make the monsoon retreat period's months the driest throughout the year. After this rather pleasant period the relatively cold winter season sets in, lasting from the middle of December until the middle of March. Much of the weather in these months is influenced by regularly passing disturbances from the west, specifically from the Mediterranean Sea. On average each winter month encounters five of such western disturbances. In the northern parts of the country, foremost Peshawar and Kashmir, these disturbances are responsible for considerable amounts of rainfall, especially in the months from January onwards. In both months March and April with the heat of the sun setting in, hail-storms, thunderstorms and tornadoes are a frequent appearance. The provinces Punjab and NWFP are particularly delivered with significant amounts of rainfall, whereas the valleys of the submontane zone receive precipitation in the form of snow. These winter water sources are most important for farmers in the summer-dry areas. Especially for Quetta and Kalat divisions, as they do not receive rainfall in the summer originating from the monsoon currents. As far as temperatures are concerned in the winter season, at Kalat for instance in the extreme cold spells temperatures around 22°C subzero have been recorded.

2.2.2 Water resources

Approximately 1% of the country's area is covered by water bodies and 2% by snow and glaciers. Pakistan has one of the largest glacial deposits in the world. The glacial area covers 13% of the northern mountainous ranges located on the upper Indus Basin. The Indus Basin contains the greatest area of perennial ice outside the Polar Regions. The glaciers and snow covers are a pivotal source for one of the worldwide largest contiguous irrigation systems. Runoff originating from the Himalayan Ranges is dependent on the heat available to melt the stored water in the ice and snow, thus a large part of the annual runoff is generated in the summer season and influences the Indus Rivers flows significantly (LUAOP, 2009).

The major water source in the country however, is the Indus River with its major tributaries – the river Kabul on the right bank, the rivers Sutlej, Beas, Ravi, Chenab and Jhelum on the left bank (Figures 2.2 and 2.5). QAZILBASH (2005) estimated a total hydroelectrical potential of 50.000 MW for Pakistan, which is mainly generated by the Indus River and its tributaries. Water flows in the Indus Basin System are, inter alia, regulated by three very large hydropower dams- Tarbela- Mangla and Chashma, with current live storage capacities of 17.89 cubic kilometers. Originally the dams were designed to store a capacity of 23 cubic kilometers; however mainly through rapid sedimentation 22 percent of the storage capacity is lost (WORLD BANK, 2005). Moreover, 19 river barrages, 12 inter-river link canals, 45 large canal commands, and more than 900.000 mostly diesel-operated tubewells are spread over the territory. An average annual flow of around 172 Billion Cubic Meter (BCM) is estimated for the Indus River System (ASLAM AND PRATHAPAR, 2006). The Basin approximately covers 65% of the country's territory and is part of a hydrological system consisting of overall 3 hydrological units, namely the Indus basin – the Karan Desert closed Basin and the Makran Coast on the Arabian Sea. Although these latter two comparatively dry basins cover a territory of 35% (17%-Karan; 18% Makran), due to the lack of a perennial supply, they only account for an inferior annual flow. A large groundwater aquifer covers a gross command area of 16.2 million ha in the Indus Basin and serves as an additional source of water (RANDHAWA, 2001). In the 26 years from 1977 to 2003 average annual river flows amounted to 141 Billion Cubic Meters at river inflow monitoring stations. The majority of inflow is recorded in the summer growing season from April to September, whereas only 1/5 of the total annual inflow is recorded in the winter farming season from October to March. More than half of the total Indus River System inflows are recorded for the Indus River, which receives approximately 86% of its inflows in the summer farming season. These summer inflows are significantly dominated by glacial melt, whereas the winter season inflows originate from winter rainfall. The Kabul River which has a share of approximately 15% in the total Indus River System inflows is another glacial melt and snow fed river with the majority received in the summer season. The remaining rivers of the Indus River System (IRS) (Chenab and Jhelum and others) are different in one characteristic. Their dominant water source in the summer season in addition to the glacial melt and snow is the significant amount of rainfall received in the monsoon season (Table 2.2) (ALI et al., 2009).

Table 2.2 Major rivers of Pakistan and their seasonal inflow distribution

River	% of IRS Inflows	% of Seasonal Distribution		Summer Inflow Water Source	Winter Inflow Water Source
		Apr-Sep (Summer)	Oct-Mar (Winter)		
Indus	44	86	14	Glacial melt	Winter Rainfall
Chenab	19	83	17	Glacial melt + Monsoon Rainfall	Winter Rainfall
Jhelum	16	78	22	Glacial melt + Monsoon Rainfall	Winter Rainfall
Kabul	16	82	18	Glacial melt	Winter Rainfall
Others	5	-	-	-	-

Source: ALI et al. (2009)

Figure 2.5 summarizes key water resources-related data for the years 2008-2009. Precipitation, especially in the northern parts of the country and foremost in the upper Indus Delta, is a vital water source. The Dependency Ratio goes to show, that the majority of the total renewable water resources are supplied externally. A substantial part of the mean annual inflow into Pakistan through the Indus River System comes from India with relatively small contributions from rivers originating in neighboring Afghanistan. The inflow from India is regulated under the Indus Water Treaty of 1960 between both countries. Central to the agreement is the distribution of the usage rights for the Indus tributaries' waters, with Pakistan being granted the right to an unrestricted use of the western rivers Chenab and Jhelum and India officially getting unbounded use of the eastern rivers Beas, Ravi and Sutlej. Through consideration of the overlap between surface water and groundwater, the internal renewable water resources (IRWR) are estimated, for Pakistan the value for 2008-2009 was 55 cubic kilometers (AQUASTAT, 2011). Almost half of the Indus Basin's drainage area is located in Pakistan, with the rest being shared by the neighbors Afghanistan, China and India. Man-made lakes and reservoirs are also plenty, with the Manchar Lake being the country's and also South Asia's largest lake. As mentioned above, although the country is home to a number of rivers its main life line is the Indus River, to which the other tributaries are attached and fed from.

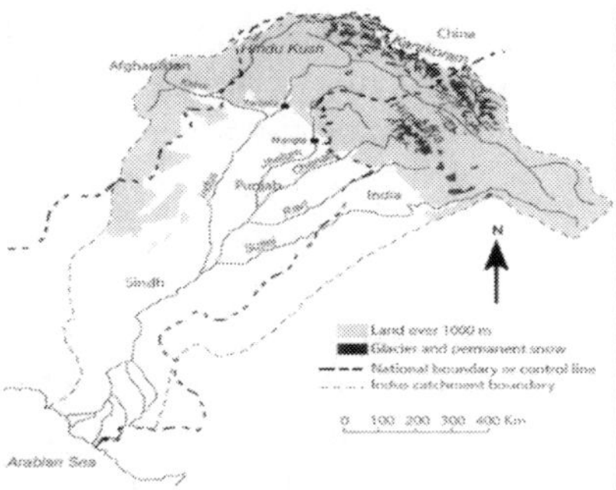

Long-term average annual precip.	Value	Unit
Depth	494	mm/year
Volume	393.3	km^3/year
Long-term average annual renewable WR		
Internal (IRWR)	55	km^3/year
External (ERWR)	175.8	km^3/year
Total (TRWR)	230.8	km^3/year
TRWR per capita	1304	m^3/year
Dependency ratio	76.17	%
Indus River system Inflow	249.08	km^3/year
Total dam capacity	27.03	km^3

Figure 2.5 The Indus Basin and the Water Resources of Pakistan, 2008/2009
Source: ARCHER et al., 2010; modified after AQUASTAT, 2011

According to REVENGA et al. (2000) the Indus in 1995 provided 17% below the United Nation's (UN) minimum standard of water required per capita per year. Their estimates for the future (2025) predict the supply to be even 50% under the minimum threshold. Water scarcity is already evident and is predicted to increase in the recent future, mainly due to the over-extraction by the agricultural sector. The Indus is highly sensitive to climate change as the majority of its water is provided by the Himalayan glaciers. No other river in Asia has a higher proportion of water provided by glacial melt (KIANI, 2005). Not only for the human population, but also for the biodiversity as flora and fauna are concerned, the Indus watershed plays a crucial role.

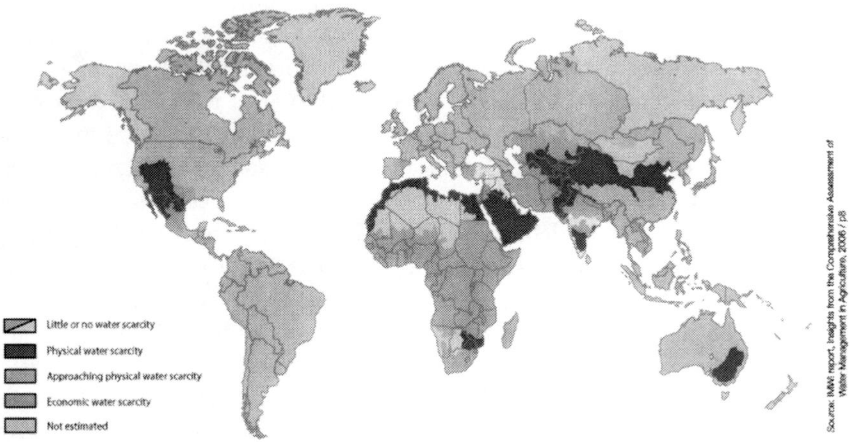

Figure 2.6 Areas of Physical and Economic Water Scarcity
Source: IWMI (2007)

As Figure 2.6 shows only some parts of the south-east (Balochistan province) are approaching physical water scarcity, the rest of Pakistan's territory is already facing physical water scarcity, referring to a situation where the water demand cannot be met with the existing water resources. As Pakistan is a developing country with irrigated agriculture as its mainstay, this situation can have devastating impacts. Countries that have sufficient domestic water resources but need heavy investments to prevent physical water scarcity are grouped into the economically water scarce areas (IWMI, 2007). This is not the case for Pakistan as it heavily relies on outer-country water supply. As far as the outflow of the Pakistani river system into the Arabian Sea is concerned, sea water intrusion into the Indus delta has been observed and is of great concern. Especially in lower flow years, the water flowing into the sea does not fulfill the minimum requirements for preventing sea water (salt) intrusion (GONZALEZ et al., 2005).

After having presented the important climatic features of the country alongside the water resources, the following considerations intend to depict the Agro-Ecological-Zones (AEZ) of Pakistan. The idea is to build a bridge between agricultural production, climate and the natural resource base and to reveal their spatial variability. Using a thermal classification approach, crop growth conditions are further elaborated on.

2.3 Agro-Ecological Zones of Pakistan

In the framework of the Agro-Ecological-Zoning methodology developed by the Food and Agriculture Organization (FAO) in a joint effort with the International Institute for Applied System Analysis (IIASA) an area of land is divided into smaller units, defined in terms of climate, soil and terrain conditions relevant to agricultural production (FAO AND IIASA, 2000). Normally the expression Agro-Ecological-Zone refers to spatial units that are related to a soil map. The level of detail to which such factors as soils, climate and land utilization types are defined may vary according to the map scale and the objectives of the study. The essential elements in defining an Agro-Ecological Zone are the growing period, temperature regime and soil mapping unit (FAO, 1996). The upcoming deliberations are devoted to the presentation of the Agro-Ecological-Zones for Pakistan with land use and water availability as additional elements for classification. According to the Pakistan Agricultural Research Council (PARC, 1980) based on physiography, climate, land use and water availability the country can be divided into ten different AEZ. Figure 2.7 depicts the main Agro-Ecological Zones of Pakistan.

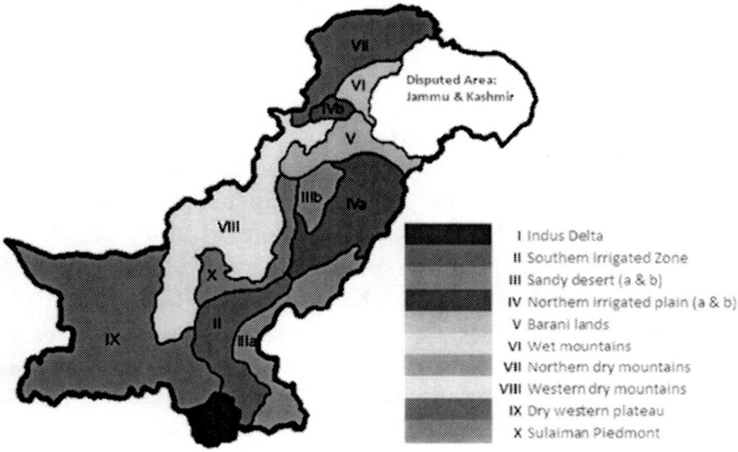

Figure 2.7 Agro-Ecological-Zones of Pakistan
Source: modified illustration after PARC, 1980

KHAN et al. (2004) provide a useful synopsis of the Agro-Ecological-Zoning in Pakistan and are therefore referred to in the following deliberations. The *Indus Delta (I)* forms the first AEZ with an arid tropical marine climate. For the mean monthly summer rainfall values around 75 mm have been recorded, whereas the mean monthly winter rainfall amounts to less than 5 mm. Summer Temperatures, measured as mean daily temperature, lie between 34°C and 40° C, whilst in winter mean daily temperature values range between 19°C and 20°C. In this particular agro-ecological setting, inter alia, rice, pulses, sugarcane, berseem and banana represent the major crops. The favorable climatic conditions make the summer season notably suitable for grazing. In autumn, the conditions for grazing are the poorest. An arid and subtropical climate characterizes the *Southern Irrigated Plain (II)*, the second AEZ of Pakistan, also referred to as the lower Indus plain. Mean monthly rainfall in summer differs for the southern and northern parts of the zone. The southern part experiences a mean monthly summer rainfall ranging from 45 to 55mm, whereas in the northern part of the plain, considerably smaller amounts of rainfall have been recorded, averaging 18 mm. Winter precipitation is scarce if not thoroughly absent, and therefore has been recorded at 0 mm. Temperatures in summer range between 30° and 50°C maximum, whereas in winter the minimum temperature can fluctuate between 0° and 12°C. Cotton, wheat, sugarcane, rice, berseem and sorghum are the major crops grown in this particular Agro-Ecological-Zone. As in the case of the Indus Delta, the summer season is considered the best grazing season.

The *Sandy desert Zone (III)* physiograpically is characterized by sandy and loamy soils. Moreover, in the zone (a) moving sand dunes, high sand ridges and salt lakes are common. Due to persisting water shortage, with maximum rainfall averaging 300 mm annually, the vegetation in this desert zone (a) is of xerophytic nature. Nevertheless, some clay fractions are also found. In the Rabi season (winter farming season) the crops grown amongst others comprise wheat, rape, millet and mustard, whereas the Kharif (summer farming season) crops include cotton, sugarcane, pulses and rice. The land is also utilized for grazing. Short trees are characteristic for the Sandy Desert Zone (b), with rainfall in the northern parts of this zone varying between 300mm and 350mm. The soils in zone (b) are relatively calcareous. The crops grown are cotton, sugarcane and rice in the Kharif season. Wheat and gram alongside maize, sorghum and millet are typical Rabi season crops. Parts of the area are also suitable for grazing. The *Northern Irrigated Plains (IV)* are located between the Sutlej and Jhelum rivers. The climate is of semi-arid (east) to arid (southwest) nature. As far as temperature is concerned, winters can have minimum temperatures of around 6°C, whereas summers are hot with 40°C. In the eastern parts of the zone rainfall varies between 300 mm and 500 mm. Significantly lower levels of rainfall with 200 mm are recorded in the southwestern parts, specifically in the valleys of Mardan and Peshawar with mean monthly rainfall averaging 20-30 mm in both seasons. The *Alluvial Valleys (Zone IV b)* are characterized by soils high in clay content (silty clays, clay loams), whereas the dominant soil types of the eastern parts of the Irrigated Plains (Zone IV a) are sand, loam and clay. Sugarcane and sugarbeet, tobacco, wheat, maize and berseem are crops of the alluvial valleys. The northern part of zone IV (a) is known for its canal irrigated agriculture and the crops wheat, sugarcane, rice, oilseed and millet. Citrus and mangoes are found in the central and southern parts of the zone IV (a). In a dry country with one of the largest irrigation networks the *Barani (V- rainfed)* zone, comprising the Potwar Plateau and the Salt Range at the foot of the mountains, is a rarity. Monthly average rainfall fluctuates around 200 mm in the summer months, whereas winters receive significantly less precipitation with 35 mm – 50 mm. Generally, the northern parts receive higher precipitation, whereas the southern parts are of a semi-arid nature. Nevertheless, rainfall is received monthly, facilitating the rainfed cultivation of crops such as wheat, oilseed, mustard, rapeseed and pulses. Agriculturally important districts of the Punjab province are located in this zone (e.g. Gujranwala and Sialkot). The soil profile of the zone is dominated by non-calcareous to slightly calcareous silt loams and calcareous loams. The area is also classified as the most suitable area for crop production as additional irrigation, bound to sufficient water availability as in the rest of the country, is not required.

The *Wet Mountains - High Mountains zone (VI)* is characterized by relatively high amounts of rainfall, ranging from monthly 235 mm in the summer season to 116 mm in the winter months. Small emphasis is put on rainfed agriculture in this zone, most of the territory is covered with forest vegetation. The soils consist of silt loams to silty clays. From a physiographical viewpoint the area includes high mountains and wide valley plains. The climate in the eastern part is classified as humid, whereas the western part is sub-humid, with a relatively dry summer. Temperatures as low as 0°C are common in the winter months. The summer temperatures normally peak at a maximum of 35°C. Silty loams and silty clays are the dominant soil types of this region paving the way for the cultivation of maize, wheat, olives and apples (high altitude areas). Land in the undifferentiated climate zone or *Northern Dry Mountains zone (VII)* is mainly used for grazing. The summers are mild and the winters are cold, rainfall levels are generally low in both seasons. Irrigation is vital, which is provided by the rivers and streams in the region. The soils are calcareous in lower altitudes and acid in higher altitudes, but generally deep and clayey. The major crops of the region are wheat, maize, rice and fruits. Barren hills up to 3000 m and steep slopes are most characteristic of the *Western Dry Mountains zone (VIII)*. Deep and loamy soils are most prevalent in the valleys, where the majority of agricultural production takes place. The major crops of this Agro-Ecological-Zone are wheat and fruits (e.g. peaches, plum, apricots, and apples). Nevertheless, the largest share of land is mainly used for grazing. In the extreme north of the zone a sub-humid climate can be classified, whereas the south is rather of a semi arid-nature. The rainfall received, does not considerably differ between the seasons, however with a slightly higher amount in summer. Temperature in the south can easily touch the 35-39°C mark, at the same time subzero temperatures in winters are not an irregular incidence either.

Inter mountain basins and plateaus are dominant in the *Dry Western Plateau zone (IX)*. The climate is of arid and tropic nature, thus where water is available and soils are good agricultural cultivation is possible. Especially the lower regions of this zone have xerophytic vegetation, whereas the higher altitudes can also have a thin strip of juniper forests. A special feature of this zone is the availability of a breeze from the coastal area, facilitating the cultivation of fruits and vegetables subject to sufficient water availability. Other crops of the region include sorghum and wheat. Winters in the coastal zone are generally warmer with values around 15°C. The northern areas, although warmer in the summer with 40°C have colder winters with temperatures as low as 3°C. The area is very dry with most of the parts receiving rainfall at a monthly average of 2 mm. An exception to the dry nature of this agro-ecological-zone is the south eastern part, receiving a mean monthly rainfall around 35 mm in the summer season, triggered by the monsoon. Moreover, grazing represents an important type of land use as grasses alongside the xerophytic desert vegetation are common. An arid and hot subtropical continental climate dominates the *Sulaiman Piedmont zone (X)*. Temperatures in summer fluctuate between 40°C and 43°C and in winter vary between 6°C and 8°C. Mean monthly rainfall in the summer season is below 15 mm. Floods originating from the hill torrents provide a vital source of water for irrigation. The streams have been responsible for shaping the alluvial landscape of this zone. In the slightly sloping areas the soils are loamy; further into the plains clays clearly dominate. Nurtured by the water from the torrents the crops wheat, gram and rice are cultivated and remain the main type of land use in this area.

After having presented the country's agro-climatic profile, in following the idea is to discuss the developments and structural changes in the agricultural sector of Pakistan, including deliberations on land tenure, land use and the technical resource base.

2.4 Development of the Agricultural Sector in Pakistan

The lion share of the area devoted to the production of crops is located in the Punjab. The main agricultural crop produced throughout the country is wheat, with a share between 60% and 80% in total cropped area. As per the productivity is concerned, for the production of wheat the provinces Sindh, Balochistan and Punjab in the period 1987-2006 have been especially suitable. Rice productivity has been better in Balochistan and Sindh, whereas the country's major cash crop cotton has been comparatively more successfully grown in the Punjab and the Sindh provinces. More than 90% of the value added in the important crop sector is contributed by wheat, rice, cotton and sugarcane (ESP, 2011). Land reforms in the country are ongoing and have historically been an area of particular attention and concern.

2.4.1 Land use pattern in Pakistan

According to the LUAOP (LAND USE ATLAS OF PAKISTAN) (2009) the country's total geographical area including water bodies and glaciers amasses to

83.5 million hectares, with the lion share being contributed by rangeland and exposed rocks. Latest figures suggest that approximately 5% of the total area is covered by forest, 27% by rangeland, 20% by agricultural land, 25% by exposed rocks and the rest by desert, water bodies and glaciers. The bulk of the Agricultural land is located in the provinces Punjab and Sindh, with the Punjab province clearly possessing the largest share. Deserts, which make up 10% of the total geographical area, are common for the Sindh and Balochistan provinces, whereas forests are dominant in the NWFP. Balochistan has the largest share of exposed rocks with almost 50% of the total territory being covered by them. Rangeland, which is a pivotal source for livestock development, is evenly distributed throughout the provinces with a slightly bigger share for the Punjab province. The highest share of waterlogged and saline land is found in the Sindh region, closely followed by the Punjab province, mainly owing to intensive agricultural practices involving irrigation. The northern areas, foremost the NWFP possess the country's glaciers with a share of 100%. As far as the historical development of land use patterns in the country is concerned, the country's geographical area (excluding glaciers and water bodies) can be divided into five different categories, including reported area, forest, area not available for cultivation, cultivable wasteland, current fallow and net sown area (Table 2.3).

Table 2.3 Decade-wise land utilization in Pakistan 1951-2002

in million hectares

Period	Reported Area [Σ1-5]	Forest [1]	Not available for cultivation [2]	Cultivable waste [3]	Current fallow [4]	Net area sown [5]	Total cultivated area [4+5]
1951/52	46.45	1.39	20.75	9.16	3.54	11.61	15.15
1961/62	50.99	1.68	18.73	12.46	4.85	13.27	18.12
1971/72	53.55	2.83	20.4	11.11	4.77	14.44	19.21
1981/82	53.92	2.85	19.9	10.86	4.89	15.41	20.3
1991/92	57.61	3.46	24.34	8.85	4.85	16.11	20.96
2001/02	59.28	3.61	24.5	9.13	5.67	16.32	21.99
Annual growth rate 1951/52 - 2001/02	0.49%	1.91%	0.33%	-0.01%	0.94%	0.68%	0.75%

Source: modified after KUROSAKI (2007)

Table 2.3 presents 51 years of data to depict the decadal land use change in Pakistan. According to the Ministry of Food, Agriculture and Livestock (MINFAL) the area reported refers to the total physical area of the villages or districts. Whereas the area not available for cultivation is that uncultivated area of the farm which is used for housing purposes, farm roads and other related purposes. The land category cultivable wasteland refers to those areas, which although available for cultivation, were not cropped in the reference period. The total cropped or cultivated area is calculated as the sum of current fallow and net sown area. As the data reveals, there has been a considerable change in the forest and cropped area. The cultivated area in the 51 year reference period has grown by more than 30%, with an average annual growth rate of approximately 0.75%. Forests on the other hand have shown the highest annual increases to the tune of 1.91% almost doubling throughout the 51 years considered. A slight increase in the conversion of wasteland into productive land has been observed. The reported area has also significantly increased.

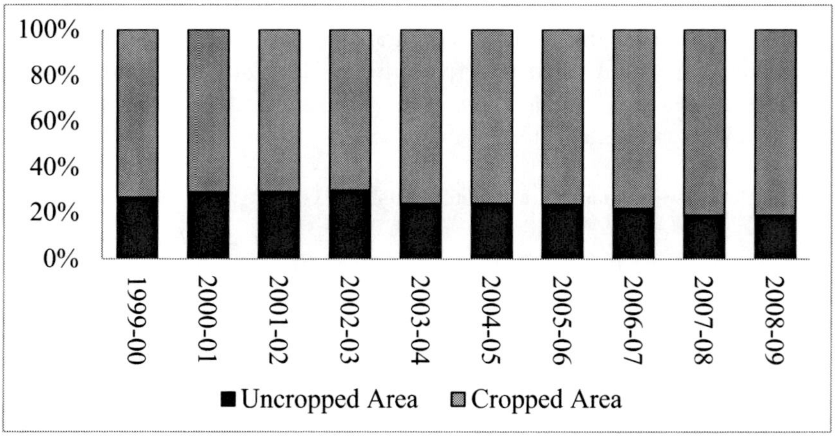

Figure 2.8 Utilization of Arable Land, 1999-2009
Source: PAKISTAN STATISTICAL YEARBOOK, 2010

Approximately 37 percent of the total area (29.45 ha) is arable land. According to the latest estimates by the Food and Agriculture Organization (FAOSTAT, 2011), 81 percent of the arable land is currently cropped (see Figure 2.8). With almost 20 million hectares of the cropped area being irrigated, the country has one of the highest proportions of irrigated cropped area in the world (PAKISTAN STATISTICAL YEARBOOK, 2010).

Growth in cropped area has been very impressive over the years, from 11.6 million hectares in 1947 to 23.85 million hectares in 2009. However, according to the UNITED NATIONS POPULATION DIVISION (2011), Pakistan is the 6th most populous nation in the world, home to 173 million people. Within the top six most populous countries, Pakistan has the highest average population growth rate with 1.8% per year. Thus for 2100 the United Nations have forecasted a total population for Pakistan numbering 261 million people. As currently most of the arable land is already cropped and utilized for agricultural purposes, land expansion most likely is not going to be an option to feed the future population. Figure 2.9 depicts the abundance of arable land in South Asia on a per capita basis. As can be seen, the arable land per capita available to a Pakistani is 0.11 ha, very similar to the immediate neighbor India with 0.13 ha per capita. Both countries, considering the population pressure, in future will also have to deal with an increased pressure on their land resources.

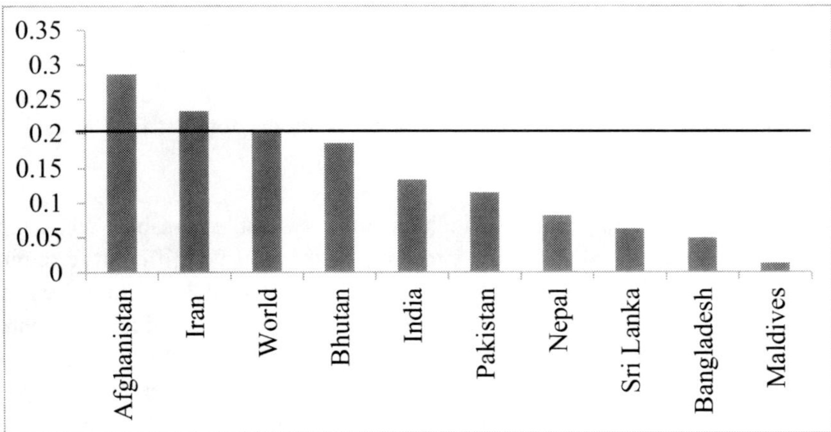

Figure 2.9 Arable Land Abundance in Southern Asia 2008 (hectares/capita)
Source: own illustration using FAOSTAT, 2011

Globally, the average arable land available per capita amounts to 0.20 ha (FAOSTAT, 2011). Pakistan has a significantly lower value, illustrating the pressure on its land resources.

2.4.2 Agricultural Production – Major Crops of Pakistan

Pakistan is home to two major crop growing seasons, namely "Kharif" (summer) and "Rabi" (winter). The summer growing season spans from April (sowing) to October (harvest), whereas the winter season commences in November and lasts until March. Crops such as rice, maize, millet, cotton, sugarcane represent the major summer crops. Wheat, mustard, rapeseed, barley and gram are typical Rabi season crops (AHMAD, 2005).

Figure 2.10 summarizes the main crop production regions of Pakistan and also depicts the dominant cropping pattern by province. The cropping pattern cotton-wheat covers the largest share of total agricultural area in the country. According to the Agricultural Statistics of Pakistan (MINFAL, 2007) on average more than 30% (7.1 million ha) of the total available area for cultivation is used for cotton-wheat cropping. The largest area for cotton-wheat cultivation is located in the Punjab province with 5.5 million ha; a much smaller share is held by the Sindh province with 1.6 million ha. Mixed crops occupy close to 28% (6.33 million ha) of the total cultivable land, with the largest area (4.1 million ha) being located in the Punjab province. Rice-wheat is another important cropping pattern used in Pakistan, with about 19% (4.25 million ha) of the agricultural land occupied by this combination. Again the highest share of land for this pattern also, is occupied by the Punjab province with 2.8 million ha. The maize-wheat cropping pattern is used in the NWFP and the Punjab province, occupying approximately 10% (2.1 million ha) of the total available land for cultivation.

The remaining cropping patterns include pulses-wheat, vegetables-wheat and peri-urban. Generally the Punjab province with its five different cropping patterns (cotton-wheat, rice-wheat, mixed crops, pulses-wheat, maize/wheat-oilseeds) represents the country's major agricultural production centre, occupying 69% (15.5 million ha) of the agricultural area. With three different cropping-patterns (cotton-wheat, rice-wheat, mixed crops) over an area of 4 million ha (18% of the total agricultural area), the Sindh province also plays an important role for the country's agricultural production. The other two provinces Balochistan and NWFP, although also important, do not play a leading role in terms of agricultural production. Balochistan possesses 5% (1.1 million ha) of the agricultural area, whereas the NWFP occupies close to 8% (1.8 million ha). The major cropping patterns in the Balochistan province comprise mixed crops, orchards/vegetables-wheat, rice-wheat and peri-urban. The NWFP has three major cropping patterns with maize-wheat, mixed crops and pulses-wheat.

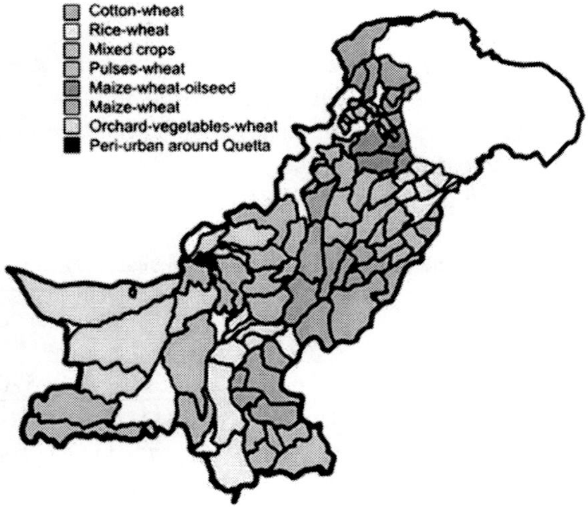

Region	Cropping Pattern
Punjab I	Cotton-wheat
Punjab II	Rice-wheat
Punjab III	Mixed crops
Punjab IV	Pulses-wheat
Punjab V	Maize/wheat-oilseeds
Sindh I	Cotton-wheat
Sindh II	Rice-wheat
Sindh III	Mixed crops
NWFP I	Maize-wheat
NWFP II	Mixed crops
NWFP III	Pulses-wheat
Balochistan I	Mixed crops
Balochistan II	Orchards/vegetables-wheat
Balochistan III	Rice-wheat
Balochistan IV	Peri-urban

Figure 2.10 Major crop production regions of Pakistan
Source: Modified after FAO, 2004

The following statistical abstract intends to address the major crops wheat, rice, cotton and sugarcane in detail with reference to the cultivated area, farm structures and yields. A short review about vegetables in Pakistan shall also be given.

Wheat

Not only is wheat the staple food of the people in Pakistan, it is also the most important crop as far as agricultural policy formulation is concerned (ESP, 2011). Moreover, wheat contributes about 45% of the daily caloric consumption of the population. Wheat is also a critical commodity in terms of total contribution to national food security (GARCIA, 2009).

Table 2.4 Development of Area, Yield & Production of Wheat - Pakistan 1961-2010

WHEAT	1961-1971	1972-1981	1982-1991	1992-2001	2002-2010
Area Harvested (ha)	5433601	63114782	7512590	8209520	8490989
Yield (Hg/ha)	9213	13871	17043	21207	25005
Production (tons)	5074335	8811824	12821720	17426790	21266222

Source: FAOSTAT, 2011

According to the latest data available from the FAOSTAT (2011), on average around 8.5 million hectares between 2002 and 2010 were cultivated with wheat; yield per hectare in the same period averaged 2.55 tons, with a total wheat production of 23 million tons. According to the latest Economic Survey of Pakistan (ESP, 2011), wheat production in the country has increased from 23 million tons in 2009-10 to 24 million tons in 2010-11. As Table 2.4 shows, the area harvested with wheat has considerably increased from 1961 to 2010 by almost 35%. From 1961 to 2001 the expansion has been very notable, whereas in the last decade a slowdown has been observed. In the foregone 50 years wheat yields have more than doubled, although they are still behind the global average of approximately 3 t/ha. Also noteworthy is the fact, that the wheat crop occupies the largest area under a single crop. Thus, according to the Agricultural Statistics of Pakistan (MINFAL, 2007) almost half of the total cultivated area is cropped with wheat. With 72%, the Punjab province has the largest share in the total area allocated for wheat production in Pakistan. The other two important provinces for wheat cultivation, as far as area is concerned, are Sindh and NWFP, with shares of 14% and 10% respectively. As far as total production is concerned the scenario is not different either, the Punjab province contributes 75% of the country's total wheat output, followed by Sindh with 14%, NWFP with 7% and Balochistan with 4%.

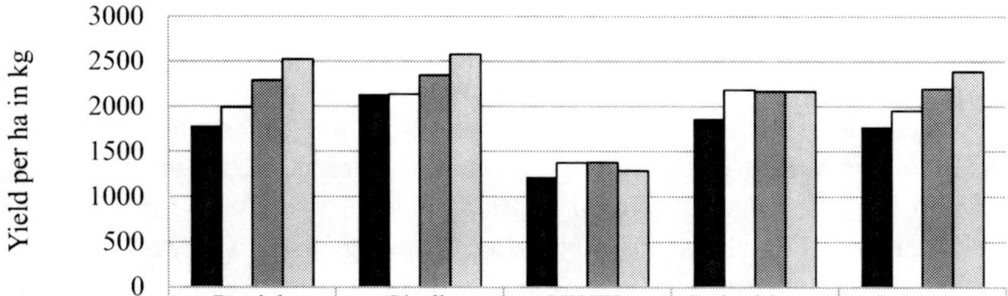

	Punjab	Sindh	NWFP	Balochistan	Pakistan
■ 1987-1991	1781	2134	1221	1867	1773
□ 1992-1996	1988	2135	1375	2181	1951
▩ 1997-2001	2287	2343	1378	2167	2196
▨ 2002-2006	2521	2576	1286	2167	2388

Figure 2.11 Provincial average wheat yield development in Pakistan 1987-2006
Source: own illustration after MINFAL, 2007

Figure 2.11 shows the development of average wheat yields in the different provinces over a time period of 20 years. Except for the NWFP, where average wheat yields have rather dropped, all other provinces have considerably improved wheat yields during the reference period. The highest improvements have been achieved by the Punjab province, which has managed to increase average wheat yields from 1.7 t/ha in 1987-1991 to 2.5 t/ha in the period 2002-2006. The other remarkable improvement has been observed in the Balochistan province, with the average wheat yield increasing from a value of 1.9 t/ha in 1987-1991 to 2.2 t/ha in the period 2002-2006. Although the Sindh province's average wheat yields from a country perspective were sufficient in the period from 1987-1991 at 2.1 t/ha, it still has achieved improvements to the tune of 2.6 t/ha in the most recent observed period. As per the country is concerned average wheat yield fluctuates between 2.4 t/ha and 2.5 t /ha. The key constraint to wheat production is generally the timing and availability of the irrigation water supply, and the efficiency of water use on-farm (GARCIA, 2009).

According to the latest Agricultural Census of the country (AC, 2000), on average 80% of the total registered farms in Pakistan cultivate wheat. Out of the total farms under wheat, approximately 58% are small farms with average land sizes varying between 0.5 and 2.0 ha, 37% are medium farms with average land sizes between 2 ha and 10 ha, whereas 5% are large farms with average land sizes around 10 ha minimum and 60 ha maximum. In 2010-2011 wheat contributed 13.1% to the total agricultural value added (ESP, 2011).

Rice

Rice is the second most important food crop, accounting for about 6 % of the daily caloric intake (GARCIA, 2009). However, at the same time it also plays a crucial role as a cash crop, earning substantial foreign exchange. The national rice crop is predominantly rainfed and it is planted from May-June during the kharif season. Punjab is the major rice growing area, whereas major high yielding varieties are produced in the Sindh province. Rice has emerged as a major export commodity and Pakistan enjoys a near monopoly status in the export of fine aromatic Basmati- rice (JUNEJO et al., 2007).

Table 2.5 Development of Area, Yield & Production of Rice - Pakistan 1961-2010

RICE	1961-1971	1972-1981	1982-1991	1992-2001	2002-2010
Area Harvested (ha)	1400118	1792306	2022510	2244510	2570356
Yield (Hg/ha)	17107	23755	24244	27720	31977
Production (tons)	2432364	4266516	4900520	6251412	8259706

Source: FAOSTAT, 2011

According to the latest data available from the FAO (2011), on average more than 2.5 million hectares between 2002 and 2010 were cultivated with rice (paddy); yield per hectare in the same period averaged 3.2 tons, with a total rice paddy production of 8.3 million tons. The rice crop has been particularly strongly affected by the floods of 2010, recording the lowest production since 2002 in the period 2010-2011 with approximately 5 million tons (ESP, 2011). As Table 2.5 shows, the area harvested with rice has considerably increased from 1961 to 2010 by almost 46%. From 1961 to 1981 the expansion has been very notable, whereas in the period from 1982-2001 a slowdown has been observed. In the period between 2002 and 2010 however, another considerable growth in the harvested area has been recorded. In the foregone 50 years rice paddy yields have increased by 47%, although they are still behind the global average of approximately 4 t/ha. The rice crop occupies approximately 13% of the country's available area for crop cultivation (MINFAL, 2007). With 62%, the Punjab province has the largest share in the total area allocated for rice production in Pakistan. The other important province for rice cultivation, as far as area is concerned, is Sindh with a share of 29%. As far as total production is concerned the Punjab province in absolute terms dominates rice output in Pakistan, producing 48% of the country's rice on 62% of the total rice area.

However, in relative terms the Sindh province outperforms the other provinces, producing 40% of the country's total rice output on half the area used by the Punjab province.

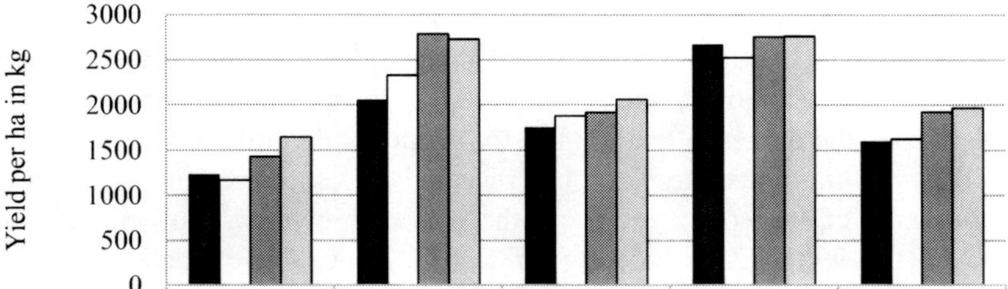

	Punjab	Sindh	NWFP	Balochistan	Pakistan
■ 1987-1991	1235	2058	1754	2676	1600
□ 1992-1996	1171	2330	1880	2524	1626
▩ 1997-2001	1428	2787	1920	2757	1922
▨ 2002-2006	1645	2730	2063	2762	1967

Figure 2.12 Provincial average rice yield development in Pakistan 1987-2006
Source: own illustration after MINFAL, 2006

Figure 2.12 shows the development of average rice yields in the different provinces over a time period of 20 years. Except for the Sindh province, where average rice yields after healthy improvements have rather stagnated in the period 2002-2006, all other provinces have considerably improved average rice yields during the reference period. The highest improvements have been achieved by the Punjab and the Sindh provinces, which have both managed to increase average rice yields by approximately 25% between 1987 and 2006. In absolute terms the Sindh and Balochistan provinces, with average rice yields around 2.7 t/ha in the period 2002-2006, have outperformed the other provinces. The Punjab province in particular has shown improvements, however yield levels at 1.6 t/ha remain lowest in the country. As per the country is concerned average rice yield fluctuates around 2 t/ha.

According to the latest Agricultural Census of the country (AC, 2000), on average 24% of the total registered farms in Pakistan cultivate rice. Out of the total farms under rice, approximately 47% are small farms with average land sizes varying between 0.5 and 2.0 ha, 47% are medium farms with average land sizes between 2 ha and 10 ha, whereas 6% are large farms with average land sizes around 10 ha minimum and 60 ha maximum. In 2010-2011 rice has contributed 4.4% to the total agricultural value added (ESP, 2011).

Cotton

Cotton is the country's major cash crop. As far as total cotton production, consumption and export are concerned, Pakistan globally ranks amongst the top five countries. Moreover, the cotton crop supports the country's industrial powerhouse, the domestic textile industry, from textile mills to knitting, dyeing and oil expelling units. Research has revealed a close correlation between the performance of the cotton crop and the overall growth rate of the economy (BANURI, 1998). Only recently in 2009 biotech field trials have been introduced in the country (OSAKWE, 2009). Studies for India have shown promising economic impacts for the farmers through adopting Bt-Cotton (MAL et al., 2011).

Table 2.6 Development of Area, Yield & Production of Cotton - Pakistan 1961-2010

COTTON	1961-1971	1972-1981	1982-1991	1992-2001	2002-2010
Area Harvested (ha)	1619682	1974120	2487790	2934810	2980289
Yield (Hg/ha)	8667	9352	15417	16737	20111
Production (tons)	1419375	1854470	3909541	4914932	5999864

Source: FAOSTAT, 2011

According to the latest data available from the FAOSTAT (2011), on average more than 2.9 million hectares between 2002 and 2010 were cultivated with seed cotton; yield per hectare in the same period averaged 2 tons, with a total seed cotton production of 5.9 million tons. Like the rice crop, the cotton crop was also severely affected by the floods in 2010, recording a 9% reduction in total output as compared to the previous production period (ESP, 2011). As Table 2.6 shows, the area harvested with seed cotton has considerably increased from 1961 to 2010 by almost 45%. In the periods from 1971 to 1981 and1981 to 1991 the expansion has been very notable, whereas in the period from 2001-2010 a slowdown has been observed. In the foregone 50 years seed cotton yields have increased by 57% and achieved levels around the global average. The cotton crop occupies approximately 7% of the country's available area for crop cultivation (MINFAL, 2007). With 80%, the Punjab province has the largest share in the total area allocated for cotton production in Pakistan. The only other important province for cotton cultivation, as far as area is concerned, is Sindh with a share of 19%. As far as total production is concerned the Punjab province in absolute terms dominates cotton output in Pakistan, producing 82% of the country's cotton. The Sindh province produces the remaining 17% of cotton in the country, whereas 1% is jointly shared between Balochistan and NWFP.

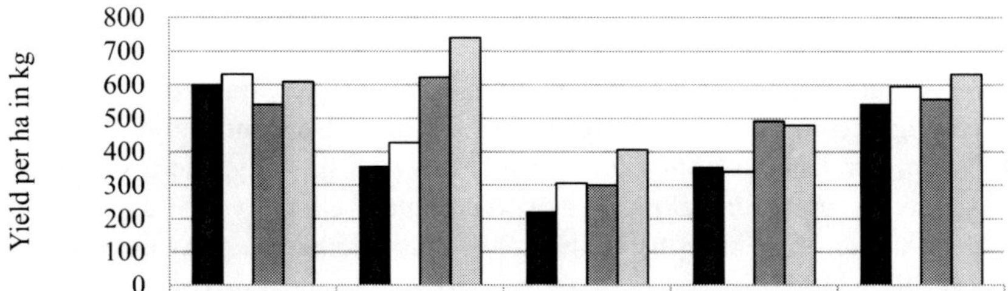

	Punjab	Sindh	NWFP	Balochistan	Pakistan
■ 1987-1991	603	358	222	356	544
□ 1992-1996	631	428	306	340	595
▩ 1997-2001	541	622	300	492	557
▨ 2002-2006	609	740	407	480	631

Figure 2.13 Provincial average Cotton lint yield development in Pakistan 1987-2006
Source: own illustration after MINFAL, 2007

Figure 2.13 shows the development of average cotton lint (export cotton) yields in the different provinces over a time period of 20 years. Except for the Punjab province, where average cotton lint yields have rather stagnated in the period 1987-2006, all other provinces have considerably improved average cotton lint yields during the reference period. The highest improvements have been achieved by the Sindh province, which has managed to increase average cotton lint yields from 358 kg/ha in the period 1987-1991 to 740 kg/ha in 2002-2006. In absolute terms the Sindh and Punjab provinces, with average cotton lint yields around 600-700 kg/ha in the period 2002-2006, have outperformed the other provinces, which are not important cotton production regions. Although improvements have been remarkable in the NWFP, the yield level at 400 kg/ha remains lowest in the country. As per the country is concerned average cotton lint yield fluctuates around 650 kg/ha, surpassing the global cotton fiber yield of approximately 620 kg/ha. A major handicap to the yield increase and cotton production in general is the existence and wide prevalence of mealy bugs and the cotton curl virus in Pakistan (OSAKWE, 2009). According to the latest Agricultural Census of the country (AC, 2000), on average 25% of the total registered farms in Pakistan cultivate cotton. Out of the total farms under rice, approximately 50% are small farms with average land sizes varying between 0.5

and 2.0 ha, 44.5% are medium farms with average land sizes between 2 ha and 10 ha, whereas 0.5% are large farms with average land sizes around 10 ha minimum and 60 ha maximum. In 2010-2011 cotton has contributed 6.9% to the total agricultural value added (ESP, 2011).

Sugarcane

After cotton, sugarcane is the country's second most important cash crop, generating income for the domestic farmers and also contributing to foreign exchange earnings. Not only domestically but also globally, the importance of the sugarcane crop for Pakistan is overwhelming. Globally, the country ranks amongst the top five in terms of sugarcane cultivated area. After the end of the British rule in 1947 the country only possessed 2 sugar mills (MEMON, 2002). Currently 81 mills with a crushing capacity of approximately 60 million tons exist in Pakistan (YASEEN et al., 2005).

Table 2.7 Development of Area, Yield & Production of Sugarcane - Pakistan 1961-2010

SUGARCANE	1961-1971	1972-1981	1982-1991	1992-2001	2002-2010
Area Harvested (ha)	535564	700998	865690	986200	1032222
Yield (Hg/ha)	359784	366516	404767	459120	499198
Production (tons)	19427164	25739180	34849120	45398150	51545211

Source: FAOSTAT, 2011

According to the latest data available from the FAO (2011), on average more than 1 million hectares between 2002 and 2010 were cultivated with sugarcane; yield per hectare in the same period averaged 49 tons, with a total sugarcane production of 51 million tons. According to the latest Economic Survey of Pakistan (ESP, 2011), sugarcane production in the country has increased from 51 million tons in 2009-10 to 53 million tons in 2010-11 (ESP, 2011). As Table 2.7 shows, the area harvested with sugarcane has considerably increased from 1961 to 2010 by almost 48%. From 1961 to 1981 the expansion has been very notable with 25%, whereas in the period from 1982-2001 a slowdown has been observed. In the foregone 50 years sugarcane yields have increased by 28%, although with 49 t/ha they are still behind the global average of approximately 65 t/ha. The sugarcane crop occupies approximately 5% of the country's available area for crop cultivation (MINFAL, 2007). With 64%, the Punjab province has the largest share in the total area allocated for sugarcane production in Pakistan. The other important provinces for sugarcane cultivation, as far as area is concerned, are Sindh and NWFP with a share of 25% and 11% respectively. As far as total production is concerned the Punjab province in absolute terms dominates sugarcane output in Pakistan, producing 59% of the country's sugarcane on 64% of the total sugarcane area. However, in terms of productivity the Sindh province outperforms the other provinces, producing 30% of the country's total sugarcane output on 25% of the total sugarcane allocated area. The remaining part is produced in the NWFP, with a share of 11%. In the NWFP also a small percentage of sugar-beet is cultivated.

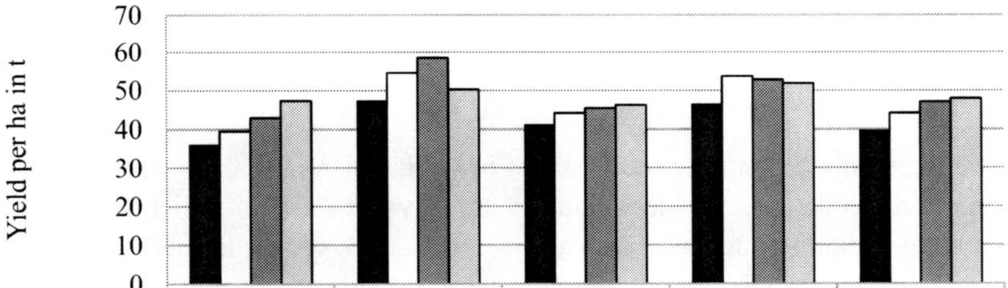

	Punjab	Sindh	NWFP	Balochistan	Pakistan
■ 1987-1991	36	47	41	47	40
□ 1992-1996	40	55	44	54	44
▩ 1997-2001	43	59	45	53	47
▨ 2002-2006	47	50	46	52	48

Figure 2.14 Provincial average Sugarcane yield development in Pakistan 1987-2006
Source: own illustration after MINFAL, 2007

Figure 2.14 shows the development of average sugarcane yields in the different provinces over a time period of 20 years. All provinces have shown improvements during the reference period. The Sindh province however, has slightly struggled in the period 2002-2006 to sustain the previously achieved increases in yield. The highest improvements have been achieved by the Punjab province, which has managed to increase average sugarcane yields from 36 t/ha in the period 1987-1991 to 47 t/ha in 2002-2006. In absolute terms the Sindh and Balochistan provinces, with average sugarcane yields around 50 t/ha in the period 2002-2006, have outperformed the other provinces, although the Balochistan province with less than 1% share in the total sugarcane harvested area is not an important sugarcane cultivating region. Although improvements have been achieved in the NWFP and the Punjab province, yield levels at 46-47 t/ha remain stagnant, this development is also observed in the other provinces.

As per the country is concerned average sugarcane yield fluctuates around 48 t/ha.

According to the latest Agricultural Census of the country (AC, 2000), on average 13% of the total registered farms in Pakistan cultivate sugarcane. Out of the total farms under sugarcane, approximately 35% are small farms with average land sizes varying between 0.5 and 2.0 ha, 56% are medium farms with average land sizes between 2 ha and 10 ha, whereas 9% are large farms with average land sizes around 10 ha minimum and 60 ha maximum. In 2010-2011 sugarcane has contributed 3.6% to the total agricultural value added (ESP, 2011).

Vegetables

Vegetables rank among the minor crops in Pakistan. As far statistical records are concerned, tomatoes, potatoes and general vegetables are distinguished. In Pakistan vegetables such as radish and cucumber are grown throughout the year, whereas other vegetables such as tomato, onion and carrot are either summer or winter crops. Generally, the cultivation of vegetables in the country is bound to a higher risk for the farmers, as the crops are more prone to disease and pest attacks when compared to the major crops cotton, rice, sugarcane and wheat. In total more than 30 different species of vegetables are grown in the country (BAKHSH, 2007).

Table 2.8 Development of Area, Yield & Production of Vegetables - Pakistan 1961-2010

VEGETABLES	1961-1971	1972-1981	1982-1991	1992-2001	2002-2010
Area Harvested (ha)	135297	155429	214408	316045	391507
Yield (Hg/ha)	88752	119987	121222	133587	128814
Production (tons)	1204794	1864210	2611805	4229883	5039373

Source: FAOSTAT, 2011

According to the latest data available from the FAO (2011), on average around 391000 hectares between 2002 and 2010 were cultivated with vegetables; yield per hectare in the same period averaged 12.8 tons, with a total vegetables production of 5 million tons. As far as vegetables in Pakistan are concerned, the major crops in this group include onions, tomatoes, pumpkins and okra, with shares of 30%, 10%, 7% and 4% respectively in total area cultivated with vegetables. Other important vegetables include carrots, peas and cauliflower. As

Table 2.8 shows, the area harvested with vegetables has considerably increased from 1961 to 2010 by almost 65%. From 1982 to 2001 the expansion has been very notable with 32%. In the foregone 50 years wheat yields have increased by 30%.

Vegetable production in Pakistan is regionalized, thus for instance onions are only grown in selected Punjab districts, whereas other vegetables are grown predominantly in other provinces. The majority of vegetables are grown in the peri-urban settings of Karachi, Lahore, Peshawar (BAKHSH, 2007). With 59%, the Punjab province has the largest share in the total area allocated for vegetable production in Pakistan. The other three provinces Sindh, NWFP and Balochistan have an equal share in cultivated area, with 16%, 15% and 12% respectively. As far as total production is concerned the Punjab province contributes 67% of the country's total vegetable output, followed by Balochistan and NWFP with 13% and Sindh with 7%.

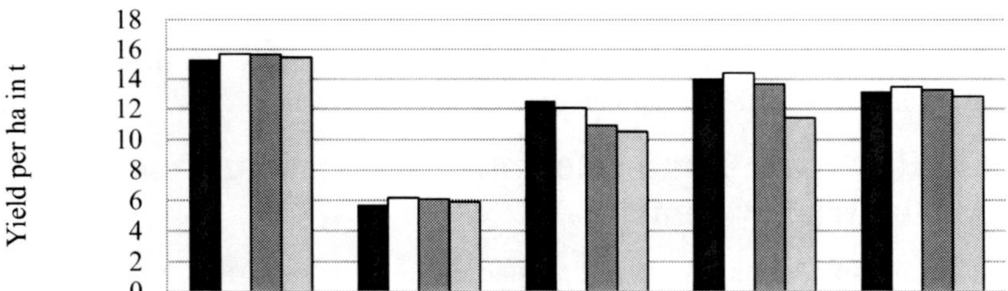

	Punjab	Sindh	NWFP	Balochistan	Pakistan
■ 1987-1991	15	6	13	14	13
□ 1992-1996	16	6	12	14	13
▩ 1997-2001	16	6	11	14	13
▤ 2002-2006	15	6	11	11	13

Figure 2.15 Provincial average Vegetables' yield development in Pakistan 1987-2006
Source: own illustration after MINFAL, 2007

Figure 2.15 shows the development of average vegetables' yields in the different provinces over a time period of 20 years. All provinces have somewhat remained at the same yield levels during the reference period. The Sindh province in particular, has the lowest yield level with 6 t/ha, which is also linked to the lower importance of the province for vegetable production and the fact that vegetable production in the country is highly regionalized. The regionalization is not very clear in the statistics regarding yields, however when considering the fact that some provinces are specialized on vegetables with higher yields such as carrots, cauliflower and onions, whereas others produce vegetables with smaller yields such as i.e. green beans, the situation becomes clear. In absolute terms the Punjab province, with average vegetables' yields around 15 t/ha in the period 2002-2006, has outperformed the other provinces, although the NWFP and the Balochistan province with yield levels at 11 t/ha have also remained important regions for vegetables cultivation. As per the country is concerned average vegetables' yield stagnates around 13 t/ha.

More than 54% of the vegetables are produced by smallholder farmers with landholdings smaller than 1 acre. As the land size does not facilitate the cultivation of the major crops cotton, rice and sugarcance, the only option is to grow vegetables that are high in vitamin content and the cultivation is of a rather short duration. Moreover, vegetables in Pakistan generally receive higher prices. Availability of irrigation water and farm management play a pivotal role in determining the success of vegetable cultivation, too often water is not available and traditional farm management practices hinder success than promote it. In 2010-2011 by growing at 9% vegetables have contributed the bulk of the growth in the minor crops (ESP, 2011).The following subchapter is devoted to the description of the different land tenure systems in Pakistan.

2.4.3 History of Land reforms, Farm structures and tenurial systems

Land reforms and land tenure in Pakistan go hand in hand and have had a long and difficult developmental path throughout the country's history, first and foremost owing to the fact that land reforms have been perceived as a daunting and complex legal issue to be tested against Islamic-Sharia compliance. After the breakdown of the British colonial rule in the subcontinent and the creation of the three independent states India, Pakistan and Bangladesh (West Pakistan until 1971), first steps were undertaken to distribute land to the people. Land reform in this context refers to the abolition of large land holdings in order to prevent land concentration in a few hands. The so generated land is then distributed to cultivators and tenant-cultivators (GAZDAR, 2007).

In 1947, after independence land in Pakistan was predominantly allocated to a few, creating the landlord system. Generally, a land ceiling did not exist, if so then the ceiling was set much too high, permitting the ownership of large land parcels by a selected elite. This left the majority of peasants landless, drawing them into poverty and misery. Many studies have revealed the positive relationship between landlessness and poverty, which is also an indicator of social acceptance and standing, thus land ownership defines the class somebody belongs to, be it superior or inferior. This situation of a society utterly being stratified into classes hinders the progress of the country and allocates power, especially political say into a few hands, as has been the case in Pakistan. Being an agrarian country, feudalism has also significantly shaped the country's political development. When after the breakdown of Pakistan into West Pakistan and Bangladesh in 1971, 60-70% of the Pakistani parliamentary seats were occupied by feudal landlords, the eastern brother Bangladesh did not have one Member of Parliament from a feudal background. This development is crucial to understand the existing land related issues in Pakistan.

By the time of independence in 1947, three major and one minor type of land tenancy or so called land revenue systems simultaneously existed in South Asia. The major revenue settlements in British India included the *zamindari*, *mahalwari* and *ryotwari* systems. Occasionally, the *jagirdari* system could also be found, which was more or less a special system not referred to as revenue settlement, as the jagirdars did not have to pay land revenue tax. The zamindari system was headed by the zamindar, who managed land claims and was also responsible for the collection of the land revenue tax or payments for the state. This empowered him with special discretionary rights over the tenants. Zamindars (and Jagirdars) were both intermediaries between sate and cultivator. In the territory today known as Pakistan and other major parts of the north-western subcontinent, the mahalwari system dominated. The entire village or so called mahalla owned the land and thus was also collectively responsible for paying the land dues to the state. Although being a collective system, the mahalwari approach allowed individual claims which were recorded in the mahalla register. The idea was to collectively share the costs and benefits from the usage of land. Anyone that enjoyed ownership rights had also responsibility to bear the costs. Occasional cultivators and tenants did not receive ownership rights, but were also freed from the responsibility towards land revenue payments. In the southern and western parts of British India, the ryotwari system was most prevalent. In contrast to the mahalla system, in the ryotwari system the individual was granted private property right to land, although the true ownership remained with the state. Jageerdars were influential people that gained access to land without restrictions and the obligation to pay a land tax. However, jageerdaris were less common in British India.

After the colonial period the mahalwari and ryotwari systems were inherited by Pakistan, with ryotwari being most prevalent in the southern parts of the country, whereas the north and north-east were dominated by the mahalwari system. After the colonial period in 1947, several steps were undertaken by the governments of the newly independent states to ensure a more balanced land distribution. The major reform interests were related to the abolishment of the intermediary system (Zamindari and Jagirdari), transformation of tenancy rights and the introduction of land ceiling thresholds. As West-Pakistan did not inherit a large number of the intermediary systems, the initial reforms targeting the zamindars and jagirdars were primarily targeted at India and Bangladesh (East Pakistan until 1971). Large portions of land were taken from the powerful landlords and transferred to the cultivators and tenant-cultivators. As far as the reforms of tenancy rights were concerned, several loopholes existed and thus the zamindars and jagirdars also made use of their right to keep land for self-cultivation, by this they prevented the transfer of the entire land to the tenants or cultivators. The issue with landlessness and unequal right to land persisted in the South Asian countries, as the laborers were not considered as lawful claimants for land, the benefits largely were vested with the tenant-cultivators and cultivators. One group of powerful owners was once again replaced by a somewhat weaker new group of owners, still upholding a strong stratification of inferior (laborers) and superior (land owners-cultivators and tenant-cultivators). In a final post-colonial reform step the idea was to put a maximum threshold on the land size to be owned, also referred to as land ceiling. Through this individual owners were not allowed to possess land with acreage above the ceiling. However, loopholes made it easy for the landholders to keep their land, with total acreage in the excess of the ceiling. This was achieved by portioning the land into smaller parcels and distributing it officially or unofficially to family members. To this effect, the reform efforts although achieving small improvements, could not succeed as they had been designed to. In Pakistan for instance approximately 10% of the cultivated land could be generated for redistribution through the land ceiling reform, accordingly the number of beneficiaries remained low (GAZDAR, 2007; GAZDAR AND QUAN, 2004).

After its independence in 1947, having recognized landownership concentration as a serious problem, different governments (military and civil) have respectively introduced land reforms in 1958, 1972, and 1977 (NIAZI, 2003). In total the major outcome of the land-reforms has been the expropriation of approximately 2 million acres, out of which close to 80% to date have been redistributed to almost 300.000 beneficiaries. The first land-reforms were introduced in 1959 under the military rule of general Mohammad Ayyub Khan. More than half of the expropriation and redistribution of land was accomplished under the 1959 land-reforms (KHAN, 2000). The 1959 reforms included the introduction of a land ceiling threshold on private ownership. The ceilings of

500 acres for irrigated and 1000 acres of un-irrigated land were introduced on individual holdings. The problem of land concentration prevailed, as family holding were not ceiled. Moreover, land productivity was also one crucial factor in implementing the ceiling. As the productivity standards of 1947 were taken into consideration to assess 1959 land productivity, large landholdings were given another instrument to circumvent the ceiling and keep most of their allotted area. Thus more than 50% of the area released by the landlords was uncultivated area, which needed substantial amounts of investments before it could be cultivated. The reforms in general had failed, as even after the redistribution the landlord on average possessed land in the range of 5000-7000 acres (KHAN, 1981; HUSSAIN, 2002). In 1972 the second land reform efforts were undertaken in Pakistan, lowering the ceiling on irrigated land to 150 acres and 300 for un-irrigated land. Effectively however, this ceiling was put much higher. The land productivity was assessed once again based on 1940 standards, clearly understating the achievements made since independence regarding yield, cropping patterns and cropping intensities. To this effect, the actual ceiling including the obsolete productivity standards amounted to more than 400 acres for irrigated land (KHAN, 1981). The 1972 reforms lead to a redistribution of only 0.6 million acres, without breaking the persisting power of larger landholdings. The third and last reform efforts were introduced in 1977, reducing the land ceiling to 100 acres for irrigated and 200 acres for un-irrigated land. Additionally an agricultural tax for larger holdings was introduced, smallholders were excluded. Only 13000 people benefited from land distribution in this third reform effort. As a novelty, religious holdings the so called waqf were exempted, causing wide ranging protests against land reforms in general. This issue was further aggravated by the military government of Zulfiqar Ali Bhutto, who introduced Islamic law into the discussion upon request of the religious leaders. By this the religious leaders and owners of land challenged the concepts of land reform and land ceiling in the court of sharia compliance, winning their case. The Sharia Appellate Bench of the Supreme Court declared the imposition of land ceilings as contrary to the injuctions of Islam. Although since then local unrest in the landless community is frequent, the issue has been put to rest; too many facets and angles have to be studied to find a solution, which the government as for now is not ready to undertake.

Farm structure and types of land tenure

Especially in developing countries the possession of agricultural land is not only an economic but also socio-cultural asset with great value for the owner. From a structural viewpoint, many small, owner-cultivated land parcels dominate Pakistan's farming sector. From independence onwards, several tenurial changes have occurred, foremost the shift from tenant cultivation towards owner cultivation. Several studies have argued that the owners had realized that with their influence and contacts, a healthy access to input markets could be somewhat facilitated. Moreover, the high almost unbearable costs for the tenants relative to the output the land generated were one factor explaining this shift. Another factor of course was the fact that through technological improvements farming was made more lucrative and access to this specific factor remained a privilege of the large holdings, thus the improvements in productivity made the owners even stronger and they started reclaiming their land and started cultivating it (NIAZI, 1997; KUHNEN, 1996).

According to the latest Agricultural Census of the country in 2000, the total number of privately operated farms in Pakistan is estimated at 6.62 million covering an area of 20.41 million hectares. The average farm size in this setting amounts to 3.08 ha. As far as the country's provinces are concerned, the Punjab province is home to 59% of the total farms in the country with a total area of 11.23 million ha. The average farm size in the Punjab province resembles the country's average of approximately 3 ha. The second highest share of farms is found in the NWFP with 20%. These 1.36 million farms occupy close to 2.3 million ha of the country's total cultivated area. However, the average farm size in the NWFP with 1.7 ha is considerably below the country's average. The largest farms are found in the Balochistan province with average farm sizes around 7.9 ha, clearly surpassing the country average. Regarding the number of farms however, Balochistan is home to only 5% of the total farms in the country, occupying an area 2.6 million ha. Table 2.9 provides an overview on the number of farms and the farm area by province. The majority of the farms (86%) are under 5 ha covering 38% of the cultivated area, whereas 7.8% of the farms are medium sized and account for 16% of the cultivated area. Most characteristically for Pakistan the large farms with a share of 6% in total farms account for a cultivated area larger than 10.8 million ha (45%) (USAID, 2010). In total, close to 60% of the total cultivated area is in holdings of more than 5 ha (NIAZI, 2003). Land tenure can be defined as a multi-dimensional process or relationship among people with a legal or customary basis that determines the usage of land by specifying the user, the subject of use, the time frame of use and the relevant conditions of use.

Table 2.9 Number of Farms and Farm Area-by Province

No. in million & area in million acres

Administrative Units	Farms		Farm Area		Avg. size of Farm Area
	Number	%	Hectares	%	
Pakistan	6.62	100	50.43	100	7.6
Punjab	3.86	59	27.76	55	7.2
Sindh	1.07	16	10.69	21	10
N.W.F.P.	1.36	20	5.59	11	4.1
Balochistan	0.33	5	6.39	13	19.4

Source: Pakistan Agricultural Census (AC, 2000)

Regarding the allocation of property rights, land tenure can be further classified into different categories such as private (i.e. agricultural or residential holding), communal, open access (i.e. common grazing) and state (i.e. ownership of forests) (COX, 2002). In the previous chapter several traditional land tenure systems were addressed. However, these rather outdated land tenurial systems such as zamindari, ryotwari and mahalwari are not of great significance for explaining the current land tenure situation in the country. According to the Pakistan Agricultural Census three tenurial types can be distinguished, namely owned, owner-cum-tenant and tenant farms. As summarized in Table 2.10, 78% of the total privately managed farms are cultivated by the owners themselves, whereas 0.56 million (8%) are cultivated in an owner-cum-tenant relationship and the remaining percentage by tenants. The owner cultivated farms account for the majority of the area with a share of 73% (36.97 million acres), followed with 15% for owner-cum-tenant operated and 12% for tenant cultivated farms. Except for the Punjab, where the tenancy types owner-cum-tenant and tenant have an equal share, all other provinces are clearly characterized by a dominance of either owner or tenant cultivation, without much emphasis on the mixed type of land tenurial system.

As the census does not explicitly provide detailed information on the specific patterns and actual managing units of the different tenurial systems HAIDER AND KUHNEN (1974) have further classified land tenure in Pakistan into seven categories. These categories include landlord, small landlord, family-owner-cultivator, marginal-owner-cultivator, tenant of better standing, tenant at-will and landless laborer.

Table 2.10 Tenure Classification of Farms and Farm Area-by Province

No. in million & area in million acres

Province	Number of Farms				Farms Area			
	Total	Owner	Owner-cum-tenant	Tenant	Total	Owner	Owner-cum-tenant	Tenant
Pakistan	6.62	5.13	0.56	0.93	50.4	36.97	7.32	6.13
Punjab	3.86	3.04	0.42	0.4	27.8	19.25	5.4	3.11
Sindh	1.07	0.7	0.04	0.32	10.7	8.11	0.74	1.84
N.W.F.P.	1.36	1.13	0.08	0.15	5.59	4.24	0.81	0.54
Balochistan	0.33	0.27	0.01	0.05	6.39	5.37	0.38	0.64

Source: Pakistan Agricultural Census (AC, 2000)

The ownership of land is highly stratified between these groups, thus landlords on average own more than 150 acres of irrigated land, whereas the other above mentioned tenurial categories on average own 25-150 acres, 7.5-25 acres, less than 7.5 acres, more than 12.5 acres, less than 12.5 acres and no land at all, respectively. The upcoming chapter is devoted to the understanding of the country specific agricultural policy.

2.5 Domestic Agricultural Policy

The ultimate goal of Pakistan's agricultural policy is aimed at maintaining a high growth rate for the sector, with the intention to cope with the growing population pressure (FAS USDA, 2009). Currently the country is home to more than 170 million people. Especially to this background agricultural policy plays a key role in determining the future development of livelihoods in Pakistan. In the recent past however, this specific area of government policies has been neglected. Trade and macroeconomic policies can have decisive indirect effects on agriculture, especially as many of major agricultural commodities are traded on an international scale. For a country like Pakistan with an overwhelming importance of the agricultural sector, agricultural policy design is of particular importance. Historically, the focus of agricultural policy has been threefold, to control the trade deficit, to ensure the supply of life sustaining and important goods and to direct investments into the preferred sectors (KEMAL, 1988).

The following historical trade policy review is mainly based on previous work conducted by DOROSH AND VALDÉS (1990). As far as the specific objectives, instruments and outcomes of agricultural trade policy throughout the years prior to the WTO accession are concerned, five different periods can be distinguished. The years between 1952 and 1959 can be classified as a period of increasing government controls. After independence in 1947 Pakistan did not devaluate its currency, but was still able to pursue a liberal trade policy, surged by the high demand from war struck Korea. Until 1952 the country had issued import licenses under the so called Open General License Agreement, allowing for imports of any product. The end of the Korean War marked the end of this initial phase of liberal agricultural policy, especially for cotton and jute which had been mainly demanded. After the demand slump, the policy of import controls was introduced, using import licenses for specific goods in the consumer goods category. Through this initiative the supply of important consumer goods and intermediate goods could not be satisfied, corruption lured and a black market for the import licenses flourished. Cotton and jute were exported, however bound to export duties. Most other agricultural products were banned from being exported, decreasing the real export value by almost 70% in the 1959 as compared to 1952.

The second period from 1959 to 1972 was marked with significant cuts in export taxes, which at the end of this period were completely abolished. Although the restriction of imports in this period also played a leading role on the policy agenda, export increases were mainly targeted. For increasing the exports a special scheme awarding value based bonuses for the exports of manufactured goods was introduced, in order to provide incentives for the producers. Depending on the stage of processing the bonus rates fluctuated between 20 and 40 % of the fare on board value. However, these bonuses were not released in cash; they were more or less transferred into vouchers, which could be used to purchase restricted items from the import commodity list. This bonus scheme led to a devaluation of the exchange rate for items purchased through the very same. The import license policy of the previous period prevailed, generating high revenues for the license holders, driven by a demand which clearly exceeded supply. As the import licenses were held by so called traditional importers under the OGL scheme, a concentration of wealth and trade benefit became most evident, as a remedy another OGL scheme was introduced in order to issue more licenses for new importers. This however only benefitted a few, the monopolistic import structure prevailed. In 1964 another remedy against the concentration of trade benefits was the introduction of a free list of goods, the import of which was allowed without the possession of an import license. Just one year later, the Indo-Pak war led to a drastic reduction of commodities on this free list. This second period was thus characterized by a greater reliance on market mechanisms and the availability of foreign cash aid. These developments led to a decrease in trade protectionism.

In the third period from 1972 to 1976 a major devaluation of the rupee against the dollar took place, from 4 Rs. for 1 $US to 11 Rs. for 1 $US. A devaluation of the dollar in the same period led to a fixed rate of 9.9Rs. for 1$ in 1973. As one major change the bonus scheme was abolished. With the end of the bonus scheme the import policies were also rectified, with the introduction of two separate lists, namely free list and tied list. The free list did not put any restrictions on the country's the imports generated from, whereas the tied list included commodities whose import was tied to a specific trading partner or country. Despite several changes in the import policies, the restrictions on the import of consumer goods remained, executed through high taxing or complete bans. The only export subsidies that remained were export tax rebates and financing, all others were withdrawn. Exports were first taxed through the collection of duties, which at the end of the period were also withdrawn.

The introduction of export subsidies and the so called Raw Material Replenishment Scheme (RMR) characterized the fourth historical trade policy period from 1978 to 1982. As domestic inflation surpassed the global levels, the rupee appreciated. Trade balance management was executed by the introduction of export subsidies through compensatory rebates on the fare on board value rather than the devaluation of the rupee. To foster exports, access to a number of raw materials that were previously banned was granted in the framework of the RMR scheme. The fifth period from 1982 to 1988 was marked by an exchange rate policy change. The rupee was no longer linked to the dollar and a gradual depreciation of the rupee was initiated. One outcome of this development was the increased liberalization of imports that was manifested in lifting the bans on the majority of imported goods. As most bans were lifted the free import list was converted into the negative import list with only some commodities on it. At the end 1989 the negative list was further decimated. Additionally a tied imports list was kept, that regulated the import from specific countries. Different studies have estimated a decline in the average rates of protection from approximately 150% in 1963 to 66% in 1981, whereas in the same time period an increase in the protection for capital goods has been observed (NAQVI AND KEMAL, 1983). Historically, the country has run a policy to restrict imports, promote exports and increase revenues. Clearly the restriction of imports has played a far more important role. In the later phases (1980's) an orientation towards export promotion has been observed. Not all industries have benefitted from the export subsidies as most of them have been allocated to traditional industries with strong lobbies, causing a concentrated growth effect of trade, especially in the textile industry. Before the country's accession to the World Trade Organization (WTO) in 1995, protection has been enacted through tariffs, quotas and bans.

On the other hand exports have been promoted through subsidies. The agricultural sector in this context has been taxed more than it has been promoted. Especially the exports of cotton and rice have been heavily taxed, while wheat has been at times completely banned from being exported. The foregone overview has revealed that from a historical perspective the agricultural sector as compared to the manufacturing sector has been a neglected area of government policies. However, with the accession to the WTO in January 1995 and the associated requirements and challenges, the country has clearly understood the need to focus on the agricultural sector. The main focus as per agricultural policy is concerned, is laid on sustaining food security, increasing productivity, commercial agriculture, income diversification and export orientation, also to the background of the challenges arising from a WTO membership (BOI, 2007). These goals of making agriculture globally more competitive and to strengthen the sector in the domestic circuit are also pursued by the Medium Term Development Framework (MTDF) of Pakistan, bringing planners and executers under one umbrella.

Within the WTO Pakistan is an active member of different negotiating country-alliances such as the G20, G33 and Cairns Group. Within the G20 Pakistan strongly advocates an offensive stance towards the three pillars of the Agreement on Agriculture (AoA). In the G20 Pakistan is especially negotiating on the reform of the Green Box criteria, as due to the current definition too many loopholes exist, that allow for a transfer of subsidies into the green box, thus circumventing the true idea of the AoA. As a regular member of the G33 group Pakistan also pursues a rather defensive stance, demanding special safeguard mechanisms for developing countries against cheap imports. Since 2005 however, Pakistan has joined the Cairns Group, which is aggressively pursuing the liberalization of trade, asking for the complete elimination of export subsidies by 2013 and the improvement of market access amongst others.

Pakistan's latest trade policy review has been completed in January 2008. However, even before the completion of the Uruguay Round, Pakistan had liberalized its trade regime, which was mainly driven by the structural adjustment programme linked to the IMF and World Bank conditionalities. The non-tariff-barriers had been completely abolished and tariff bindings were used. According to the latest data from the WTO, Pakistan today has almost completely bound all its agricultural tariff lines, whereas the overall tariff binding coverage amounts to 98.7%. On an average agricultural tariffs are bound at 95.6%, whereas the actually applied tariffs in 2010 averaged only 17%. In comparison the tariffs for non-agricultural goods were on average bound at

54.6%, whereas in 2010 only 13.4% were actually applied. For all imported goods the bound tariff measured 59.9%, whereas only 13.9% where applied in 2010. In 2009 34.8% of the agricultural imports under the AoA were duty free, whereas 36.1% in non-agricultural goods were free of import duty. Over the time period from 2003 to 2005 the share of the collected import duties in total tax revenue amounted to 15.8%. No tariff quotas and special safeguards were used. The only agricultural tariffs not completely bound at 100% are for a few products or product groups including animal products, beverages and tobacco and agriculture related transport equipment. On average the highest tariffs are applied for dairy products and beverages & tobacco, with values of 30%, 52.5%, respectively. Foremost animal products, FVP (fruit, vegetable and plants), cotton and oilseeds were not taxed, with duty free shares of 45.3%, 73.1%, 99.8% and 20.3%, respectively. According to the latest Trade Policy Review (TPR, 2008) self sufficiency based food security has been declared a government priority. The food security policy is primarily targeted at the food grains wheat, rice and maize; additionally edible oils and sugar are also included. Domestic support measures are directed towards low-value crops such as sugarcane, with the result that domestic prices exceed world levels by up to 60%. Direct and indirect subsidies are used in the form of on-freight and income tax concessions for certain agricultural exports. The state owned Trading Corporation of Pakistan is engaged in the trade of the commodities wheat, sugar and cotton. The ongoing reforms since 2001 in the agricultural sector are foremost directed towards an increasing involvement of the private sector in marketing, input supply, wheat procurement, storage facility construction, establishing cooling facilities and the maintenance of international quality standards. Credit disbursement by banks to small farmers has been another focal point of the current reforms.

Additionally, a scheme to increase the use of certified seeds has been discussed. The introduction of internationally accepted quality controls and management according to international standards is another government reform priority. In total 41 commodities are already handled according to international quality standards. A scheme named ‘Water Vision 2025’with a total investment volume of 36 billion Rs. has been initiated to boost the water sector and improve the irrigation infrastructure. The trade instruments can be generally distinguished into border measures and domestic measures. The border measures include nominal tariff protection, non-tariff measures such as import authorization and general bans, export measures such as bans (e.g. wheat) and registered export contracts and other measures such as the involvement of state owned enterprises in the foreign trade of commodities such as cotton, wheat and sugar. The domestic support comprises market price support, subsidies and corporate agricultural farming. Pakistan's average level of nominal protection has decreased. Especially the nominal protection for agriculture has shown a significant decline, with current tariffs remaining below the level for manufactured goods (Figure 2.16). Most of the country's domestic support has been focused on the "Green Box" measures of the AoA and with 1% represents only a small fraction of government expenditure. Foremost wheat, cotton, rice and sugarcane are market price support targets, after the latest WTO review however market price support has been more or less limited to wheat and cotton. Mainly based on the production-costs the Agricultural Prices Commission recommends commodity prices. For wheat and rice the domestic price is generally lower than import parity or respectively exports parity, whereas for cotton it almost equals export parity.

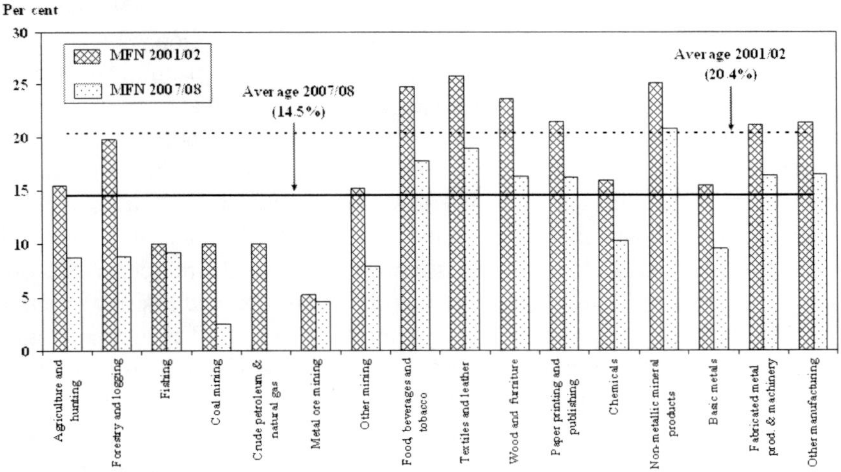

Note: Excluding specific rates.

Source: WTO Secretariat calculations, based on data provided by the authorities of Pakistan; and Pakistan Customs online information. Viewed at: http://www.cbr.gov.pk/newcu/TARIFF/Tariff-2007-2008.asp.

Figure 2.16 Tariff averages 2001/02 & 2007/08

In the recent past 2007/08 the domestic sugar price has exceeded import parity by almost 10%, somewhat assisting farmers (CBR, 2004). The government pays fertilizer subsidies to farmers to the tune of maximum 250 Rs. per 50 kg bag of Phosphorus fertilizers. In order to keep the domestically produced fertilizer prices below import parity an indirect natural gas subsidy is paid to the producers (DOROSH AND SALAM, 2007). Moreover, most of the canal irrigation in the Punjab province is subsidized. The main subsectors of the agricultural sector include crops, livestock and fisheries. The major policy relevant crops comprise wheat, cotton, tobacco, sugarcane and edible oils. As wheat is concerned, government influence is still enacted through price support and directing subsidies towards mills. Generally the dual objective of the policy directed towards the wheat crop is to achieve food security and also guaranteeing minimum producer prices in form of a price band. Within the price band producers can move freely and are also free to make a choice to sell to the government or privately. The new wheat policy, which is being implemented progressively since the latest trade review in 2008, is directed towards decreasing the government influence in the wheat market, thus i.e. already in 2006 the transport restrictions on inter-provincial trade had been lifted (DOROSH AND SALAM, 2006). For cotton the main domestic policy goal is to ensure high quality standards, therefore the mills have to provide a premium to producers for high quality cotton, with standards being set by the country's cotton standards institute. In the case of tobacco the intervention is limited to the negotiation of pre-sowing contracts between farmers and manufacturers which are based on minimum prices fixed by the tobacco board that in turn is paid for its service by a fee that is levied on the production of tobacco. Sugar cane is indirectly subsidized through rebates, however in turn the mills are forced to export certain quantities, if they fail to export mandatory quantities a fine is levied on them. Sugar imports are subject to a tariff of 15% on the value. The prices for edible oils from canola, sunflower and palm are set on the open market. A board promotes oilseed production in Pakistan and assists farmers on a non-profit basis with its small sized oil processing plants. The oilseed-board is also financed through a cess coming from the imports of edible oils. As far as livestock is concerned, the government exercises support through financing the work of the recently (2007) established Dairy Development Board and Pakistan Dairy Development Company "Dairy Pakistan". Both pursue goals related to promoting important livestock sectors, if required through financial assistance. The agricultural policy of the country for the livestock sector is primarily aimed at increasing the involvement of the private sector and a higher orientation towards markets. For the fishing subsector strict licensing schemes are in place to prevent overfishing, whereas laws on protecting turtles have also been implemented. Pakistan has not utilized much domestic support, as its total aggregate measure of support reveals (Table 2.11).

As tables 2.11 and 2.12 reveal the country's AMS has always been negative. Having a look at the recent AMS structure, there seems to be enough room for Pakistan to allocate more resources to the agricultural sector, which is of a pivotal importance to the country. By first achieving positive protection, Pakistan can further utilize the *de minimus* provisions under the AoA. Especially the Green Box can provide a strong area to concentrate on in this regard, as no ceilings on support exist (BADAR et al., 2007).

Table 2.11 Aggregate Measure of Support for Pakistan 1986 -2004

Year	Product Specific AMS	Non Product Specific AMS	Total AMS
1986-88	-640	87.1	-552.9
1995-96	-203	10.8	-192.2
1996-97	-72	15.5	-56.5
1997-98	-143	22.5	-120.5
1998-99	-191.29	26.7	-164.59
1999-00	-257.37	12.6	-244.77
2000-01	-337.01	161.55	-175.46
2001-02	-218.43	0	-218.43
2002-03	-163.27	1.89	-161.38
2003-04	-80.9	58.7	-22.2

Source: BADAR et al. (2007)

Table 2.12 Aggregate Measure of Support for Pakistan 1986 -2004 by Commodity

Product	1986-88	1995-96	1996-97	1997-98	1998-99	1999-00	2000-01	2001-02	2002-03	2003-04
Wheat	-252	-172	-72.4	-143	-191	-257	-377	-218	-163	-80.9
Seed Cotton	-187	0	0	0	0	0	0	0	0	0
Rice (Basmati)	-117	-20.1	0	0	0	0	0	0	0	0
Rice (Coarse)	-48.6	-10.4	0	0	0	0	0	0	0	0
Sugarcane	24.2	0	0	0	0	0	0	0	0	0
Onions	-0.2	0	0	0	0	-0.1	0	0	0	0
Potatoes	-57.8	0	0	0	0	-0.05	0	0	0	0
Gram	-1.1	0	0	0	0	0	0	0	0	0
Soybean	-0.4	0	0	0	0	0	0	0	0	0
Sunflower	-0.1	0	0	0	0	-0.02	0	0	0	0
Safflower	0	0	0	0	0	0	0	0	0	0
Total	-640	-203	-72.4	-143	-191	-258	-377	-218	-163	-80.9

Source: BADAR et al. (2007)

The sector contributes significantly to the overall economy of the country, thus the upcoming chapter aims at unveiling these sector specific contributions with respect to GDP, Employment and Trade.

2.6 Agriculture in the National Economy

As in many if not all developing countries, the agricultural sector in Pakistan too plays a vital role for feeding its rapidly growing population. Ever since independence in 1947 the agricultural sector has substantially contributed to economic activity in the country, currently providing nearly 21% of the country's Gross Domestic Product (GDP), offering employment to 45 % of the labor force and bearing responsibility for almost 65% of the livelihoods of the population, directly or indirectly linked to the sector (ESP, 2011). A look at the structural evolvement of the sector reveals diverse periods of development, ranging from periods of rapid growth to such of stagnation and significant productivity declines. By using different indicators Figure 2.17 illustrates the decreasing importance of the agricultural sector as a whole.

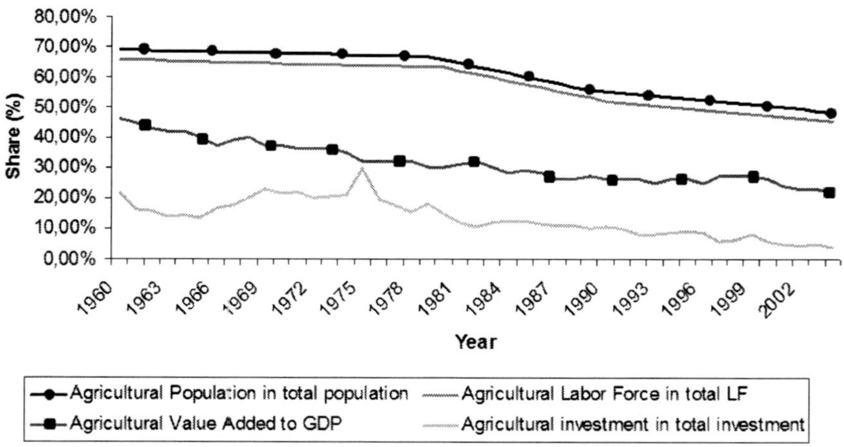

Figure 2.17 The importance of Agriculture- measured by key indicators 1960-2004
Source: own illustration, AHMED (2010)

As illustrated in Figure 2.17, in the past 20 years the agricultural labor force has constantly lost its share in total labor force. Relatively, investments in this sector have been on higher levels during the phase of the green revolution (1966-1976), thereafter a decrease is indicated from formerly 30% in 1976 to less than 10% in 2004, this does not mean that 20% less investment is taking place in the agricultural sector, it can be a shift from large scale government investments to more small scale private investments; moreover a "share" in this regard only is a relative measure. The share of agriculture in GDP has been decreasing since 1960, following the same pattern like investments and vice versa, one reason is that GDP growth in Pakistan from 1960-2004 has outperformed the growth rate of the agricultural sector, leaving the decrease to be a mere issue of ratios; not to forget that the growth of output has been mainly carried by agriculture, at least up to the end of the 1990's.

2.6.1 Agriculture and the GDP

More recent observations show, that over the course of the last decade the agricultural sector's contribution to the overall GDP has clearly declined, making room for the more competitive sectors such as Industry and Services, with shares of 27.1% and 52% in 2010-2011, respectively. The share of agriculture in GDP in 2010-2011 was recorded at 20.9%. In comparison, in 1999-2000 the agricultural share in GDP amounted to 25.9%. Although a decline in value added is observed, the relative importance of the sector for the country cannot be ignored, as it still represents the mainstay of the country's population. Based on the World Development Indicators (WDI, 2011), Figure 2.18 presents the sectoral contribution to GDP from 1990 to 2009.

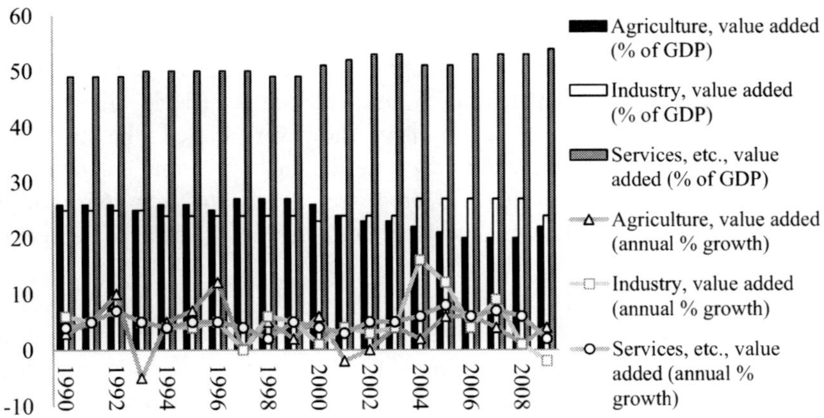

Figure 2.18 Sectoral contribution to GDP, 1990-2009
Source: own composition using WDI, 2011

Due to structural problems and natural calamities the agricultural sector has lost significant growth momentum as its growth slowed down to 2.7 % in the last decade as against 4.4 % in the 1990s and 5.4 % in the 1980s (ESP, 2011).

Natural calamities have especially hit the important crop sector of the country. An overview on the recent growth performance of the sector and its subsectors is provided in Table 2.13.

Table 2.13 Agricultural Performance, 2004-2011

in percent

Year	Agriculture	Major Crops	Minor Crops	Livestock	Fishery	Forestry
2004-05	6.5	17.7	1.5	2.3	0.6	-32.4
2005-06	6.3	-3.9	0.4	15.8	20.8	-1.1
2006-07	4.1	7.7	-1	2.8	15.4	-5.1
2007-08	1	-6.4	10.9	4.2	9.2	-13
2008-09	4	7.8	-1.2	3.1	2.3	-3
2009-10	0.6	-2.4	-7.8	4.3	1.4	2.2
2010-11	1.2	-4	4.8	3.7	1.9	-0.4

Source: ESP, 2011

The agricultural sector consists of the subsectors crops, livestock, fishing and forestry. Major crops include wheat, cotton, rice, sugarcane and maize, whereas the minor crops comprise potatoes, pulses, onions and chilies. The crop subsector is of paramount importance for the overall agricultural sector and approximately accounts for 37% of agriculture in Pakistan, influencing the growth performance of intertwined other subsectors. Rising incomes and changing patterns of consumption have however shifted the traditional importance of the crop sector in favor of the more vibrant and demanded livestock sector. In the 1990's the crop sector used to stem 65% of the agricultural sector, today it is the livestock sector, making up 56% of the sector.

2.6.2 Employment

As far as employment is concerned, no other sector is as important as the agricultural sector which employs close to 45% of the country's total labor force. However, compared to the period from 1990 to 1995 a gradual decline in line with the sectoral share in GDP can be observed. From 48.5% in 1990-1995, the agricultural share in total employment has declined to 43.2% in the period from 2002-2008, nevertheless maintaining a high labor absorption capacity. Table 2.14 presents the distribution of the total employment by economic activity throughout the period from 1990 to 2008.

Table 2.14 Distribution of the Total Employment by Economic Activity, 1990-2008

Economic Activity	1990-1995	1996-2001	2002-2008
Agriculture, Hunting, Forestry and Fishing	48.54%	47.09%	43.20%
Mining and Quarrying	0.14%	0.12%	0.09%
Manufacturing	11.39%	10.73%	13.63%
Electricity, Gas and Water	0.79%	0.76%	0.72%
Construction	6.67%	6.32%	6.12%
Wholesale and Retail Trade and Restaurants and Hotels	13.15%	13.96%	14.71%
Transport, Storage and Communication	5.20%	5.30%	5.68%
Financing, Insurance, Real Estate and Business Services	0.79%	0.85%	1.09%
Community, Social and Personal Services	13.24%	14.83%	14.73%
Activities not adequately defined	0.09%	0.03%	0.06%

Source: own composition using ILO, 2011

The rural sector in Pakistan is home to more than half of the country's population. Thus agriculture in these areas takes the vital role of job provision and subsistence. A relatively stable source of employment is pivotal for the livelihoods of the rural people. To this effect, the agricultural sector also adds value to the general economic development of Pakistan and through economic activity provides the means for the rural people to access basic life and health sustaining services (DOROSH AND VALDÉS, 1990). A somewhat disaggregated view on the employment capacity of the rural sector is presented in table 2.15.

Here the rural labor force is distinguished into the two categories rural agricultural and rural non-agricultural. The first two columns depict the different (sub-) sectoral contributions to employment and GDP. With a total share of 67.5% in total employment, the rural sector substantially absorbs labor in Pakistan. However, in comparison to the 35.1% share of the rural non-agricultural sector, the contribution of the rural agricultural sector to the country's employment amounts to 32.4%. The remaining 32.5% of the employed population are urban workers. Although the rural sector contributes the lion share to employment with 67.5%, the urban sector with its employment absorption capacity of 32.5% has a significantly higher share in the country's

GDP with 56%. Out of the 44% share of the rural sector in GDP, 20% are added by the rural agricultural sector. The third column depicts the elasticity of employment with respect to output growth. RAO (1975), in field studies, estimated a value of 0.3 for the agricultural sector, capturing the increased labor productivity associated with yield-increasing innovation. Through weighting both the informal and large-scale sectors, the urban employment elasticity of is calculated at 0.5. According to GAVIAN et al. (2002) the employment elasticity for the rural non-agricultural-sector at 0.9 captures the development that increased farmers' demand raises production (for details see USAID, 2009).

Table 2.15 Sectoral Employment and GDP Shares in Pakistan

Sector	Employment Share	GDP Share	Employment Elasticity
Rural Agricultural	32.4%	20%	0.3
Rural Non-Agricultural	35.1%	24%	0.9
Subtotal	67.5%	44%	-
Urban	32.5%	56%	0.5
Total	100%	100%	-

Source: USAID, 2009

The average agricultural value added per worker in Pakistan has increased from 800$ in 1994 to 900$ in 2009 (FAO, 2011).

2.6.3 Trade

According to the FAOSTAT (2011) agricultural trade in Pakistan is broadly composed of non food products, food products, cereals, dairy products plus eggs and meat. Food products take the largest share in agricultural imports, with values around 50%. These are followed by non food products with 25%. With a share of 24% cereals are also a very important import-commodity group. A very small share (2%) in agricultural imports is taken by meat and dairy products. Agricultural exports are dominated by cereals, occupying a share close to 67% in total agricultural exports. Approximately 10% of the exports are non food products, whereas 20% are food products and 3 % of the exports are contributed by the meat sector.

In terms of export value the top commodity is milled rice, generating 1.6 billion dollars from an exported quantity of 2.5 million tons in 2009. The other major contributors to agricultural exports were hydrogenated oil and refined sugar, with export values of 120 million dollars and 82 million dollars, respectively. In terms of export value, other important commodities include broken rice, cotton lint, crude materials, molasses, wheat flour, tangerines and cotton waste. As far as the export quantities are concerned, the ten most important commodities include milled rice, molasses, refined sugar, wheat flour, tangerines, broken rice, potatoes, hydrogenated oil, dates and mangoes. Generally, quantity wise the most noteworthy contributions to the agricultural imports are made by wheat, palm oil, cotton lint, rapeseed and cake of soybeans, with quantities of 1.8 million tons, 1.76 million tons, 89 million tons, 80 million tons and 21 million tons, respectively. The highest import values are recorded for palm oil, cotton lint, wheat, rapeseed and tea. The evolution of key agricultural trade indicators from 1961 to 2009 is presented in Figure 2.19.

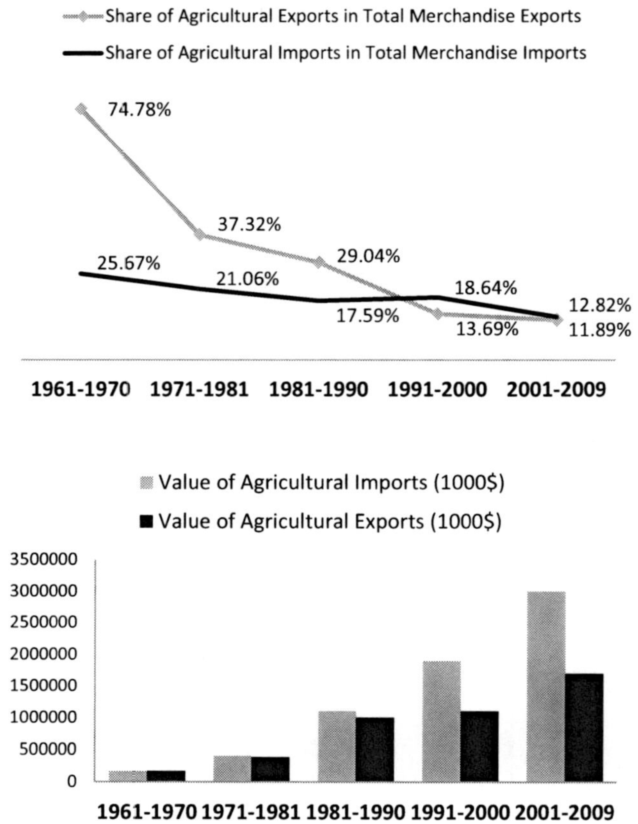

Figure 2.19 Evolution of Agricultural Trade in Pakistan and its importance 1961-2009
Source: own illustration using Data from FAOSTAT (2011)

The value of agricultural exports has increased from approximately 194 million dollars in the period 1961-1970 to 1.75 billion dollars in the latest considered period between 2001 and 2009. In the early years of the green revolution in Pakistan, the share of the agricultural exports in total merchandise exports amounted to vaguely 75%, whereas imports held a share of 25% in the total value of merchandise imports. Thus trade in Pakistan in the 1960's was highly agriculture-driven and focused. In the 1960's the country was a net-exporter of agricultural commodities, with an agricultural trade surplus of 32 million dollars. With other sectors such as industry and services starting to gain momentum in the 70's, the share of the agricultural exports in total merchandise export value dropped to 37%, although still contributing a considerable amount to foreign exchange earnings in the country with a total agricultural export value of 437 million dollars. In the same period in terms of share the agricultural imports almost remained constant at 21%, however with a total value of 447 million dollars which in comparison to the previous ten years had increased by 64%, indicating the growing pressure on domestic production through i.e. a very fast population growth that needed a healthy food supply. This period also marks the shift of the country from a net-agricultural-exporter to a net-importer, with an agricultural trade deficit in the respective period of 10 million dollars. Another decline in the relative importance of agricultural exports came in the period between 1981 and 1990, in agricultural exports accounted for a share of 29% of total merchandise exports. The share of agricultural imports in total merchandise trade amounted to 17.6% in the same period. The agricultural trade deficit in this period was 50 million dollars. In the 1990's, nothing changed significantly in terms of export value, which remained around 1 billion dollars, contributing 13.7% to the total merchandise exports. The agricultural imports' value however grew at a rapid pace (42% compared to the previous period). In the most recent considered period between 2001 and 2009, the agricultural trade deficit amounted to 125 million dollars; with an agricultural export value around 1.75 billion dollars and import value around 3 billion dollars. According to latest available estimates the sector contributed around 12% of the total merchandise export value and 13% of the total merchandise import value.

Food accounted for 18% of the agricultural exports and 10% of the imports, whereas non-food commodities contributed 72% and 90%, respectively. Major markets for exports were the United Arab Emirates, Afghanistan, Iran and the United States. The major import partners in 2009 were Malaysia, India, United States, Russian Federation, Indonesia and Canada (UN COMTRADE, 2011). The role of productivity in determining the sector's performance is subject of the following subchapter, with a particular focus on the technology related aspects.

2.7 Investment and Agricultural Credit

In the recent past the farming sector's requirements for inputs have increased, especially the demand for fertilizers, plant protection detergents and new seed varieties has gained significant momentum. Where on the one hand the demand for important inputs has increased, their prices have also skyrocketed. As the previous chapters have revealed, the relative importance of agriculture in GDP has declined in Pakistan. Nonetheless the importance is still vital as the majority of the population is employed in the sector and depends on it for the sake of their livelihoods. In Pakistan the sector's neglect in terms of policies has also led to an outflow of resources previously directed towards the sector's development.

Savings in the small farming sector, that forms the backbone of the total agricultural sector, are hard to accumulate. People in the rural sector have to feed large families and also finance health support when needed. Savings in such circumstances are very rare, thus the only way these small scale farmers can finance their farming operations is via credit. The country's credit system can be classified into traditional informal and formal. The traditional system includes relatives, friends, traders and rotational credit unions. The rotational credit unions are formed by people who intend to lend money to each other and also people who need money that they cannot raise themselves. A needy person who cannot arrange a larger sum e.g. to pay for a daughter's wedding or for the purchase of an agricultural asset pays a monthly sum into the rotational credit pot, organized by one person in the union. The amount of the credit sum is fixed at the beginning. Thus the first person, who is collectively assessed as the neediest gets the full amount at the beginning of the rotation. Once the rotation is completed, everyone will have received the amount once, this regardless of inflational fluctuations (for a detailed overview on rotational credit see Bardhan, 1989). As informal sources are non-institutional in nature, the time taken by them to issue the loan or credit and the involved procedures are rather short, however this ease of access translates into rather high interest rates (HUSSAIN AND THAPA, 2012). Although rural households and farmers can easily access informal credit systems, not always the required cash volume is available, as the resources of the informal lenders are mostly much smaller than those of the formal credit lenders such as the ZTBL (Agricultural Development Bank) and other commercial banks. Moreover NGO's are also involved in providing credit to rural farmers (IQBAL et al., 2003).

In general the development of the rural credit market in Pakistan has never taken off, which can be linked to several reasons such as the acceptance of land as the only collateral and the overwhelming power of landlords who in the development of the rural small scale sector see a circumcision of their own powers. Understandably, the transaction costs for credit agencies in the urban areas are clearly lower, thus the formal credit institutions have traditionally preferred the urban over the rural regions. On top of these issues a limited understanding of agricultural requirements has led to a shift of the attention of formal credit institutions and thus private sector investments towards policy promoted other sectors such as real estate and industry.

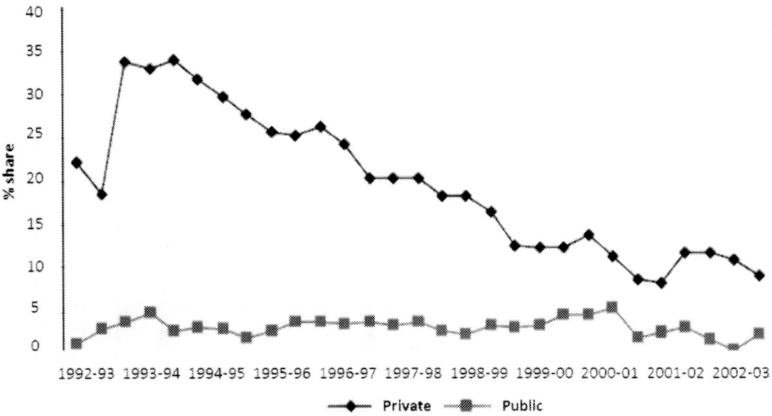

Figure 2.20 Structure of Investments in the Agricultural Sector of Pakistan 1992-2003
Source: MALIK AND RANDHAWA (2006)

As Figure 2.20 shows the share of the public sector investment in agriculture is on a very low level, additionally it has been declining in the last decade. Thus infrastructure has been neglected and the private sector has also pulled out its financial resources.

The largest credit institution of the country is the ZTBL that besides receiving funds from the World Bank and Asian Development Bank, predominantly borrows from the State Bank of Pakistan and thus contributes the lion share of the total credit disbursed to the agricultural sector of the country. In contrast the commercial banks as the other credit institutions, more or less are dependent on their own funds. From 1987-2005 on an average the ZTBL has contributed 55% of the total credit to the sector (Figure 2.21).

Figure 2.21 shows that in the period from 2005-2011 the ZTBL's share in total disbursed credits has decreased, slightly indicating a paradigm shift towards a greater attention from the commercial banks and NGOs towards funding agricultural production and development in the country. This increased attention has also been promoted by the agricultural policy after the WTO accession and surely the input price hikes in the recent past. The paradigm shift is manifested in an agricultural policy that requires banks to fulfill a lending target, with fines as an instrument of enforcement. Being the country's major agricultural regions, the Punjab and Sindh provinces have both received the largest share of the total credit disbursements.

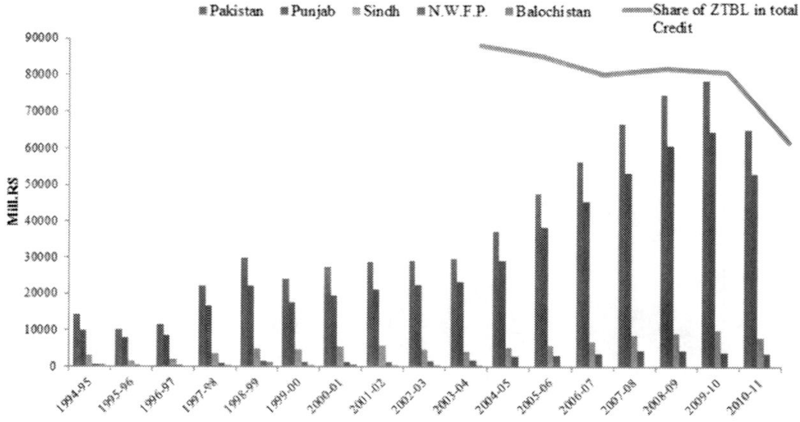

Figure 2.21 Agricultural credit advanced by Zarai Taraqiati Bank Limited (ZTBL)
Source: own illustration based on Agricultural Statistics of Pakistan (AC, 2011)

IQBAL et al. (2003) report that the institutional credit as the percentage of agricultural GDP, after initially growing from 1971-1986, at the end of the 80's up to the mid 90's constituted a decreasing and smaller portion of the agricultural GDP, before growing again from 1997 onwards (compare also Figure 2.20).

Table 2.16 Growth of Institutional Agricultural Credit in Pakistan 1971-2002

Period	Compound Growth Rate (Percent)	
	Nominal Credit	Real Credit
1971-72 to 1975-76	86.48	57.41
1975-76 to 1980-81	20.91	10.93
1980-81 to 1985-86	25.64	18.14
1985-86 to 1990-91	3.64	-4.43
1990-91 to 1995-96	7.22	-3.07
1995-96 to 2001-02	15.87	8.84

Source: IQBAL et al. (2003)

Many small scale farmers and landless laborers are also targeted by the ZTBL in special credit programmes (AC, 2011). Table 2.17 covers the period from 1987 to 2011 and presents an item-wise break-up of the loans disbursed by ZTBL. On a country-wide scale the majority of loans have been paid to the farmers to finance fertilizer and farm mechanization. Thus, 31% of the total loans have been used for buying fertilizer, whereas 19.5% have been acquired to buy tractors and other machinery.

Table 2.17 ZTBL loans disbursed by purpose (averages 1987-2011)

Province	Seed	Fertilizer	Pesticides	Dairy Farming	Poultry Farming	Fisheries	Tubewells	Tractors	Others	Total in Rs. Million
Pakistan	15.3%	31.0%	13.8%	5.3%	1.0%	0.2%	4.0%	19.5%	9.9%	17917
Punjab	16.0%	30.9%	14.9%	4.3%	0.8%	0.0%	2.6%	23.1%	7.4%	13258
Sindh	14.3%	36.1%	14.0%	5.7%	1.2%	0.7%	4.6%	7.5%	15.9%	3236
N.W.F.P.	14.7%	21.9%	4.1%	19.2%	3.0%	0.0%	3.7%	13.1%	20.2%	986
Balochistan	1.8%	17.5%	0.9%	1.7%	0.5%	2.7%	42.9%	14.7%	17.3%	436

Source: Agricultural Statistics of Pakistan (AC, 2011)

In the Balochistan province the majority of loans (42.9%) have been passed by the ZTBL to finance tubewells. In the N.W.F.P. dairy farming has been another major field of investment with 19.2% of the province's loan flowing into the dairy sector. The Punjab and Sindh provinces have resembled the country's pattern of an emphasis on fertilizer, pesticides and machinery.

2.8 Total Factor Productivity

There is a general consensus on the definition of productivity; however there is no single measure of productivity (NAKAMURA AND DIEWERT, 2002). Economists agree that productivity is a measure of the goods and services produced per unit of input. The most comprehensive measure of aggregate or sectoral productivity is Total Factor Productivity (TFP). TFP measures the ability to combine the factors (labor and capital) more effectively over time. This can be due to changes in qualities (more appropriate skills or embedded technologies), better methods of organization and acquisition of efficient management techniques and know-how, enhancements in worker's education and expertise, gains from specialization, introduction of new technology, innovation or upgrading of existing technology and enhancement in Information Technology (APO, 2008). Additionally, TFP considers multiple inputs at a time and is often measured residually, as that change in output that cannot be accounted for by the change in combined inputs or in other words, is the portion of output not explained by the amount of inputs used in production (OECD, 2001). Mostly TFP is used to assess the contribution of technology in growth.

The TFP calculations by AHMED (2010) reveal an average annual output growth of 3,9% in the agricultural sector over the last four decades, which has exceeded the aggregate population growth rate of about 2.7%. This rate of growth in agriculture has been sustained by the technological progress embodied in the high-yielding varieties of grains and cotton with supporting public investment in irrigation, agricultural research and extension (R&E), and physical infrastructure. Agricultural growth, in turn, has made significant contribution to the overall economic growth of 5.4% per year during this period. Despite rising per capita income, food demand is likely to grow rapidly given the low level of current per capita income. Recent projections for future food supply and demand, call for sustained efforts for increasing production of essential items (wheat, edible oils, etc.) (CHAUDHRY AND CHAUDHRY, 1997). Faced with limits to further expand the cultivated land and slowing returns to further input intensification, productivity growth assumes a central role in meeting the challenges of the future, and moreover as shown in Table 2.16, in agriculture TFP explains a substantial part of growth. The periods of high/low agricultural growth have generally coincided with periods of robust/poor performance of the national economy. Over the entire period Pakistan's agricultural TFP has grown with an annual average of 1.8%, very similar values were computed by other studies for a shorter time period (EVENSON et al., 2004). The calculations by AHMED (2010), found a share of 44.8% in growth that is explained by TFP. These values computed for the Agricultural sector are very similar to the aggregate values for the total economy, this clearly shows that although decreasing in importance in relative terms, in absolute terms throughout the entire period at study (1960-2004) Pakistan's Agriculture has driven the aggregate economy. Of course, with increasing amount of per capita income, increasing levels of education, consequently rising living standards a general shift from the agricultural society to the industrialized society will most probably follow, not to forget the service sector, which is increasingly becoming important in Pakistan. So the last decade slowly has been indicating this "decreasing" trend, however as poverty prevails in Pakistan and hunger is prevalent over large parts, the role of the agricultural sector will be a detrimental one. Table 2.18 summarizes the relative contributions of capital, labor and TFP (technology proxy) to GDP growth in the agricultural sector. TFP growth has significantly contributed to the overall agricultural GDP growth in the country. Technology has played a key role in determining the direction of growth in the sector. To tackle expected future shortages in food supply, resulting from adverse impacts from environmental change, technological solutions can be expected to play a leading role.

Table 2.18 Productivity and GDP growth in Agriculture 1960 – 2004

Period	GDP growth (%)	Minus share of Capital (%)	Minus share of Labor (%)	TFP growth (%)
1960-1965	4.38%	0.80%	1.87%	1.71%
1965-1970	5.13%	0.96%	2.25%	1.92%
1970-1975	3.53%	1.02%	2.38%	0.13%
1975-1980	3.52%	0.87%	2.04%	0.61%
1980-1985	6.35%	0.44%	1.03%	4.88%
1985-1990	4.52%	0.44%	1.02%	3.06%
1990-1995	3.47%	0.56%	1.32%	1.59%
1995-2000	3.01%	0.49%	1.14%	1.38%
2000-2004	-0.46%	0.38%	0.89%	-1.73%

Source: own illustration, 2008

To what extent the country is exposed to threats from global climate change shall be discussed in the subsequent chapter. In this chapter the vulnerability of the country to various climate change related factors shall also be studied.

3 The role of climate in agricultural production and local vulnerabilities

The upcoming chapter is devoted to the discussion of the natural and scientific mechanisms through which climatic variables affect crop production. Although the literature on these mechanisms is rather incomplete as yet, there seems to be sufficient evidence to suggest, that climate change will certainly impact on agricultural productivity and lead to different possible outcomes (CALZADILLA et al., 2010). The upcoming deliberations also analyze the country's vulnerability to climate change by analyzing historical climate related records.

3.1 Agricultural Production under Climate Change and the response of crop growth

Climate change is very likely to affect future crop growth and consequently crop management (KERSEBAUM et al., 2010). In order to understand the mechanisms associated with climate change and crop growth and subsequently crop or agricultural production, it is necessary to first analyze the principle abiotic requirements for plant growth. Besides the gas associated with global warming "carbon dioxide" (CO_2) three other factors are primarily important for crop or plant growth, these include sunlight, nutrients and water (WOLFE AND ERICKSON, 1993; HOUGHTON et al., 2001).

Figure 3.1 Climate and soil interaction – impacts on crops and livestock
Source: MENDELSOHN AND DINAR, 2009

Figure 3.1 reveals that every species, be it crop or livestock, has an optimal range in which its growth is maximized. In this regard, the climate ranges sup-optimal and optimal can be distinguished. Certain so called limiting factors such as soil and water can impact the climate-crop relationship and either in case of the existence of a limitation lead to lower growth rates and in the absence of such to a higher growth rate, respectively. The existence of limiting factors such as degraded soils and water scarcity lead to overall lower levels of growth and production, this throughout all climate ranges. Most importantly though Figure 3.1 reveals that higher temperatures above the optimal threshold in combination with the existence of the aforementioned limiting factors are most harmful to crop and animal production.

CO_2 is the essential substrate for photosynthesis, the process through which water and carbon dioxide are transformed into sugars driven by the energy from sunlight (ROTTER AND VAN DE GEIJN, 1999). These sugars or carbohydrates are used by the plant for growth (ROLLAND et al., 2006). Most of the studies that have been conducted to date are clear in one aspect that elevated carbon dioxide levels increase the rate of plant growth (WARRICK 1988; ALLEN (JR.), 1991; TUBIELLO et al., 2007), however other factors that interact with carbon dioxide in this process have been clearly studied too narrowly. Most of the understood ripening processes associated with higher carbon dioxide availability are related to the leaf level (KIMBALL et al., 1993; GHANNOUM et al., 2000). The findings from the leaf level analysis are then extrapolated to the plant, which itself has a much more complex system of processes. Thus for instance, elevated CO_2 levels could also affect the plant roots as increased levels of plant growth have been found to be associated with a higher demand for nutrients. In order to saturate this increased nutrient demand, the plant could react through intensified root growth to extract the nutrients from the soil. Various studies have been carried out, though inconclusive about the plant level effects of increased carbon dioxide availability. The findings have shown that the developmental rate of the plant could be altered, making it grow slower, faster or at a similar rate. The understanding of the linkage between elevated CO_2 concentrations, plant physiology and plant development is an important requirement, as the choice of the most CO_2 responsive crop or plant culture is closely associated to this understanding (for a detailed see MORISON AND MORECROFT, 2006). Clearly more dimensions of the CO_2 impact have to be considered, when drawing any conclusions. The scales that have to be taken into consideration should amongst others comprise biochemistry, plant physiology, plant cell, plant organs (i.e. stems, roots, leaves), individual plants and entire cropping regions (ROGERS AND DAHLMAN, 1993). Although studies in general have confirmed the beneficial effects of additional carbon dioxide availability for plants, certain distinctions have to be made regarding the mode of CO_2 fixation in different plant types with respect to the chosen photosynthesis pathways. Plants can have different photosynthesis pathways. In the majority of plants or crops (approximately 96%) the primary product of a carbon fixation is 3-phosphoglycerate, a chemical compound with three carbon atoms. Such crops and plants are thus named C_3-plants, which amongst others comprise rice, wheat, soybean and cotton. Another group of plants named C_4-plants save water by first producing the C_4 compound called oxaloacetate through carbon dioxide fixation. These C_4 plants amongst others comprise corn, sorghum and sugarcane (HELDT AND PIECHULLA, 2011). According to a detailed review by REDDY AND HODGES (2000) on climate change and global crop productivity, elevated carbon dioxide concentrations as compared to current levels could be beneficial to and stimulate growth of C_3 plants, whereas little effects are expected for C_4 plants, that are already optimized for current carbon dioxide concentrations. Not only photosynthesis for the two groups C_3 and C_4 behaves differently, plant growth also tends to respond differently to higher concentrations of CO_2 (WARRICK, 1988; KRAMER, 1981). More recent research by AINSWORTH AND LONG (2005) has come to find that in case of a doubling of the atmospheric CO_2 concentration significantly higher (double) increases in leaf photosynthesis for C_3 plants as compared to C_4 plants can be expected. Thus higher yields for C_3 crops in the range of 10-20% seem realistic, whereas they are marginal for C_4 crops and plants, if occurring at all (GIFFORD, 2004).

As has been noted earlier, a full understanding of the effects of elevated CO_2 concentrations on plant or crop growth and yields requires a closer look at the interaction of this so called "carbon fertilization effect" with other variables that are important for crop development. Many limiting factors such as nutrients, soils, pathogens and diseases etc. exist, their omission from the analysis of a carbon fertilization effect can lead to overestimations of the positive effects (FUHRER, 2003; TUBIELLO AND EWERT, 2002). Thus, early experiments carried out in laboratory settings overstated the carbon fertilization effect. After Free Air Carbon Dioxide Enrichment-FACE studies were carried out, a lower impact was revealed (CLINE, 2007). As the atmosphere in future is expected to alter in different ways concurrently, an analysis considering only one component such as CO2 for crop impact assessment does not seem sufficient (UNSWORTH AND HOGSETT, 1996). The upcoming deliberations intend to present the interactions between elevated carbon dioxide concentrations and several abiotic and biotic factors of production such as water (precipitation), nutrients, temperature, soil quality, diseases, weeds and pests.

Carbon-dioxide interactions with other limiting factors are difficult to study, not only because of the associated complexity but also the relatively expensive manipulative experiments (LUO et al., 2008). The interactions between increased levels of CO_2 with precipitation and temperature were studied with regard to ecosystem carbon and water dynamics in different climatic zones. In their study LUO et al. (2008) found that none of the considered interactions between Temperature, Precipitation and CO2 (three-way) was substantial for carbon or water processes related to biomass production. However, two-way interactions proved positive for temperature and carbon dioxide and temperature and (doubled) precipitation, revealing the amplification of one factor by the other. Between temperature and halved precipitation levels a negative interaction was found. Squire and UNSWORTH (1988) found that determinate crops with specific developmental stages would generally grow faster in higher temperatures, however certain stages of the life cycle could be shortened and thus limit the gains from an increase of carbon dioxide concentrations. Accordingly, MORISON (1996) notes that if temperatures and carbon dioxide concentrations rise in tandem, initially this might favor a faster plant development process, however as this would imply a shortening of the growing period, overall yield production could be hampered by a constrained potential growth. KOTI et al. (2005) over an entire season exposed different soybean genotypes to treatments of elevated CO_2, temperature and UV-B radiation and found that the damaging effects of temperature and UV-B increases could not be ameliorated by a doubling of the CO_2 concentrations. In some cases negative temperature-CO2 interactions have been found for crops such as wheat (XIAO et al., 2005) and rice (MATSUI et al., 1997). According to HANSEN et al. (2008) the interactions between increased warming and carbon dioxide levels and their impact on photosynthesis are disturbed by a misconception about the uncertainty related to the carbon fertilization effect on photosynthesis after a certain threshold or so called inflection point, this specifically in open field experiments (Figure 3.2). They state that, as the response of photosynthesis to rising CO_2 concentrations is nonlinear, after an initial slope, which has been the main interest of past research, the gains in carbon acquisition are significantly reduced, whereas after this slope or inflection point increases in nitrogen and water use efficiency are expected (Figure 3.2). Thus FACE experiments that prove certain positive interactions between CO_2 and photosynthesis should consider higher levels of uncertainty in their experimental designs.

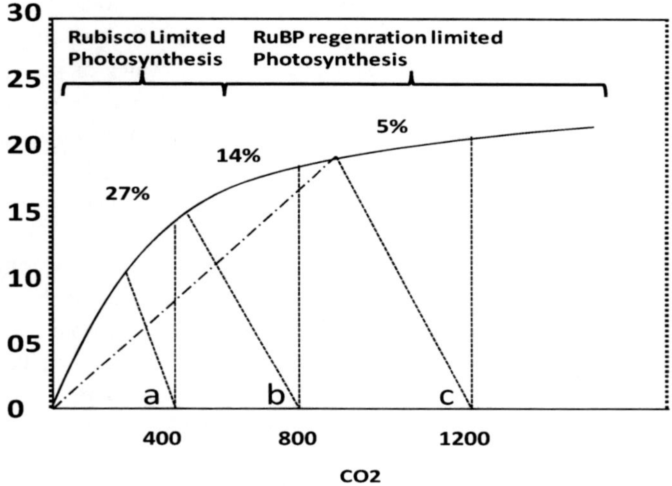

Figure 3.2 The response of photosynthesis to internal CO_2
Source: AINSWORTH AND ROGERS (2007)

As far as the interaction between photosynthesis and temperature is concerned, a hill-shaped relationship has been found (Figure 3.3)

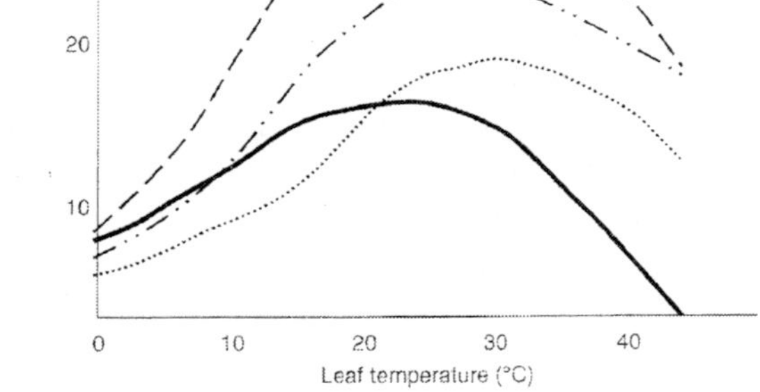

Figure 3.3 Photosynthesis-Temperature Relationship
Source: MENDELSOHN AND DINAR, 2009

Besides the interactions between carbon dioxide concentrations and temperature rise, there is another important variable to consider when analyzing the impact of climate change on crop growth and yield, namely water or precipitation. A temperature rise would normally imply an increased demand for water, however from a plant physiological point of view, increasing temperatures and higher CO_2 levels reduce stomatal conductance, this leads to less water requirement by the plant (HANSEN et al., 2008). This contributes to an increase in water use efficiency (WUE), the ratio of carbon fixed to water transpired (ROGERS AND DAHLMAN, 1993). These WUE increasing effects are primarily confirmed for single plants or crops, the response of entire ecosystems is hardly known (CENTRITTO AND LORETO, 2005). As the majority of global Agriculture is rainfed, projected changes in precipitation patterns will determine the direction of the impacts from climate change. For now, the Global Circulation Models' (GCM) simulations have to be relied upon to draw conclusions, which at times might not reflect the true nature of expected impacts and their magnitude (TUBIELLO et al., 2007). MENDELSOHN AND DINAR (2009) show that irrigation can help to reduce plant sensitivity to warming provided that financial and natural resources are available to the farmer making the switch.

Another key requirement for plant growth is the availability of nitrogen (N) and other soil nutrients such as phosphorus (P). A simultaneous availability of important soil nutrients and elevated carbon dioxide levels could benefit plant growth (KIM et al., 2003). Experiments in which the availability of these important soil nutrients was constrained did not show any significant carbon fertilization effect on plant development (SCHNEIDER et al., 2004). For a thorough understanding of the climate change-crop growth nexus, the focus has to be also turned to the behavior of weeds and pathogens under altered atmospheric conditions. A detailed report prepared by ARCHAMBAULT et al. (2001) found that the competition between weeds and crops shows significant changes with elevated CO_2 concentrations, although responses can differ substantially between C_3 and C_4 plants and weeds and thus also alter their interactive behavior. Another finding of the report is that herbicide efficacy changes can be expected to occur approximately 40 years from present, with both temperature rise and co2 increase affecting it significantly. However, in their economic analysis they also find that the CO_2 fertilization effect can ameliorate the losses from herbicide efficacy loss, provided that the likewise CO_2-benefiting weeds are controlled by the farmer. The weed control seems overwhelmingly important, as in the case of plants (weed or crop) with the same photosynthetic pathway, clearly weeds are favored by the carbon fertilization effect (ZISKA AND BUNCE, 2006). Insect outbreaks and pests are also prevalent in the present; however the timing as a result of climate change is expected to change. First incidences have been recorded already, where earlier insect activity due to warming has occurred. Pest outbreaks and diseases can be expected to be aggravated by future climate change (GAN, 2004).

The foregone discussion has revealed that although certain single-factor relations between climate change and crops are important and in most experimental settings significant with respect to the results, a more sophisticated analysis is clearly needed. To this background the single-factor relationship between elevated carbon dioxide levels and plant growth in general has been positively confirmed. Plant growth will be most likely enhanced through an extended CO_2 availability. Nevertheless, when interactions between other climate and growth relevant factors are considered, results do not appear as clear anymore. Temperature and carbon dioxide interaction in certain experimental settings has proven to be beneficial for crop growth, whereas studies exist that have found contrasting results, at times even negative interactions have been revealed; as for instance in the case of rice and wheat, when considering relative yield growth. Another revelation has been the difficulty with the theoretical and empirical setting in which the results have been found, much too often under ideal conditions only, beneficial climate change impacts for crop growth have been confirmed. However, it is very difficult to find ideal conditions in real a world setting, such as growing conditions with adequate water and nutrition supply and the absence of weed pressure. Despite these drawbacks, the general literature seems to have a much clearer understanding of single leaf or plant level effects as compared to the understanding of ecosystem responses to climate change. Equipped with this scientific background on climate change and crop growth, the following considerations are devoted to the analysis of the vulnerability of Pakistan to adverse effects from climate change.

3.2 Vulnerability to Climate Change in Pakistan

Pakistan is a developing country facing several obstacles to development. Thus in a nutshell vulnerability to climate change can be framed as follows: the climate is arid and hot, regularly the country is exposed to risks from extreme weather events (droughts and floods), socio-economic conditions are deteriorating, major income sources of the population depend on climate sensitive sectors, awareness for environmental protection is missing, warning systems are not in place or are outdated, political instability and capacity to react to abrupt and long-term environmental changes is low. Specifically the agricultural sector is at risk as it is the mainstay of the population, with a total contribution of 25% to overall GDP. The industry of the country is heavily agro-based, with agro-based exports weighing as heavy as 80% in total exports. Moreover, almost two thirds of agriculture is irrigated, requiring sufficient water supplies. On top of the export contribution, agriculture in Pakistan has to fulfill a significant product contribution. As two thirds of the population are engaged in agriculture (direct and indirect), the sector also provides a substantial factor contribution. More than 30% of the population lives below the poverty line.

Great uncertainties in the magnitude and direction of the impacts from future climate change exist; therefore in following from a historical point of view climatic key indicators shall be studied, in order to be able to judge on future developments and vulnerabilities in the course of this study. Pakistan as an individual country came into existence in 1947. Climate records have been well documented since the country's existence. Due to persisting political instability and the associated risk of losing these valuable records, the United States - NOAA data rescue program has provided a valuable platform for preserving these data records digitally. The rescued data ranges from 1947 to 1971, covering a wide range of meteorological information.

3.2.1 Temperature

In order to understand future climate change an analysis of the past climate is highly valuable. On the basis of past climatological data, foremost temperature and precipitation records, this analysis can be carried out. Data records for the independent country Pakistan are generally available starting from 1947 (see NOAA CENTRAL LIBRARY, 2011). The years starting from 1947 provide reports on a general weather outlook rather than specific data. A complete dataset covering detailed climatic information is available only from 1951 onwards. Considering that world climate refers to the mean or normal condition over a long period, a 50 years stretching dataset fulfils the requirements for analysis (TREWARTHA, 1954). In the framework of the United Nations Development Programme (UNDP) initiated Project "Climate Systems and Policy" long term historical data has been studied to assess climatic trends over a period of 40 years (1960-2000). In line with the general global trend the results clearly reveal that climate in Pakistan is also changing. The analysis reveals that throughout the 40 years at study the mean annual temperature has increased by 0.35°C, averaging 0.08°C per decade (MCSWEENEY et al., 2010). A detailed study conducted by the (Pakistan) Global Change Impact Studies Centre (GCISC) for the period from 1970 to 2000 based on observations from 54 meteorological stations found significant increases in both, daily minimum and daily maximum temperatures in 80 percent of the sampled stations (SHEIKH et al., 2009). According to the final report of the country's Task Force on Climate

Change (TFCC) published in 2010, area average mean annual temperature in Pakistan over the period of the last century, in accordance with the global trend, has increased by 0.6°C. The slope of this warming for the period 1901 to 2000 has been estimated at 0.06 °C. Alarmingly, the last fifty years clearly indicate an increased rate of decadal warming, averaging 0.24°C (see Figure 3.4). The dataset considered spans from 1901-1947 (India-Pakistan's current territory) to 1947-2001 (Pakistan).

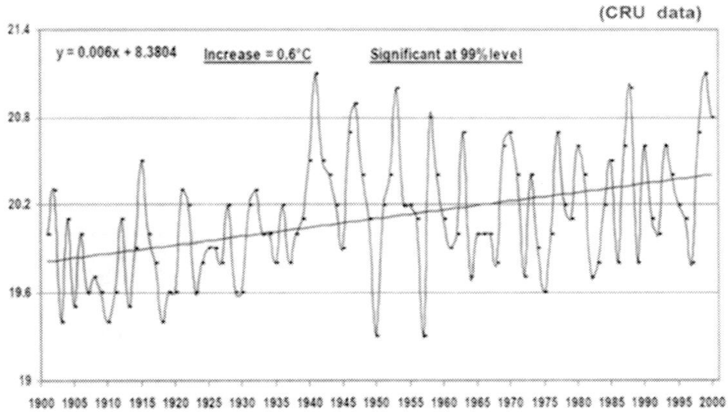

Figure 3.4 Annual Temperature (°C) Trend 1901-2000 Pakistan
Source: TFCC, 2010

According to the Pakistan Meteorological Department (CHAUDHRY et al., 2009) during the period 1901-2000, the increase in mean annual temperature in the Northern half of Pakistan was found to be higher with a value of 0.8°C as compared to the country as a whole with 0.6°C. Furthermore, data from the Climate Research Unit (CRU) in the United Kingdom relative to the national scale indicate a higher increase in mean annual temperature for Northern Pakistan. In addition, based on data from 1951 to 2000 the TFCC report highlights the general warming trend in mean and maximum temperatures for the summer season (April and May), this throughout the country. For the same time period, the Monsoon Season that spans from July to September has generally shown a decreasing trend in temperatures (TFCC, 2010). According to MCSWEENEY et al. (2010) the warming trend is primarily observed for the months from October to December, with precisely 0.19°C per decade. Additionally, the frequency of hot days and nights (days or nights where temperature exceeds a certain threshold by 10%) in Pakistan has increased, whereas the frequency of cold days and nights has decreased respectively. As can be seen in Figure 3.5 the temperature anomalies relative to the 1970 to1999 mean climate have also increased. The black solid curve shows the mean of observed data from 1960 to 2006 and the marked curve shows the median, whereas the shading represents the range of model simulations of recent climate across an ensemble of 15 models (MCSWEENEY et al., 2010).

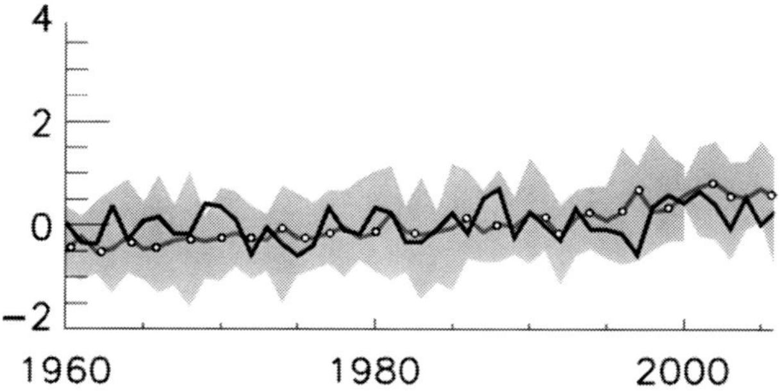

Figure 3.5 Mean Annual Temperature Anomaly – Pakistan (°C) 1960-2006
Source: modified after McSWEENEY et al., 2010

When analyzing selected stations in the country a general rising temperature trend can be observed. The Balochistan province for instance throughout the time period from 1960-2009 has shown an increasing temperature trend; this is revealed by having a look at important stations in the province, comprising Quetta, Dal Badin and Zhob. Using a linear regression temperature increases ranging from 0.73°C in Zhob to 2.43°C in Dal Badin and 2.84°C in Quetta are revealed. However, the coastal areas of the province have shown decreasing temperatures in the range from -0.5°C to -1.5°C. In the observational period from 1960-2009 the monsoon months from June to September reveal a general decrease in temperatures, except for the Balochistan region including the Sulaiman and Kirthar ranges. The lower Indus Plains and Western Highland zones have both followed decreasing temperature trends, also in the winter season stretching from December to March. The hottest season of the year from April to May has shown a warming trend throughout all zones and seasons. Only in the Greater Himalayas and the Western Highlands the temperatures for the monsoon-retreat period from October to November have dropped, in all other zones increases can be observed (see Table 3.1)

Table 3.1 Mean Temperature Trends in different AEZ of Pakistan (°C) 1951-2000

Regions/Seasons	Annual	Monsoon (Jun-Sep)	Winter (Dec-Mar)	Apr-May	Oct-Nov
I (a): Greater Himalayas	0.04	-0.80	0.32	1.09	-0.06
I (b): Sub-montane / monsoon dominated	-0.19	-0.57	0.00	0.13	0.12
II: Western Highlands	-0.72	-1.48	-0.65	0.17	-0.47
III: Central & Southern Punjab	0.11	-0.25	0.03	0.83	0.31
IV: Lower Indus Plains	-0.08	-0.55	-0.07	0.35	0.15
V (a) : Balochistan Province (Sulaiman & Kirthar Ranges)	0.11	0.46	0.63	0.79	0.50
V (b): Balochistan Plateau (Western)	1.17	1.3	0.43	2.17	1.80
VI: Coastal Areas	0.00	-0.18	0.05	0.03	0.30

Source: SHEIKH et al., 2009

Although in certain areas of the country temperatures in the past have shown a decreasing trend, the overall picture clearly reveals the increasing warming of the country. This has been underlined by different studies that have been undertaken to reveal the historical temperature trend over Pakistan (AHMAD et al., 2003; FAROOQI et al., 2005; SHEIKH et al., 2009). The foregone deliberations have shown the vulnerability of the country to future warming, which could affect all important agricultural regions and further exacerbate the need for adaptation strategies to cope with the change. In following the other major determinant of climate shall be addressed from a historical perspective, namely precipitation.

3.2.2 Precipitation

The following deliberations based on the two studies conducted by the Global Change Impact Studies Centre (SHEIKH et al., 2009) and the Pakistan Meteorological Department (CHAUDHRY et al., 2009) aim to reveal historical precipitation observations for Pakistan. By using data records from the Climatic Research Unit (CRU) SHEIKH et al. (2010) found that the mean annual precipitation over Pakistan averaged by area, has increased by a significant percentage. The data used comprises the 20th century, in those years where Pakistan's territory accounted to India (1901-1946), the data-points relevant to Pakistan's territory have been extracted. As Figure 3.6 shows an increasing trend for precipitation over the last century is revealed. Although the frequency of fluctuations is large the extremes are most evident.

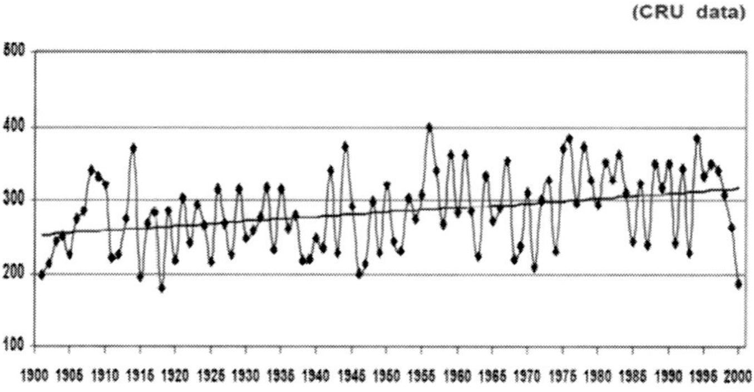

Figure 3.6 Annual Precipitation (mm) Trend 1901-2000 Pakistan
Source: TFCC, 2010

As the majority of the annual precipitation is received in the monsoon period, this general rising trend can be explained by the increased variability of the monsoon. According to the Pakistan Meteorological Department the monsoon precipitation has increased over the entire country except for the southern coastal region, for which a significant drop has been observed (Table 3.2).

Table 3.2 Precipitation trends (ΔP %) in different AEZ of Pakistan (1951-2000)

Regions/Seasons	Annual	Monsoon (Jun-Sep)	Winter (Dec-Mar)
I (a): Greater Himalayas	0.49	1.73	-0.04
I (b): Sub-montane / monsoon dominated	0.3	0.38	0.53
II: Western Highlands	-0.02	0.22	0.00
III: Central & Southern Punjab	0.63	0.57	0.99
IV: Lower Indus Plains	0.22	0.45	-0.27
V (a) : Balochistan Province (Sulaiman & Kirthar Ranges)	1.19	1.16	1.14
V (b): Balochistan Plateau (Western)	0.1	-0.2	-0.4
VI: Coastal Areas	-0.82	-1.34	0.00

Source: SHEIKH et al., 2009

For the period from 1951-2000 the highest annual increases have been recorded for the Central & Southern Punjab region and the Balochistan province. Especially in the monsoon season significantly higher amounts of precipitation have been observed in the greater Himalayas. Except for the Central & Southern Punjab Region the precipitation increases in other zones remained marginal in both directions. In accordance with the global trend these observations seem plausible. Globally, land precipitation over the course of the last century has increased by approximately by 2% (HULME et al., 1998). When having a look at regional or zonal patterns, for the zones 30N to 85N precipitation hikes of up to 12% in comparison to long run normals have been found (DORE, 2005).

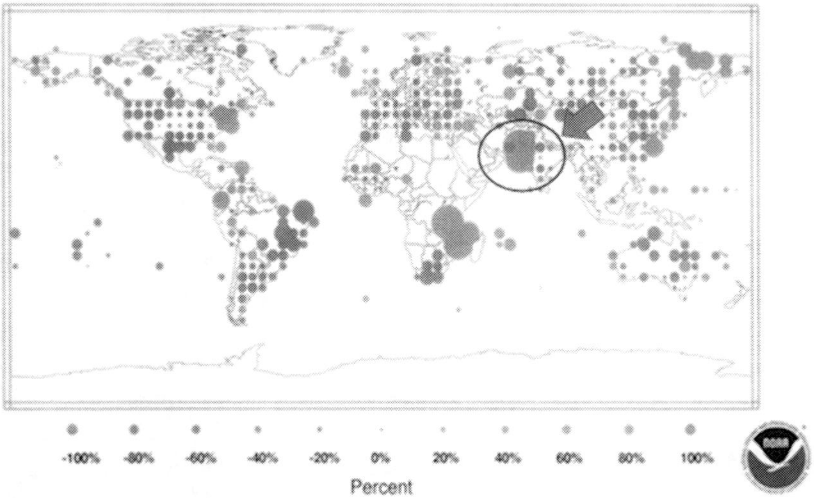

Figure 3.7 Precipitation Anomalies in September 2011
Source: NOAA, 2011

In the recent past the high temporal variability has been confirmed for Pakistan. In Figure 3.7 the National Oceanic Atmospheric Administration's (USA) observations for the monsoon period relative to a 1960-1990 base period have clearly revealed the magnitude of departures from normal for Pakistan, which have effectively doubled and are amongst the highest precipitation anomalies in the world. Generally rainfall in most parts of the Asian peninsula is low but with high temporal variability, specifically owing to the monsoon season. According to CHAUDHRY (1994) in Pakistan 7 out of 10 considered stations have clearly shown an increasing trend during the monsoon season. Research concerning the historical precipitation patterns over Southern Asia and Pakistan specifically, has shown that there is clear consensus on an increasing precipitation trend. Especially the monsoon season has been found to be responsible for this development. Interestingly, the main question to be answered in future will be the interaction between increasing temperature and precipitation and whether the positive precipitation trend will compensate for the negative warming effects or cause damage itself through higher variability and consequently extreme events such as floods. To this background the upcoming subchapter shall highlight the country's vulnerability to extreme events from a historical perspective.

3.2.3 Extreme Events

In 2009 the World Bank Environmental Department published a report on Ecosystem based approaches to Climate Change. In the framework of climate risk assessment the World Bank identified the five most daunting climate threats and ranked those 12 countries most at risk (Table 3.3).

Table 3.3 Five Climate related threats, and the 12 Countries Most at Risk 2009

Low Income | Middle Income

Rank	Drought	Flood	Storm	Coastal 1m	Agriculture
1	Malawi	Bangladesh	Philippines	All low-lying Island States	Sudan
2	Ethiopia	China	Bangladesh	Vietnam	Senegal
3	Zimbabwe	India	Madagascar	Egypt	Zimbabwe
4	India	Cambodia	Vietnam	Tunisia	Mali
5	Mozambique	Mozambique	Moldova	Indonesia	Zambia
6	Niger	Laos	Mongolia	Mauritania	Morocco
7	Mauritania	Pakistan	Haiti	China	Niger
8	Eritrea	Sri Lanka	Samoa	Mexico	India
9	Sudan	Thailand	Tonga	Myannmar	Malawi
10	Chad	Vietnam	China	Bangladesh	Algeria
11	Kenya	Benin	Honduras	Senegal	Ethiopia
12	Iran	Rwanda	Fiji	Libya	Pakistan

Source: WORLD BANK, IRIN (2009)

The countries appearing on these twelve ranks were either middle or low income. The identified threats from climate change include drought, flood, storm, coastal sea level rise (1m) and greater agricultural uncertainty. When having a first look at Table 3.3, one can see that some of the world's poorest nations top the list of countries at the highest risk. As per droughts are concerned, most alarmingly 11 out of 12 countries leading the ranking are low income countries. In the case of flood threat 9 out of 12 countries are low income countries and as far as agricultural uncertainty is concerned 10 out of 12 countries are low income states. This table has been created for the period up to

2009, thus some extreme events of unprecedented nature like the epic floods in Pakistan in 2010 that devastated the country in large parts could not be considered. Millions were left homeless and important harvest was destroyed, exacerbating issues of food security. According to the (North American) National Center for Atmospheric Research (NCAR) the abnormal increase in the monsoon rain in Pakistan and the rest of the subcontinent originated from the warming of the Indian Ocean's surface waters by two degrees Fahrenheit over the last forty years. The heated air carried more moisture which added approximately eight percent more water vapor into the monsoon systems over land, stirring up storms and causing them to pull additional water, which then can caused devastation (TRENBERTH, 2010). The inclusion of this flood disaster in 2010 would have certainly lead to a higher ranking for Pakistan in terms of flood threat. Nevertheless, the above table clearly reveals that Pakistan is highly vulnerable to threats from climate change, foremost originating from flood and agricultural uncertainty. Moreover, although the country does not appear in the drought threat ranking it has a long stretching history related to it. In the period from 1871 to 2006 more than 14 major drought events of severe nature have been recorded (CHAUDHARY et al., 2001).

From 1998 to 2002 the province of Balochistan was hit by severe drought conditions, affecting 84% of the population directly, killing 30% of the province's livestock and causing mass migration due to widespread hunger and disease. La Niña resulted in abnormal weather conditions, causing meteorological drought in most parts of the country. Intense heat in combination with precipitation shortage throughout the winter period and parts of the summer season caused this 4 years of severe drought (SAJJAD et al., 2009). CRUZ et al. (2007) report evidence on serious increases in the frequency and intensity of extreme weather related events such as extended drought periods, tropical cyclones, flash floods and severe dust storms in the Asian region. In this context, both the frequency and the intensity of climate related extreme events in Pakistan have increased in the recent past. Pakistan in its 65 years of existence has been hit by numerous climate related extreme events. Not only climatic extremes have been responsible however, geological disturbances have also been very intense in this region, causing severe earthquakes and affecting nearly all parts of the country. The severe impact of natural disasters in Pakistan can be judged by the fact that in the period from 2000-2009 about 79000 people were killed and 28 million affected (IFRC, 2009).

Table 3.4 Climate related natural hazards in Pakistan (1935-2011)

Type of Disaster	Date	Death Toll	No Total Affected	Damage (000 US$)
Earthquake	31.05.1935	60000	-	-
	27.11.1945	4000	-	-
	28.12.1974	4700	-	-
	08.10.2005	73338	5128309	5200000
Flood	1950	2900	-	-
	Aug 73	-	4800000	661500
	02.08.1976	-	5566000	505000
	Jun 77	848	-	-
	Jul 78	-	2246000	-
	15.07.1992	-	6184418	-
	08.09.1992	1334	6655450	1000000
	02.03.1998	1000	-	
	22.07.2001	-	-	246000
	09.02.2005	-	7000450	-
	10.08.2007	-	-	327118
	02.08.2008	-	-	103000
	28.07.2010	1985	20359496	9500000
	Aug 11	-	5800000	-
Storm	15.12.1965	-	10000	-
	26.06.2007	-	-	1620000
Drought	Nov 99	-	2200000	247000

Source: EM-DAT VERSION 12.07 (2012)

Table 3.4 based on the CRED's (Collaborating Centre for Research on the Epidemiology of Disasters) data on emergency events presents a timeline of documented major extreme events in Pakistan, dating back as far as 1935 when Pakistan was known as Western India. Climate related natural hazards amongst others can comprise tornados, hurricanes, floods, cyclones, drought, landslides and heat or cold waves (CBSE, 2006). Pakistan however, has been specifically vulnerable to floods, storms and droughts (AHMED AND SCHMITZ, 2011a). As shown in table 3.4, although the highest death toll was recorded from earthquakes, the highest number of people indirectly or directly affected was estimated for floods and droughts. Regarding the economic losses incurred by natural disasters, floods top the list, specifically owing to the magnitude and country wide impact. For instance in 2010 during the monsoon floods, in total 84 out of 121 districts were affected.

In another study conducted by Germanwatch the Global Climate Risk index 2012 (HARMELING, 2011), which indicates the level of exposure and vulnerability to climate related extreme events such as flood, storms and drought, was estimated using 2010 data. In addition to the latest data, the 20 year time period from 1991-2010 has also been considered for the calculations.

For 2010, Pakistan clearly ranks first as being the most severely affected country, followed by Guatemala, Colombia and Russia. For the period from 1991-2010 Pakistan ranks eighth.

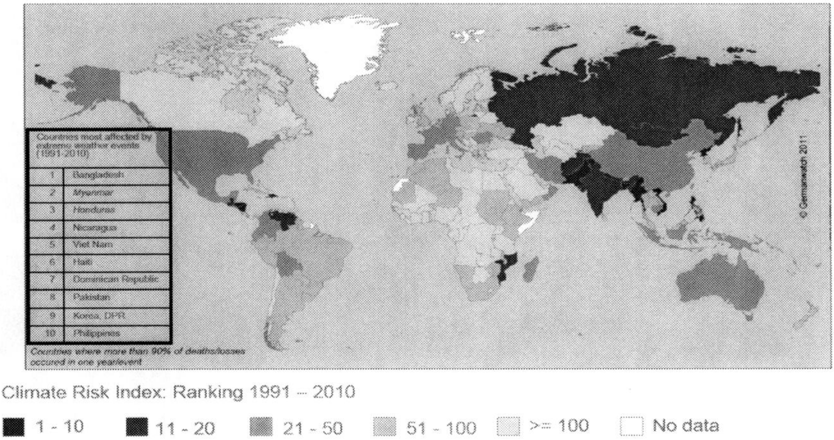

Figure 3.8 World Map of the Global Climate Risk Index 1991-2010
Source: HARMELING, 2011

The ranking in Figure 3.8 is calculated as an average for the years considered, however some countries are continuously affected by extreme events and some are only ranking high because of exceptional catastrophes in a certain year (HARMELING, 2011). That Pakistan throughout this time period of 20 years has been continuously affected has been shown in Table 3.4 based on data from the CRED.

The foregone discussions in chapters 2 and 3 have presented a picture on the importance of the agricultural sector for the country and its steadily growing population. A high dependence on the sector also translates into a high dependence on climate. Therefore, from a historical perspective climate change was analyzed. By elaborating on temperature trends, precipitation trends and the history of extreme events these deliberations have revealed the country's vulnerability to climate change in the past. After understanding these interdependencies, how these impacts can be quantified and modeled, shall ensue in the upcoming chapter.

4 Modeling the Impacts of Climate Change on Agriculture

Where the previous chapters have introduced the interrelationship between agricultural production and climate change, the following considerations intend to cover the aspects related to the assessment of the economic impact of climate change on agriculture. The economics of climate change is not yet well understood; especially the impacts for the agricultural sector are plagued by great uncertainty (KELLY AND KOLSTAD, 1999). The idea behind this review is to supplement previous deliberations made by various authors regarding the same issue such as ADAMS et al. (1998), KURUKULASURIYA AND ROSENTHAL (2003), CLINE (2007) AND MENDELSOHN AND DINAR (2009). A comparative approach will be followed in order to address the strengths and weaknesses of the respective methods.

As the majority of scientists today agree, climate change is a fact and also most likely to occur in future. In an extensive study for the USA, 97% of the scientists agreed that climate change is a reality (ANDEREGG et al., 2010). The idea behind economic assessment or modeling is to uncover the effects that a change will have, with the intention to unveil and understand the costs that will arise from such a change with reference to adaptation and mitigation. Being a multifaceted phenomenon climate change demands multidisciplinary research attention. This multidisciplinary approach has been termed integrated assessment in the literature. Integrated Assessment Models (IAM) have been defined in different ways. WEYANT et al. (1996) for instance have opted for a broad definition, namely 'any model using knowledge from research in multiple disciplines'. KELLY AND KOLSTAD (1999) define an IAM as 'any model which combines scientific and socio-economic aspects of climate change primarily for the purpose of assessing policy options for climate change control'. IAMs have been the primary tool of choice for studying climate change impacts, thus several studies have evolved throughout the last 20 years, only to name a few: NORDHAUS AND YANG (1996); DOWLATABADI AND MORGAN (1995); LEMPERT, et al. (1996); CHAN et al. (1999); Watanabe et al. (2005). As the focus is to study the impacts of climate change on agriculture the following deliberations will concentrate on the agricultural module of a general IAM-framework.

Starting with the question, whether the intention is to measure the impacts of environmental change on a multiplicity of sectors, broadly Agriculture, Industry and Services, or the idea is to analyze and study one particular sector only, two types of economic impact assessment models can be found in the literature. The question of sectoral impacts forms one module or aspect in IAM. Models that study all sectors in one framework are termed economy-wide or Computable General Equilibrium models (CGE), whereas models that focus on one particular sector in a less aggregated way are known as partial-equilibrium models (Figure 4.1).

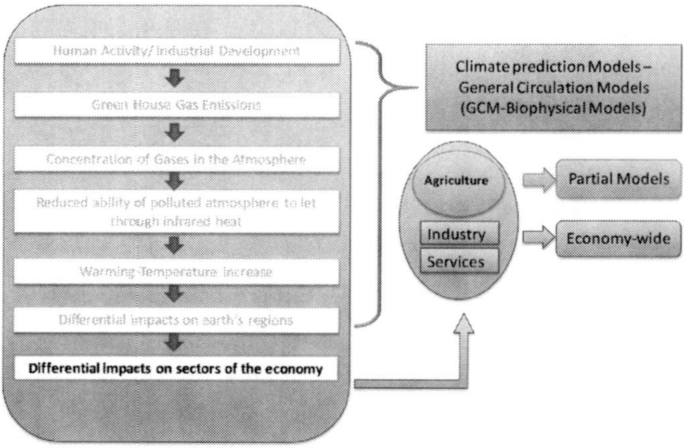

Figure 4.1 Stylized Integrated Climate Change Impact Assessment Framework

Source: own illustration (2012)

In following, with particular reference to the agricultural sector, different types of partial-equilibrium models and economy-wide models shall be discussed. Partial models can be further classified into crop growth simulation models and econometric approaches.

4.1 Crop Simulation Approach or Agronomic models

Crop simulation models intend to capture the behavior of agricultural yields based on a number of factors including climate, soil and farm management. The production function approach or empirical yield model, which will be conceptually exemplified in detail later, is the primary model in this class. A general feature shared by the crop simulation class models is that they are calibrated to a specific location and thus are primarily used to assess the vulnerability to climate change (= expected yield losses) of local agricultural production systems (PADGHAM, 2009). The effects for a specific location are then extrapolated to a much wider area (ROSENZWEIG AND IGLESIAS, 1998). Detailed agronomic science provides the basis for this class of models. Their advantages are clearly rooted in the fact that they are scientifically calibrated and provide the possibility to incorporate carbon fertilization and hydrological variables. A major advantage of these models lies clearly in their high flexibility to include different modules such as economic, hydrological and climatological, to represent detailed local conditions (MENDELSOHN AND DINAR, 2009; LEE, 1998). Many studies have been carried out using crop simulation models, thus for instance the maize crop has been studied by RITCHIE et al. (1998) amongst others, MEARNS et al. (1999) have studied wheat, soybean has been studied by LAL et al. (1999) and rice by SASEENDRAN et al. (2000) amongst others. More recently, the WORLD BANK (2008, 2009) has conducted studies in India and Morocco using integrated assessment approaches including agronomic models. The Indian study used a combination of a regional climate model, a hydrological model and an agro-meteorological simulation model with the aim to understand the impacts of climate change in India's drought and flood affected agricultural areas. Farm economic models were then fed with the estimates of the single modules to better understand the economic effects for agriculture. A similar procedure was applied for the Moroccan study. Crop simulation models have been formulated into software solutions and have been extensively used for making management decisions on the farm level. An overview of selected crop simulation models is presented in Table 4.1.

Table 4.1 Crop Simulation Models

Model	Crop	Studies
CERES	Maize	Mearns (1999)
ICASA/IBSNAT	multi-crop	Rosenzweig and Iglesias (1998)
EPIC	Erosion, Soil, multi-crop	Sharpley and Williams (1990)
CROPWAT	irrigation requirement	FAO (2004)
DSSAT	Software-including multiple models	Rosenzweig and Iglesias (1998)

Source: own illustration (2011)

The limitations of the crop simulation approach are primarily related to the neglect of autonomous adaptation when assessing future climate change. Historical relationships between inputs and outputs are considered for modeling future impacts. The agronomic relationship is captured but the farmer's management practice is believed to be exogenous or fixed. Only if farmer behavior remains constant over the time considered, also with regard to warming, the results can be plausible. However, as the farmer adapts to a changing environment, the results clearly tend to overestimate damages from climate change. Some studies have given attention to this limitation and included possible adaptations into the model (e.g. changing the crop variety), however these adaptations are only valid for a specific location and exogenously introduced (YOU, 2001; STUCZYINSKI et al., 2000). Another caveat is the possibility to only model one crop at a time, without allowing for crop switching. Additionally, it is not clear, whether one specific location is representative of a much wider area, thus extrapolations can be plagued by inaccuracy and might not be reliable. As a final caveat it has been noted that only major grain crops have been studied, neglecting other possible crops which might be suited to warming (e.g. fruits and vegetables). These crop simulation class models have been characterized as impractical due to their user-unfriendliness stemming from a demanding requirement of data and high level of expertise needed to run them (MENDELSOHN AND DINAR, 2009).

4.2 Economy-wide models: Computable General Equilibrium Models (CGE)

Large enough changes in climate could lead to economy-wide impacts, affecting price structures. CGE models can capture global and economy-wide changes and serve as a remedy for the caveats of the partial models. CGEs can help to identify changes in prices and value trade effects. One great advantage of these CGE models is that while they capture the economy-wide effects of a change in the environment they also reveal effects for the micro-sectors of the economy, however provided that the model is properly selected and calibrated. Whether the CGE model is correctly specified is hard to test, as the corresponding statistical tests hardly exist (GILLIG AND MCCARL, 2002). The greatest challenge in constructing global CGEs for climate impact assessment is the estimation of sector specific climate sensitivity and inter-and intrasectoral interactions. To this effect a high level of aggregation is used, sacrificing essential detailed information on land and agricultural uses. To counter the aggregation issue and the loss of important information the models can be also constructed on a regional basis with detailed underlying country studies. Studies that have been conducted using CGE models amongst others include OURAICH AND TYNER (2012) for Morocco, JORGENSEN et al. (2004) for the United States, DARWIN et al. (1995) and DEKE et al. (2001) for various regions of the world.

4.3 Econometric Approach: The Standard Ricardian Model

To the backdrop of the other discussed models' failure to account for a full range of adaptation the so called "Ricardian Approach", which will be conceptually described in detail in the following chapters, has been developed. The Ricardian Model was named after 19th century economist David Ricardo, because of his observation that land rents would reflect the net revenue of farmland (RICARDO, 1817). SCHLENKER et al. (2005) have identified the Ricardian model as a hedonic approach of farmland pricing. Climate change is a shift in the long-run pattern of weather over time, whereas weather occurs at a point in time with great fluctuation. Reference is made towards two different phenomena (weather and climate), thus the respective economic implications amongst them will also differ. To this background the system of measurement, dependent variable or economic metric will also have to be chosen accordingly. Especially for analyzing long-term phenomena like climate change, a measure is needed that reflects the permanent income from farming. "Land value" in this regard comes in handy as it reflects long-term profitability of using the land for farming (FISCHER et al., 2007). Generally, the hedonic relation between land values and climate as an environmental variable arises due to the heterogeneity in the different qualities of land (ROSEN, 1974). These land quality attributes include climate amongst others and being hedonic in nature the model covers a wide range of substitution options (GOULDER AND PIZER, 2006). Farmers in different regions are exposed to different climatic conditions, thus the Ricardian method by making use of cross-sectional data and spatial comparison assumes that they adapt to their local circumstances. As far as adaptation to climate change in this context is concerned, the core assumption of the Ricardian Approach is that if growing rice is more profitable than growing wheat, and if the climate becomes more suitable for rice than wheat, then those farmers will adapt to the changed climate by drawing on the experiences of rice farmers elsewhere and switching from wheat to rice (POLSKY, 2004). Further elaborated, in a scenario where climate change (while controlling for other factors) leads to a shift from a base climate (Current Climate - B) to a new changed climate (Future climate change → climate B shifts to climate A) for farmers in a specific district or region (Farms facing climate B), these farmers will adapt by switching to the production pattern chosen by farms in climate state A. These adaptation costs can then be estimated using the Ricardian climate response function by either studying net revenue changes resulting from switching the production or by comparing changes in land rents or values to a base climate scenario (see Figure 4.2).

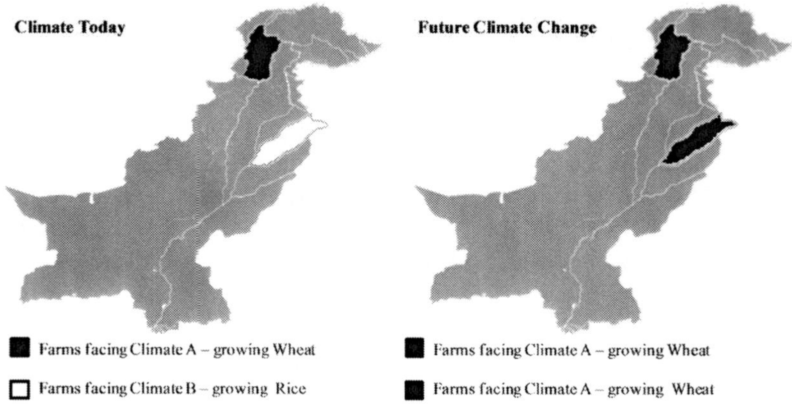

Figure 4.2 Farmer adaptation to future Climate Change in Ricardian Analysis
Source: own illustration, 2011

Departing from this observation, the idea is to compare the land rents (land values) or net revenues of farmland in different climatic zones in order to examine how these values shift with climate and other considered control variables (KURUKULASURIYA AND AJWAD, 2007). The initial approach named Ricardian Method for climate impact assessment was developed by MENDELSOHN et al. (1994) who analyzed the impact of climate change on land values and consequently farm incomes in the United States. The rationale behind the model is that responses and relations between climate and different agricultural production factors are not homogenous. This is to say that land and labor have different interlinkages with climate and thus dissimilar income responses. For instance (hired) workers with a choice to move to favorable areas and sectors, regardless of the prevailing climate, can more or less expect to be paid the same amount. Land as an input and a substantial part of the household income, however, is inflexible and cannot move, thus areas with higher productivity can be expected to have higher land values and consequently also higher net revenues (MENDELSOHN et al., 2007). The model starts from the observation that farmer decisions are dependent on uncontrollable exogenous factors including climate. Thus given these exogenous constraints a farmer has to choose a certain mix of inputs to maximize his profits. In a cross-sectional framework a comparative steady state analysis of long-term climate impacts on land values is carried out. While controlling for other influences such as varying soil conditions, land values are regressed on climate to assess the marginal contribution of each exogenous variable to the value of farmland and thus farm incomes, thus conducting a hedonic pricing of environmental attributes (LIPPERT et al., 2009). In net, the concept states that the land values are correlated with environmental variables including climate.

Being a static comparative model, changes in prices and technology cannot be modeled without specific considerations. Various other limitations have been discussed in the literature. While the approach has proven to be a practical tool for predicting the agricultural consequences of global climate change it has also generated an unusual amount of critical attention (TIMMINS, 2003; POLSKY, 2004). One limitation with particular relevance to the present study is the assumption of long run equilibrium of factor markets. Especially in developing countries the use of land values for the explanation of agricultural performance becomes problematic without the necessary considerations. Another limitation that has been subject to great attention is that the relative crop output prices and input prices remain fixed despite the impact of climate or other influences; this relation is implicitly captured in the model specification (DARWIN, 1999). Thus, the approach has been criticized because of a possible bias towards an overestimation of the benefits of adaptation due to the assumption of constant prices (CLINE, 1996; QUIGGIN AND HOROWITZ, 1999). In detail this implies that changes in consumer surplus are omitted, thus lost consumer surplus is not accounted for and so the value of the increased supply is magnified, leading to a possible overestimation of the benefits of implicit adaptation (CLINE, 1996). Earlier studies (MENDELSOHN et al., 1994) have been criticized for their neglect of additional water sources in the model, thus in general concerns regarding the considered water variables' ability to approximate the real situation sufficiently have been raised (CLINE, 1996; DARWIN, 1999). In addition, effects resulting from flooding on farm incomes have also been neglected (MENDELSOHN, 2000a). Further concerns have been directed towards the specification of climate and thus questions regarding the validity of temperature and precipitation as the lone climate proxies have been raised. Generally the approach's rationale of farmer adaptation taking place is questioned as is the validity of the GCM (Global Circulation Model)-based simulations (MENDELSOHN et al., 2007). Other drawbacks of the model include the omission of transition costs from the analysis and the effect of variables that do not vary across space such as carbon dioxide levels. The model assumes that as a measure of adaptation the farmer will pay the costs (e.g. seed costs, equipment costs) associated with the introduction of the new crops; however the "learning costs" that are associated with the cultivation of the new crop species in the initial years of adaptation incurred by crop failures for instance, are not reflected in the model. Finally, one other valid concern about the approach is that it only reflects the current policy situation (KURUKULASURIYA AND MENDELSOHN, 2007).

The limitations, although significant, are not fatal however (MENDELSOHN, 2001) and many of the raised questions have been addressed to date. The issue with the limited applicability and reliability of the land values will be addressed by examining alternative income specifications; this shall be addressed in detail in chapter 6. The restrictive assumption of constant prices is difficult to tackle. The prices for staple foods and agricultural commodities are in general determined on a global market and not only locally. REILLY et al. (1994; 1996) in one of the few existing global analyses come to find, that the gains in one region have to be set against the losses in another to quantify the net impact of climate change on agricultural output and consequently prices. At the same time they show that the expected global supply-side effects of climate change for the agricultural sector will be marginal, so they conclude that the resulting bias from the assumption of constant prices will be relatively inconsequential. Related to this aspect MENDELSOHN AND NORDHAUS (1996) find that even if total supply would change by 25%, the resulting bias from the constant price assumption would be only 7%. Thus, they conclude that until and unless the scenarios developed and modeled in accurately calibrated global models don't suggest any deleterious consequences for agricultural supplies, the assumption of constant prices may be less of an issue for the Ricardian approach (MENDELSOHN, 2000). The goal in the Ricardian approach is to assess the possible impacts on agricultural productivity, thus a first order approximation is used in the response function, what will happen to the prices in the future is not known (FISCHER et al., 2007). The limitation regarding the explicit inclusion of relevant additional water sources and irrigation has been addressed in the literature (MENDELSOHN AND NORDHAUS, 1999; MENDELSOHN AND DINAR, 2003; FLEISCHER et al., 2008). In their influential paper MENDELSOHN AND NORDHAUS (1996) have also shown that the dimensions of the modeled variables also determine whether an explicit inclusion of irrigation affects the results and renders them unbiased. They show that an inclusion of percent-irrigated land (county level) in the model does not change the results. SCHLENKER et al. (2007) find that water availability clearly capitalizes into farmland values and thus should be explicitly accounted for in the model. A thorough discussion on adaptation is presented in chapters 5.3 and 5.4. Regarding the criticism about the non-inclusion of transition or learning costs in the model, MENDELSOHN AND NORDHAUS (1999) argue that due to a gradual progress of climate change and the short lifetime of farm capital these costs will be negligible. As carbon dioxide is uniformly distributed across the globe, the carbon fertilization effect incurred by higher carbon dioxide levels cannot be measured in a cross sectional model setting. In principle every farm is exposed to the same CO_2 levels. Nevertheless, as a remedy for a possible bias resulting from the omission of the carbon fertilization effect, information from controlled laboratory experiments, resembling the different climatic regions' carbon dioxide levels, can be added to the welfare calculation exogenously (REILLY et al., 1996; MENDELSOHN AND DINAR, 2009). As a remedy for the valid criticism regarding the consideration of current agricultural policies only, country, county or district dummies can be added, however after a distortion has been sufficiently proven (KURUKULASURIYA AND MENDELSOHN, 2007). The foregone deliberations have revealed that no model is perfect for climate change impact assessment on the agricultural sector. Clearly, some methods are more suitable for certain tasks than others. Based on specific research questions and whether the intention is to measure the effects of short term weather or the impacts from a long-term climate change on a country's single sector or the entire economy and whether adaptation is to be accounted for, the appropriate model can be selected, this however by keeping in mind the strengths and weaknesses of the respective models and their practicality and accuracy.

5 Ricardian Studies- Literature Review

As mentioned earlier the Ricardian technique for climate impact assessment has experienced a steady rise in popularity since its first application in 1994 for the agriculture of the USA (MENDELSOHN et al., 1994). Alongside the rise in popularity due to its proven practicality as a tool for predicting the agricultural consequences of global climate change, it has also generated an unusual amount of critical attention (TIMMINS, 2003; POLSKY, 2004). Various countries have been studied using the Ricardian Approach (Figure 5.1).

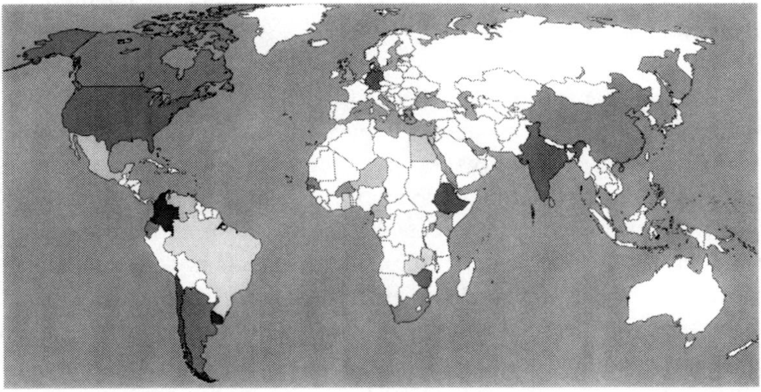

Note: only peer reviewed publications have been considered following MENDELSOHN AND DINAR (2009)

Figure 5.1 Global coverage of the Ricardian Approach 1994-2011
Source: own illustration (2012)

The first application of the Ricardian technique for climate impact assessment on agriculture was developed by MENDELSOHN et al. (1994) and applied to a developed country, namely the USA. The intention was to measure the economic impact of climate on land prices, thus farmland values were regressed on climate, soil and socioeconomic variables in order to estimate the so called Ricardian response function (best-value function) across a wide range of US counties. In a cross-sectional setting 1982 US-Census Data averages for approximately 3000 counties in the lower 48 states were used, covering important geographical, socio-economic, geophysical and agricultural factors. The effect of climatic and other non-climatic control-variables on farm revenues and land values was examined. The effect of climate on land values is confirmed by the estimations. Moreover the analysis also revealed a significantly non-linear (hill-shaped) relationship between climate and land values with additional variations by season. Based on different weighting schemes (cropland weights or crop revenue weights), the study found a gain from warming for the USA to the tune of 20$-$35$ billion (crop revenue weights), whereas losses of around $6-$8 billion (cropland weights) were also estimated. As the crop revenue weights put more emphasis on the irrigated counties also covering minor crops, the results differed, possibly offsetting the losses incurred in the major grain or crop regions. Generally, these effects are based on the finding that higher temperatures in all seasons expect autumn lead to a reduction in farmland values while increased precipitation increases them. Climate simulations revealed a gain for the US agriculture regardless of the omission of the carbon fertilization effect, also stressing the importance of the considered warming scenario.

In a following paper for the USA (MENDELSOHN et al., 1996) an improvement of the initial modeling framework was introduced. Where earlier studies had regressed farm values per acre on environmental and socio-economic variables to determine the marginal impact of each climatic variable and thus only examined active existing farms; the improved approach used aggregate farmland value per acre of total land in a county (Arable land * value per acre) as the dependent variable in order to capture the possibility that climate may also affect whether land is farmed or not, additionally analyzing the impact of climate change on the fraction of farmed land (MENDELSOHN et al., 1996). This study also found a hill-shaped relationship between temperature and land values. An optimum temperature for farming was calculated at 63°F, after which land values would decline. For the USA on average mean temperatures were found at 54-57°, leaving room for warming between 6°-8°F. Thus, where small warming was found to be beneficial for US farming increases above 63°F would clearly be damaging.

In 1999 (MENDELSOHN AND NEUMANN, 1999) examined the importance of interannual variance, finding higher interannual-variance in temperature and precipitation to be associated with lower farmland values, however this only for the summer farming season. To tackle criticism raised regarding the omission of outer-country water sources against the 1994 study by MENDELSOHN et al. (1994), an influential study by MENDELSOHN AND DINAR (2003) included available surface water by county as an additional independent variable. However, as a result only marginal differences could be observed, suggesting the issue to be rather small for the USA. Nonetheless, a following study by SCHLENKER et al. (2005) proved the impact of climate to be different for rainfed and irrigated farms, confirming the legitimacy of the concerns raised against the omission of additional water sources. Irrigated farms behaved less sensitive to climate. Subject to data availability SCHLENKER et al. (2005) suggested separate analyses for different farm types. Studying several years of data (1978, 1983, 1988, 1993, 1998, 2003) DESCHENES AND GREENSTONE (2007) found the Ricardian best-use-value functions to be instable over time, this due to certain omitted variables, which unfortunately were not further specified and pinpointed by the authors but rather assumed. Since its first application in 1994 the model has been subject to constant improvement, thus in 2006 SCHLENKER et al. (2006) introduced growing degree days instead of temperature as the dependent variable. However, without separating degree days for several differences tend to render the results biased (MENDELSOHN AND DINAR, 2009). By the example of Israel FLEISCHER et al. (2008) repeatedly showed the importance of including additional sources of water in Ricardian models as in their analysis this step significantly improved land values and also altered the climate coefficients. The developed countries that have been studied to date using the Ricardian approach include: England and Wales (MADDISON, 2000), Germany (LIPPERT et al., 2009), Canada (REINSBOROUGH, 2003; WEBER AND HAUER, 2003), USA (aforementioned) and Israel (FLEISCHER et al., 2008).

Using data of 400 separate farmland transactions in England and Wales for the year 1994, MADDISON (2000) calculated the marginal value of particular farmland characteristics. The study clearly confirmed the previous studies' findings that environmental characteristics (e.g. climate, soil) are highly important factors in determining land values. REINSBOROUGH (2003) estimated the agricultural costs of climate change scenarios by, as a novelty, introducing more sophisticated climate change scenarios, departing from the previous studies' weak approach of using uniform climate change scenarios. The study found slightly positive impacts of climate change for agriculture in Canada in future. In the same year, in another countrywide study for Canada WEBER AND HAUER (2003) analyzed the impacts of climate change on land values and subsequently a wide range of crops; they found clear benefits for Canada from climate change. LIPPERT et al. (2009) used land rental prices for Germany for the year 1999 with the dataset covering 439 districts and estimated the impacts of climate change on German agriculture. Except for eastern Germany the study found a positive impact of climate change, measured as mean temperature increase and mean temperature decline, for agriculture. Future climate change simulations stretching to 2040 were made using a regional climate model (REMO). However, the authors clearly noted that if scenarios would become too harsh, surpassing the assumed values, then losses could also become realistic. Owing to good data availability the first studies for developing countries were carried out for Brazil and India. Later, after revisions and modifications of the Ricardian model for developing country application (MENDELSOHN AND DINAR, 2003; MENDELSOHN et al., 2007a) several other developing and emerging nations were studied, including: Morocco, Burkina Faso, Cameroon, Ethiopia, Kenya, Senegal, South Africa, Ghana, Niger, Zimbabwe, Zambia, Africa (including 11 single country studies), Argentina, Chile, Ecuador, Venezuela, Sri Lanka, Mexico and China (MENDELSOHN AND DINAR, 2009). As a fundamental change in methodology the developing country studies except for South America, all used net crop revenue as the dependent variable instead of land values. The alterations made to the approach to make it applicable to developing countries will be thoroughly discussed in chapter 6.2.

The Ricardian model for India used net crop revenue as the dependent variable and found all climate variables to be significant in every season. Again non-linear but quadratic climate effects were confirmed. In comparison to smaller farms, large farms were found to have lower net revenues with climate change taking effect. This however, is believed to have been caused by the non-inclusion of the costs of household labor for the small farms, which heavily rely on it for agricultural production. Socioeconomic control variables such as literacy rate were also considered and proved to be significant. Strangely population density did in fact reduce net crop revenues or consequently land values. Results proved to be unstable over time, mainly owing to the use of interpolated weather station data. The calculations (based on 1966-1975 data) reveal an annual marginal impacts for India from a 1°C warming to the tune of -243 US$/ha. India is found to be very sensitive to precipitation with a 1mm/month increase in precipitation leading to a loss of -60 US$/ha/year. The monsoon season is found to be one key factor for the explanation of India's precipitation sensitivity (KUMAR AND PARIKH, 2001; MENDELSOHN et al., 2001; SEO et al., 2005). The Brazilian regressions used land values as the dependent variable and four different seasons to describe climate. Several control variables including population density and twelve soil dummies were also modeled. Again the sensitivity of agricultural production measured by land values was confirmed. For the dataset from 1970-1985 the marginal impact of a 1°C increase in temperature on agricultural production was estimated at -41 US$/ha/year. Precipitation with additional 1mm/month was found to have a beneficial marginal effect to the tune of 1.6 US$/ha/year. The Brazilian regressions also included the period from 1980-85, results for Brazilian agriculture proved to be more stable over time than for the Indian case, but nevertheless showing half the damage from a marginal 1°C warming as compared to the previous study period. Precipitation effects remained stable with benefits of 1.1 US$/ha/year. The African study was conducted under the project "Climate Change and Agriculture in Africa" (CEEPA, 2002) involving 11 countries. As mentioned above several single country studies were carried out. In following the results of the 11 country studies shall be presented.

As in the aforementioned Ricardian studies in the African case the impact of climate change on agriculture was to be at heart of study. The dependent variable was the estimated net value of agricultural land (farmer's perceptions). Table 5.1 summarizes the findings of the African multi-country study for the year 2050.

Table 5.1 Impact of Climate Change on farm productivity (%) by 2050

Country	Percent Change
Burkina Faso	-19.9
Cameroon	-16.8
Egypt	-5.2
Ethiopia	-1.3
Ghana	-13.8
Kenya	-9.8
Niger	-30.5
Senegal	-18.8
South Africa	-3
Zambia	-6
Zimbabwe	-4.9

Source: MADDISON et al. (2007)

Amongst the 11 countries at study those countries that were already exposed to hot and dry conditions were found to be most severely affected by future climate change. In line with this general finding the results also reveal the negative impacts to be closely associated with the initial environmental conditions in the country. Thus countries located in hot and dry areas like Niger and Burkina Faso and Senegal are estimated to suffer most from climate change. Productivity losses ranging from -18% to -30% are estimated, respectively. Ethiopia is found to be least affected from the adverse effects of climate change, mainly owing to the cooler climate it is presently facing. The African studies have been criticized for not including the effects of climate change on additional sources of water such as runoff, particularly against the background that a hydrological model for Cameroon revealed a 21% decrease in annual runoff from an increase in temperature in the range of 2°C (MOLUA AND LAMBI, 2006).

For the South American continent a 7 country study was conducted (MENDELSOHN et al., 2007b). Two seasons were used (winter and summer), with land values as the dependent variable. As a novelty climate interaction terms between temperature and precipitation were also included. Additional control variables such as 'owning a computer', age, electricity and sex of the household head, were also incorporated into the modeling framework. Country dummies were also included. The regressions tested the climate sensitivity by scale (farm size) and type (irrigated, rainfed). Temperature sensitivity was confirmed across all farm types and scales, with negative marginal impacts. Scale and type differences were not detected. Precipitation sensitivity was not proven. Overall, the effects were found to be beneficial only for the southern and Pacific edges of the continent, for the rest negative impacts are revealed. The inland region is expected to suffer losses around 100-250 US$/ha/year. Future climate simulations revealed higher losses for irrigated farms as compared to rainfed farms in the mild warming scenario, especially small farms were found to be most vulnerable. In the harsh warming scenario the losses calculated were estimated to become exceptionally high (Table 5.2).

Table 5.2 Impact of Climate Change on farm productivity (%) by 2100

Farm Scale/Type	Percent Change
Small Rainfed Farms	-35
Large Rainfed Farms	-31
Small Irrigated Farms	-48
Large irrigated Farms	-87

Source: own illustration after MENDELSOHN AND DINAR (2009)

Although being a relatively small country two studies were carried out for Sri Lanka (SEO et al., 2005; KURUKULASURIYA AND AJWAD, 2007). The first study was conducted in 2005 and included aggregate agricultural data for 25 districts of the country. Due to a serious lack of temperature variation, the study used the month of May as the proxy for temperature. The months of September, November and January were used to define precipitation in Sri Lanka. Where higher temperatures were found to be harmful for net revenues, increases in September and January rainfall were estimated to have beneficial effects. Irrigation in Sri Lanka paradoxically as compared to other studies did not increase net revenue. Future climate simulations revealed a loss of -50 US$/ha/year resulting from a 1°C temperature increase and a benefit of 0.7 US$/ha/year from a 1 mm/month increase in precipitation. A subsequent study in 2007 used farm-level data to conduct the Ricardian analysis, however did not find sufficient explanation of net revenues by climate. The control variables explained the lion share of variation in net revenues.Nonetheless, the authors concluded that the effects of climate change on the monsoon will be detrimental for the future of Sri Lanka and thus were recommended to be studied in more detail.

Another individual farm level study was carried out by WANG et al. (2008) for China. The dataset covered close to 8500 farms across 28 provinces. The dependent variable was net revenue as land values could not be obtained. By including four seasons the net revenues were regressed on climate and other control variables. As an improvement to the previous studies, control variables such as township access and road access were also included in the estimations. The study found significant negative effects of climate change on rainfed land located in southern China, whereas impacts on irrigated land were found to be marginal. Higher precipitation was estimated to have a positive effect on the value of land across both farm types. Climate change simulations using a uniform scenario with + 2.5°C and + 8% precipitation revealed losses for the farming sector to the tune of -75 US$/ha/year.

5.1 Conceptual understanding of the Ricardian Approach

To depict the fundamental differences between the traditional production function approach and the hedonic Ricardian approach, the following considerations will present both methods in their conceptual understanding. The following economic theory is obtained from MENDELSOHN et al. (1994; 1996) and SANGHI (1998).

In order to explain the specifics of the Ricardian approach the standard production function approach has to be looked upon first. The production function approach predicts the impact of differential weather patterns on crop yields by formulating a relationship between crop yield and exogenous environmental factors including climate. The simulations of future "climate"-or more realistically weather patterns with their crop yield effects are based on agronomic field experiments. The findings from these simulations are then incorporated into different economic farmer behavior models. This in turn yields a partial equilibrium model of the entire farm sector. Based on the idea that crop yields are also dependent on climatic and soil conditions amongst other inputs, the production function approach predicts yield impacts and changes in crop shares. Thus based on exogenous environmental inputs such as temperature, precipitation and site specific soil conditions and other inputs such as labor and capital the yields are calculated for different locations. These relations are empirically captured in the following four equations for a set of well-behaved production functions.

EQ (1) $$Q_i = Q_i(K_i, E) \qquad i = 1, \ldots \ldots .., n$$

With Q_i representing the yield of crop *i*, the vector *Ki* denoting the purchased inputs for the production of a specific good *i* and the vector *E* standing for external environmental inputs (location specific temperature, precipitation and soil conditions). To estimate the response of yields to climate, the yields of different locations are compared.

EQ (2) $$Q_i = \widehat{Q}_i(K_i, E) + u_{li}$$

Where $\hat{Q}i$ represents the predicted yield for crop *i* and u_{li} is an error term. The relative profitability of the chosen crop and constraints such as water availability determine the frequency with which the crop is planted. In this framework the share of the land devoted to a certain crop is understood to be a function of climate, market access and yields.

EQ (3) $$H_i = H_i(E, Z, Q_i) + u_{2i}$$

Where H_i denotes the crop share and u_{2i} an error term. The share of land devoted to the production of a specific crop is dependent on environmental factors (E), market share (Z) and on the respective crop yields per hectare Q_i. As the error term u_{li} is included in Q_i, the predicted yields $\hat{Q}i$ should be used instead of Q_i for the estimation of the crop shares. Subject to the environmental conditions faced by the farmer, income is maximized. As the profit maximizing input and yield is calculated, the net revenue for each crop can be predicted. Using a predefined crop-mix the net revenues per hectare for every country, region or district at study can be calculated (ROSENZWEIG AND IGLESIAS, 1994).

EQ (4) $$max\, N = \sum_i H_i \cdot \left(P_i \cdot Q_i - \sum_j w_j \cdot K_{ij}\right)$$

With *N* representing the net income per unit of land, P_i standing for the market price for crop *i*, K_{ij} denoting the quantity of the jth input used for the production of the ith crop and w_j representing the unit cost of the jth input. To obtain an aggregate net impact of environmental factors on yields the net impacts across all studied districts or locations have to be combined.

Although the traditional production function approach is closely linked to agronomic science and capable of predicting the impact of environmental factors on agricultural yields, it does not account for farmer adaptations (change of crop mix, different irrigation scheduling etc.) and thus creates a bias towards an overestimation of the harmful effects of climate change (Figure 5.2).

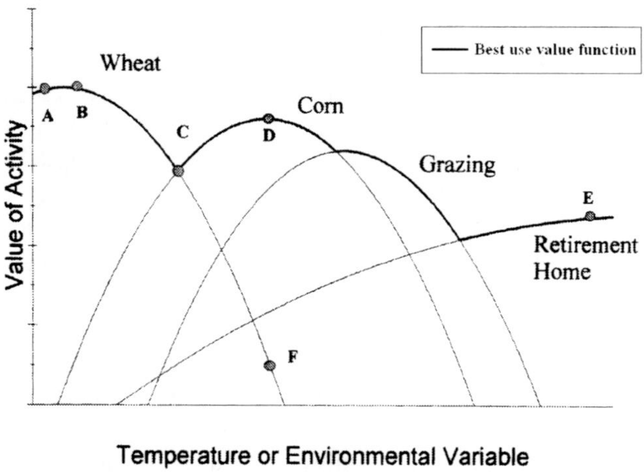

Figure 5.2 Bias of the production function approach to Climate Change analysis
Source: MENDELSOHN et al. (1994)

Figure 5.2 shows four different sectors (wheat, corn, grazing, retirement home) with their respective output values as a function of a single environmental variable, in this case temperature. In this simplified example the illustrative wheat production curve for instance shows how the value of wheat production or output continuously drops after initially rising from a colder temperature state A to a warmer temperature level B (optimum). After state B, the temperatures for wheat production have risen too high to make a production of wheat possible.

The misconception of the production function approach is best explained by having a look at point C and F. With rising temperatures the production function approach calculates a decline in wheat production to the point F, however wheat is no longer produced in reality as certain temperatures do not suit particular crops, therefore other crops are cultivated in those areas. Where the production function calculates the value at F, the reality is different, showing much higher values at point D, this to the background of an adaptation that has taken place at point C, where profit maximizing and rational farmers have realized the change in environmental conditions and taken steps to switch to corn production. In principle this is the Ricardian understanding of climate adaptation, where the switch from wheat to corn production at point C is facilitated by drawing on experiences from other farmers in other regions already facing similar environmental (temperature) conditions. Continued, this implies that at slightly higher temperatures than at point D, farmers will switch to grazing and finally at a certain point with temperature levels reaching point E, farming will not be possible, a finding shared by all models predicting future climate change. At point E farmer's will retire from agricultural production and find alternative uses for their facilities and land, such as building retirement homes, all this of course in an illustrative way to explain the issues related to climate change modeling in agriculture. To this background, as far as the Ricardian Approach is concerned, farmers in different regions are exposed to different climatic conditions, thus the method assumes that they adapt their behavior to their local climate circumstances (MENDELSOHN et al., 1996). Departing from this observation, and in order to correct for the production function bias, which is the overestimation of damages from climate change by not considering the full set of farmer adaptations, the idea is to compare the net revenues or land values (capitalized net revenues) of farmland in different climatic zones, instead of the yields. One great advantage of this approach is that it accounts for the direct and indirect effects on crop yields as well as for a wide range of adaptations that farmers make in response to a change in climate. By a spatial comparison, the model implicitly captures farmer adaptations. The method captures how farmers modify their production practices in response to environmental change embodied in water shortage, higher temperatures and precipitation changes. The approach thus tries to understand current adaptations not necessarily future adaptations. However, by understanding how farmers in different areas and regions have adapted to their local climatic conditions measured by a change in land value, an attempt is made to depict possible future adaptations. As the current land values have evolved over a long period of time with climatic and other environmental factors as their determinants, they serve as a practical fundament for the analysis of the long term phenomenon "climate change". This is a fundamental advantage over the yield based production function approach that can only model short term weather changes.

The initial approach named Ricardian method for climate impact assessment was developed by MENDELSOHN et al. (1994) who analyzed the impact of climate change on land values and consequently farm incomes in the United States. The model starts from the observation that farmer decisions are dependent on uncontrollable exogenous factors including climate and that the 'naïve farmer' or 'dumb farmer' scenario, in which the farmers regardless of occurring changes doggedly continue with their known practices, does not hold. HELM et al. (1996) argue with the slow nature of climate change, where time is given to the farmers to react and adapt. Thus given these exogenous constraints a farmer has to choose a certain mix of inputs (e.g. crops species, pesticides, irrigation) to maximize his profits. In a cross-sectional framework a comparative steady state analysis of long-term climate impacts on land values is carried out. By using a different response variable, namely net revenues or capitalized net revenues (land values), the cost structure is also affected (Figure 5.3).

Figure 5.3 shows a comparison between the sensitivity of crop yield and net revenue to climate. The graph on the left hand side (crop yield-climate), being a simplification of figure 5.2, shows the importance of including farmer adaptations, where changes in climate from state C* to C_1 without adaptation lead to a decrease in returns to R_{w0} from initial levels at R*. However, an inclusion of adaptation only reduces returns to the level R_w. Losses are incurred in both cases as farmers in the initial climate C* are hypothetically assumed to be already producing at the optimum. In the graph on the right hand side (net revenue-climate), net revenue replaces crop yield as the dependent variable, with clearly lower benefits from adaptation. This is reasoned in the fact that adaptations will be associated with certain costs which have to be included, diminishing returns. Here, the farmer is assumed to only adapt if the measure is beneficial.

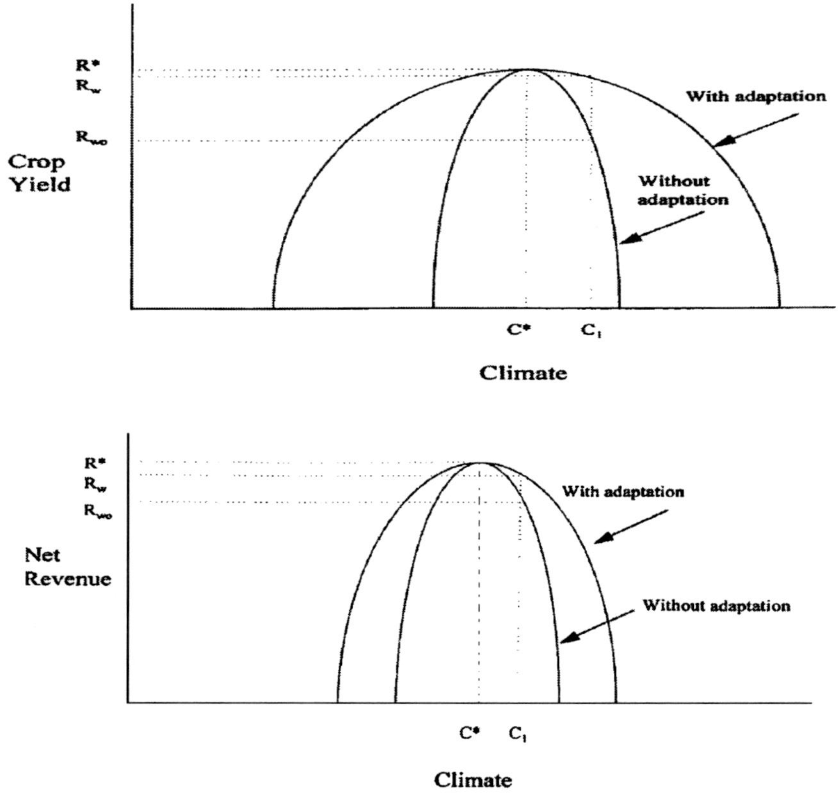

Figure 5.3 Sensitivity of crop yields and net revenues to Climate Change
Source: HELM et al. (1996)

The Ricardian model starts with the assumption of a set of well-behaved production functions.

EQ (5) $Q_i = Q_i(K_i, E) \qquad i = 1, \ldots \ldots, n$

Where, as in the previous considerations on the production function approach, the vector *Ki* denotes the purchased inputs for the production of a specific good *i*, K_{ij} is the purchased input *j* for the production of good *i* and the vector *E* stands for external environmental inputs including climate (location specific soil characteristics, precipitation and temperature). As the inputs, environmental factors and the output Q_i have prices w_j, cost minimization leads to a cost function of the following form:

EQ (6) $C_i = C_i\ (Q_i, w, E\ )$

With C_i denoting the production costs of good *i* and the vector *w* representing the factor prices. Now, the assumption of utility maximizing consumers with well behaved utility functions and linear budget constraints, who take prices as given, leads to a system of inverse demand functions for the outputs *i*=1,……n of the following form (SANGHI, 1998):

EQ (7) $P_i = D^{-1}\ (Q_1, \ldots Q_i, \ldots Q_n, Y\ )$

Where, *Y* is the aggregate income, P_i denotes the price of good *i* and *Q* the quantity of the good. Profit maximization yields:

EQ (8) $\max_{Q_i} P_i\, Q_i - C_{i-}(Q_i, w, E) - p_L L_i$

With p_L as the annual cost of land at a site and $\mathrm{C_{i-}(Q_i, w, E)}$ as the cost function of all purchased inputs except land; with $\mathrm{C_{i-}(Q_i, w, E) = C_i + p_L L_i}$.

Assuming perfect competition in the land markets drives profits to zero yielding the following equation:

EQ (9) $P_i Q_i^* - C_{i-}^*(Q_i^*, w, E) - p_L L_i^* = 0$

If under the given environmental conditions use *i* is the best use of land, the net revenues from the production of good *i* will equal the observed market rent. In other words, land rent per hectare is equal to net revenue per hectare; this is captured in the following equation, which is solved for p_L.

EQ (10) $p_L = [P_i Q_i^* - C_{i-}^*(Q_i^*, w, E)]/L_i$

Land value reflects the present value of the future stream of income from a parcel of land. A long term accumulation of net revenue determines farmland value (MENDELSOHN et al., 1996).

EQ (11) $V_L = \int_0^\infty p_L\, e^{-\varphi t}\, \mathrm{d}t = \int_0^\infty [P_i Q_i - C_{i-}(Q_i, w, E)]\, e^{-\varphi t}/L_i dt$

or

$$V = \beta_0 + \beta_1 F + \beta_2 F^2 + \beta_3 Z + \beta_4 G + u$$

Where t is time, φ is the discount rate and L_i denotes the amount of land in farm i. Land Values (V) are affected by the set of exogenous variables F, Z, and G. F is the vector of climate variables, Z is the sets of soil classification, G is the set of socio-economic indicators. EQ (11) has been termed the 'essence of the Ricardian model' by MENDELSOHN et al. (1996) (Figure 5.4).

Figure 5.4 The essence of the Ricardian Model
Source: own illustration after MENDELSOHN et al. (1996)

As the main goal is to value climate change, the change in climate from an initial level E_A to a new level E_B also reveals the change in annual welfare, which is given by:

EQ(12) $$\Delta W = W(E_B) - W(E_A) = \int_0^{Q_B} \Sigma\, D^{-1}(Q_i)dQ_i - \Sigma C_{i-}(Q_i, w, E_A) - \int_0^{Q_A} \Sigma\, D^{-1}(Q_i)dQ_i - \Sigma C_{i-}(Q_i, w, E_B)$$

Assuming constant market prices over both climate levels leads to a reduced equation:

EQ (13) $$\Delta W = W(E_B) - W(E_A) = PQ_B - \Sigma C_{i-}(Q_i, w, E_A) - [PQ_A - \Sigma C_{i-}(Q_i, w, E_B)]$$

Figure 5.5 illustrates the welfare impact of a change in the environmental variable from an initial state A to a new level B, where the supply curve (marginal cost curve) shifts outwards. A substitution of the land per hectare equation into equation (13) gives:

EQ (14) $$\Delta W = W(E_B) - W(E_A) = \textstyle\sum_i (p_{LB}\, L_B - p_{LA} L_A)$$

Where the land rent per hectare in the initial climate state A (E_A) is subtracted from the land rent per hectare in the new climate state B (E_B). The following equation (EQ 15) gives the present value of this welfare change.

EQ (15) $\int_0^{\infty} \Delta W e^{-\varphi t}\, dt = \sum_i (V_{LB} L_B - V_{LA} L_A)$

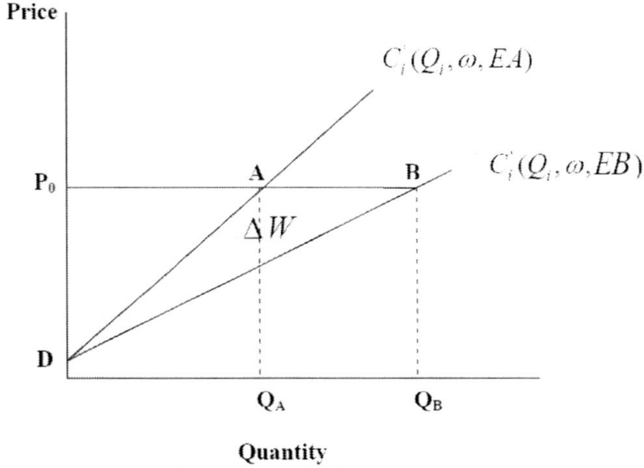

Figure 5.5 Climate Change and its welfare impacts
Source: AMIRASLANY (2010)

The present value of the stream of welfare change is captured in the difference between land values in the new environmental state and land values in the initial environmental state. Dependent on the choice of the dependent variable, whether it is annual net revenue or farmland value (capitalized net revenues) the Ricardian Model can take the form of either EQ (10) or EQ (15). Thus, by using cross-sectional observation a change in environmental conditions (normal climate) is reflected in land values and production. As land values or net revenues vary with location and climatic conditions, farmer adaptation is implicitly captured.

A beneficial climate change is embodied in higher land values, whereas if farmland values decrease, the climate change is harmful. The aggregate effect is found by summing up all effects over all observations and applying weights (e.g. amount of land in a farm, region or district). The model can thus be estimated for districts, regions and Agro-Ecological Zones. After estimating the so called Ricardian climate response function with current climatic conditions, one can replace the current climate situation with any other climate change scenario and simulate future impacts. However, as the ability of farmers to adapt besides environmental changes is also bound to many other factors that may well change over time (e.g. human capital, government support structure, technology),

(MENDELSOHN AND DINAR, 2009) have clearly called for a cautious approach when interpreting the results of the simulations and future predictions based on the Ricardian estimates of climate change impacts on agriculture.

To sum up, the Ricardian valuation model's key hypothesis is that through a change in climate the production function (crops or livestock) will be shifted. As farmers observe these changes in climate, they manage to adapt to their local conditions by adjusting their inputs and outputs accordingly. This implies that in the current situation, the economy has well adapted and land prices have reached a long-run equilibrium in the different districts (with climate as a determining factor).

5.2 Climate Change Adaptation in Agriculture

This chapter's focus is on presenting a literature review on climate change adaptation studies and to depict the most likely adaptation scenario for the developing country context. The understanding of adaptation is an important goal in itself in order to assist planning by policy makers and private individuals. Climate change impacts can be quantified using different economic simulation techniques; however the magnitude of the impacts highly depends on the consideration of adaptations while modeling.

Although many different definitions exist, adaptations in the climate change context can be understood as measures undertaken to reduce the adverse impacts from global warming on natural and human systems. Individuals, firms, communities and other entities take these actions mainly for two reasons, either to reduce the damages or increase the benefits from climate change (IPCC, 2007; BARRY et al., 2000). Adaptations however, have always been the need of societies, particularly in the agricultural sector. Climate has changed ever since the existence of the human being, thus with time civilizations have adapted to occurring changes that have impacted their lives. Societies have always adjusted, either by introducing new technologies and structural changes or by changing behavior. Over the years knowledge regarding different adaptation options has been collected through different societies. This historically accumulated knowledge has then been used by following civilizations and societies, which then have continuously enhanced this knowledge by their own climate related experiences, also regarding extreme events (OECD, 2009). From a theoretical perspective a vast body of literature has been concerned with different aspects related to climate change adaptations. To pinpoint the main directions of these literature surveys, adaptation can be classified on the basis of: intent, temporal scope, timing and agent (see Table 5.3).

Table 5.3 Classification of Adaptation to Climate Change

Basis	Type	Description
Intent based	*Autonomous adaptation*	spontaneous; not a conscious response to climate change but triggered by ecological changes in natural systems and welfare changes in human systems
	Planned adapation	based on a political decision making process to sustain a desired state; triggered by the awareness of an expected change
Timing based	*Proactive adaptation*	takes place ex-ante: before climate change impacts have been observed
	Reactive adaptation	takes place ex-post: after climate change impacts have been observed
Agent Based	*Private adaptation*	initiated and implemented by individual agents, households or firms for the sake of the own benefit
	Public adaptation	initiated and implemented by governments at several levels for the sake of the collective benefit or need
Temporal Scope based	*Short-Run adaptation*	decisions are constrained by a fixed capital stock, thus principal available options are confined to variable inputs to production (STERN, 2007)
	Long-Run adaptation	capital stock is adjustable by the decision maker

Source: modified after MALIK et al. (2010)

As far as the classification on the basis of intent or purposefulness is concerned, according to EITZINGER et al. (2007), the changing risks for agricultural production due to climate change will have several income, food and input-related implications for the sector; therefore planned adaptation clearly should be preferred over forced last minute adaptation. Another key argument in favor of the planned adaptation option is its underlying political support framework and thus its extended range of strategies which can also benefit and improve autonomous adaptations (SMIT AND PILIFOSOVA, 2001). However, the fact that even autonomous adaptation takes place in the context of given current government policies should not be neglected (WORLD BANK, 2008a).

From the "timing" perspective two dimensions of adaptation have been identified, namely *proactive (ex-ante)* and *reactive (ex-post)*. These proactive or so-called ex-ante adaptations are made in the anticipation of an expected future change, thus they heavily rely on the decision maker's ability to accurately predict future change; whereas the reactive adaptations are likely to be made ex-post, after the occurrence of change (see also KURUKULASURIYA AND ROSENTHAL, 2003). In order to evaluate the effectiveness of the above mentioned approaches to adaptations several authors have emphasized the need to have a close look at the resources available to the agents. These resources can be of financial, technological or institutional nature. If a farm or firm has larger stocks of these resources reactive adaptation might evolve as the efficient option, whereas a lack of these resources could easily render effectiveness in favor of the ex-ante option (DINAR et al., 2008). According to MENDELSOHN (2000) however, the agricultural sector due to its capability of implementing rapid adjustments in terms of technology, management and resources, is more or less well suited for adaptation to climate change as it unfolds. Moreover, especially in developing countries the agricultural sector is dominated by small-holder farmers that do not possess long-lasting capital such as new technologies which would compromise flexibility in terms of the reactive adaptation. Thus, these farmers could also adapt more or less easily using the reactive approach. In addition to these rather short-term autonomous adaptation options in developing countries, the long-term vulnerability associated with climate change can most effectively be dealt with by implementing ex-ante strategies, which should therefore not be sidelined.

For gaining additional insights into the adaptation discussion, it is necessary to distinguish between the adapting agents. The literature generally distinguishes between private and public adaptations. A private adaptation in this sense can be seen as an adaptation made by an individual for the sake of his own benefit. Public adaptations on the other hand, are made in order to benefit many people or groups of people. The construction of a coastal protection system for instance can be one such public adaptation measure, as it has many beneficiaries. The issue arising from such public adaptation schemes is that they can't discriminate between the beneficiaries, especially in the agricultural sector unequal benefit from climate change protection mechanisms can raise serious issues. Moreover, if the beneficiaries of a public adaptation scheme are very heterogeneous, this would complicate the issue further. Thus markets would have trouble in allocating the benefits to the people in a just way, whereas just would be another relative concept to be reconsidered in this context. In addition, the decision over the distribution of benefits would have to be taken at some institutional or market-institutional level, which could prove extremely difficult in politically fragile countries (see also MENDELSOHN, 2006). Bad governance and consequently the absence of private property rights would hamper incentives to protect the environment.

As the review of literature has revealed there is a certain amount of inconclusiveness in the international discussion on the focus of adaptation, whether i.e. autonomous measures outperform proactive ex-ante adaptation or vice versa. Although these different classifications in table 5.3 might suggest a distinct separation between the alternatives, the interrelations and complementarities are most evident. For example in the real world context, where the initiative to adapt might at times originate from the public level and then through relevant measures reach farmers and other autonomous decision makers. Table 5.3 shows the extremes by subdividing adaptation into "planned" and "autonomous", however between these two several other mixed forms of adaptation can and do exist, suggesting that adaptation is best described as a continuum rather than a discrete undertaking (see MALIK et al., 2010). What can be certainly concluded is the need for both approaches, as the one complements or at times substitutes the other, especially proactive solutions can benefit reactive adaptations through knowledge and experience transfer. A review on the nature of the relation between autonomous and planned adaptation, whether it is complementary or substitutional, is presented by FRANKHAUSER et al. (1999).

The need to develop site-specific vulnerability assessments and consequently site-specific adaptation measures seems most obvious for efficient outcomes (FELDMAN AND KAHAN, 2007). Which adaptation option is best or most realistic in the case of developing countries is a complicated question to answer. That the separation between these classifications of adaptation is blurry is most evident, however for the purpose of analysis and for framing the issue one scenario has to be depicted which seems to fit to the real world example of the country at study. This issue shall be addressed in the following deliberations, which stress the importance of local traditional knowledge for facilitating successful adaptation to environmental change. A detailed discussion on both types of adaptation options for the developing country perspective is carried out by COLLIER et al. (2008), using the African continent as an example. They find that the majority of adaptations will be of autonomous nature because climate changes occur gradually. Thus, the naturally forward looking behavior of private agents would protect them from being overwhelmed by the gradual character of environmental changes. Furthermore as far as planned or ex-ante adaptation is concerned, they find that without adequate information through the public sector an appropriate response would not be possible. To this background, they question the capacity of weak governments and institutions in developing countries to deliver to the demand of the changing conditions. In another study, AGARWAL (2008) based on the finding that in developing countries maintaining the livelihood is peoples' main concern, discusses possible coping strategies in the context of risks to livelihoods resulting from climatic or environmental change. Nonetheless, in proactive adaptations actors are required to plan for how changes in climate affect planned projects such as dams, roads and coastal defenses amongst others. Therefore, as the prediction of local climate change is a difficult task to achieve, reactive or autonomous adaptations, especially in the developing country context, can be expected to be the primary mode of adaptation (MENDELSOHN AND DINAR, 2009). In this context the upcoming subchapter is devoted to unveiling existing adaptations made by locals to cope with changes in environmental conditions. Particular emphasis is laid upon the importance of traditional or indigenous local knowledge, which has been inherited and nurtured over generations.

5.3 Adaptation and the importance of Local Traditional Knowledge (LTK)

Local traditional knowledge has ever since the existence of human civilizations played a key role in guaranteeing a successful adaptation to ever changing conditions that humans have been faced with. Not only restricted to South Asia, however extremely important in this particular region, which once was home to one of the great ancient civilizations of the old world, namely the Indus valley civilization. According to archaeological data the civilization was established 3300 B.C. and disappeared around 1400 B.C. As it flourished, the civilization stretched over almost the entire territory of present-day Pakistan. The Indus civilization managed to construct cities using a grid pattern and ensured water supply and the sanitary disposal of thrash by designing complex drainage and sewage systems. Moreover, ancient records hint towards the existence of affluent agricultural knowledge. As the Indus river was and also today is seasonally flooded, rich soil appears for farming, which was made use of by taking flood control measures. Archaeological records i.e. document the construction of irrigation canals for the dry season. As far as the construction of cities was concerned, plaster coated bricks were used to prepare for the floods. Like in modern day systems storage was also practiced, thus i.e. grain was collected in large sized granaries in the city centers (KENOYER AND MEADOW, 2000; POSSEHL, 2002; JAVONILLO, 2010). Although not much is known about the Indus valley civilization, the facts that are known today clearly reveal the importance of local traditional knowledge that the human being has used in ancient times and by adapting it to the demands of the local situation, even today benefits from it. WARREN et al. (1995) define the local traditional knowledge as knowledge, which is institutionalized and orally transmitted and modified from generation to generation. Today as well, many communities rely on this rich source of accumulated knowledge to assist decision making processes that can be related to but are not only restricted to agriculture, clothing, hunting, and environmental change (SILLITOE, 2007). Local traditional knowledge mostly refers to agricultural practices and rural experiences. The notion "local" is of great relevance for understanding the overall concept, as it more or less restricts knowledge to a certain community or region with its nuances and specific environmental or climatic conditions. Thus it does not indicate transfer of knowledge, but binds people to their local experiences and environments (WOODLEY, 1991; KEMPTON, 2001). A study by SUNDARAMARI (2001) found 1203 recorded traditional practices in a district of India. A recent report under the umbrella of the CGIAR documents several local adaptation measures that famers undertake in the Indo-Gangetic Plains (RIVERA-FERRE et al., 2012). Table 5.4 documents the major uses of the indigenous knowledge that have been identified by the expert groups.

Table 5.4 Major uses of local traditional knowledge in the Indo-Ganges Plains

Management	Soil management
	Water management
	Terrace management
	Rain fed cropping
	Crop diversification
	Changes in cropping pattern
	Resource efficient cropping systems
Seeds	Seed saving
	Knowledge about what crop varieties to plant
	Local communities choice of varieties
	Local seed selection process, storage, exchange system
	Changes in crop varieties
Weather	Climate prediction (local weather forecasting system)
Food	Food preservation
	Food preparation/consumption

Source: RIVERA-FERRE et al., 2012

As the illustration shows, a wide range of agriculture and food related activities are covered. From the climate context, a very interesting finding is the existence of local weather forecasting systems. Importantly, farmers or communities also practice crop diversification, change cropping patterns and varieties, and manage water systems. Various traditional systems exist in Pakistan that are still used today, one of them is the so- called "Warabandi" water allocation system. It was traditionally designed to allocate irrigation water in an equitable manner in times of water shortage by using the size of the landholding as the distributional criteria. It is understood to be of great relevance for surviving droughts (SRINIVASAN, 2004). Another water related traditional concept is the "chat" (roof), which is constructed to store water. The "chats" are traditional water reservoirs and are e.g. found in the district of Chitral. Traditional systems for infrastructure maintenance are also in place in many regions of the country, examples are the so-called "Mirzhoi" system and the "Mone" systems. In the "Mirzhoi" system, the community hires personnel to maintain the irrigation channels. Mostly smaller maintenance is delivered under "Mirzhoi" system. For larger maintenance duties the "Mone" system is used, where each beneficiary of the channel or other infrastructural facility has to contribute individually on a rotational and by turn basis. In the "Nirwalu" system an equitable water distribution is ultimately aimed at. It is a low-tech solution, where in front of a pond or small stream a piece of wood with equal sized holes is installed. Cooperation between farmers is also common, where farmers either pool and distribute their available resources or engage in a system of borrowing and swapping (NADEEM et al., 2009; ICIMOD, 2009). Some traditional practices for weather forecasting include experiences of elderly people, who forecast on the basis of indicators such as the blow of a cool breeze (poora), indicating the imminence of rain or a lull signaling a wind storm and clouds in the shape of a turkey-feather indicating approaching rainfall (RIVERA-FERRE et al., 2012).

The foregone deliberations have discussed the importance and types of adaptations and accentuated the role of local traditional knowledge when addressing adaptation to climatic or environmental change. In this context it is understood that, as adaptation at the micro-spatial scale has ever since taken place, it can be expected to continue to play a key role when the climate changes. Communities are highly dependent on their local environment and therefore have been in many instances found to be guardians of the same (depending on the duration of the dependence or interdependence); they have thrived within the systems by adapting to stresses due to drought, flood and long term weather fluctuations. In the spirit of the discussion on adaptations, the following considerations aim at presenting a remedy for the standard Ricardian model's restrictive characteristic of treating adaptations as implicit in land values, without uncovering the specific measures and the value of these measures that farmers undertake. The following subchapter presents the advanced Structural Ricardian model and its application for examining the economics of adaptation processes such as inter alia crop switching.

5.4 Black Box of Adaptation- The Structural Ricardian Method

As described in the theoretical considerations on the standard Ricardian method, the traditional model measures the change in the value of land across a set of exogenous variable such as climate and soils. Thus the standard model assumes that farmers adjust their inputs and outputs to match the conditions they are locally exposed to, in this way adaptation is implicitly captured. Adaptations in the standard setting are not explicitly modeled. Therefore they are more or less treated as a black box. How the farmers specifically adjust to match their local conditions is not revealed. To this backdrop, a handful of studies have endeavored to uncover the explicit adaptations that farmers make, when faced with different climatic conditions. The approach used for unveiling the black box of adaptation has been termed "Structural Ricardian Method" (KURUKULASURIYA AND MENDELSOHN, 2008). This advanced approach is a blend of the standard model and a choice model. It is a multiple stage model that first measures the choices (crop, livestock, and farm type) that the famers make and then estimates the conditional income for each choice. In this way structural model tries to estimate the underlying choice model that leads to the standard Ricardian locus of profit maximization.

The Structural Ricardian Studies have so far been limited to South America and Africa. MENDELSOHN AND SEO (2008) first studied the choice of farm type for South America. In subsequent studies using the same extended farm level survey, analyses of livestock and crop choice were carried out (SEO et al., 2010). For the African continent studies analyzed crop, livestock and irrigation choice (KURUKULASURIYA AND MENDELSOHN, 2008a; SEO AND MENDELSOHN, 2008A; KURUKULASURIYA et al., 2011). Crop choice sensitivity was confirmed by all studies. The African results i.e. found farmers to favor fruits, millet and vegetables over maize in case of warming. South American farmers with the incidence of temperature increase were found to opt for growing fruits, vegetables, soybeans and rice, depending on their specific location (country). The studies showed that spatially the choices clearly differ amongst farmers, reasoned in the fact that they are exposed to differential climatic conditions. This study is the first of its kind for South Asia.

Theoretical considerations and application to Pakistan

A general assumption of the choice model is that farmers make a choice amongst a set of alternative crops, livestock species or farm types to maximize the net revenue of their farm. Net revenue is calculated by subtracting all costs except household labor from the gross revenue of the farm. The choice that will earn the farm the highest net revenue will be marked as the primary choice. In this study the focus is on crop choice only, because the livestock sector could not be included due to data constraints. Thus, farmers in Pakistan can choose between seven crops, namely wheat, cotton, sugarcane, rice, maize, fruits and vegetables. Farmers in Pakistan have reported a number of different crops, thus they can actually choose more than one crop at a time. Different approaches exist that can be used to model the crop choice of the famer, either only a primary crop can be modeled or an entire crop portfolio that treats combinations of crops as a choice can be studied. Although the portfolio approach has the advantage that it covers the full range of possible adaptations, the computational complexity with an increasing number of choices available tends to restrict its use. An increasing number of choices (n) increases the number of potential combinations by 2^{n-1}. Another issue that arises with the portfolio approach is that the choices might not be mutually exclusive anymore, which would raise issues regarding the independence of irrelevant alternatives assumption in models for categorical outcomes. As farmers in this study have reported more than 10 crops, the number of combinations is large and complicates modeling. To these drawbacks, in this study only the primary crop choice is considered, defined as the crop that earns the farm the highest net revenue. Thus in cases where farmers have reported multiple crops, the crop choice is nevertheless defined as the single crop with the greatest revenue. An additional advantage of the primary crop choice model is that the choices are mutually exclusive. To estimate the choice that the farmer makes, different econometric models such as probit, logit and multinomial logit can be used. Conditional on the crop choice a second stage model estimates the net revenue for the farm. The model output can be used to forecast adaptations and impacts for different future climate scenarios and the value of adaptation can be computed by comparing adaptation and non-adaptation (MENDELSOHN AND DINAR, 2009).

The theoretical considerations follow SEO AND MENDELSOHN (2008, 2008a), MENDELSOHN AND DINAR (2009). As farmers are assumed to select the desired crop to yield the highest net farm revenue, the net farm revenue of farmer j in choosing crop i from a set of mutually exclusive choices j (j=1,2…,N) is

EQ (16) $$\pi_{ij} = V_i(K_j, S_j) + \varepsilon_i(K_j, S_j)$$

with π denoting net revenue, K representing a vector of the exogenous farm characteristics (e.g. climate, soil types) and S standing for a vector of the farmer's characteristics (household characteristics such as age, education, household size). The observable component of the net revenue function is V, whereas ε represents the error term. The model assumes that the farmer is well aware of the error term and therefore selects the crop which generates the highest net farm revenue. Thus defining $Z= (K, S)$, the farmer will choose crop i over all other crops if:

EQ (17) $$\pi_i^*(Z_j) > \pi_k^*(Z_j) \text{ for } \forall\, k \neq i.$$

$$\left[\text{or if } \varepsilon_k(Z_j) - \varepsilon_i(Z_j) < V_i(Z_j) - V_k(Z_j) \text{for } k \neq i\right]$$

Briefly, farmer j's problem can be defined as:

EQ (18) $$\arg \underset{i=1 \ldots Ii}{max}\left[\pi_1^*(Z_j), \pi_2^*(Z_j), \ldots, \pi_I^*(Z_j)\right]$$

The probability P_{ji} for crop i to be chosen by farmer j is thus given by:

EQ (19) $$P_{ji} = \Pr\left[\varepsilon_k(Z_j) - \varepsilon_i(Z_j) < V_i - V_k\right] \forall\, k \neq i \text{ where } V_i = V_i(Z_j)$$

Assuming that the error term ε is independently and identically Gumbel distributed (see GREENE, 2006) and V_k can be written linearly in the parameters such as:

EQ (20) $$V_k = Z_{kj}\gamma_j + \alpha_k$$

The probability that farmer j will choose crop i among I crops is given by (CHOW, 1983; MCFADDEN, 1981):

EQ (21) $$P_{ji} = \frac{e^{Z_{ji}\gamma_j}}{\sum_{k=1}^{I} e^{Z_{jk}\gamma_j}}$$

This is the standard derivation of the multinomial logit model, with γ_j as the parameters of the model.

From a technical point of view the issue that arises is that the errors in the selection equation (α_k) and conditional income equation(ε_i) might be correlated, because the profit described in the selection equation is only observed for the chosen crop. To this background, for the estimation of the conditional revenues, which are conditional on the choice of a certain crop, selection bias has to be corrected for in order to receive consistent estimates (HECKMANN, 1979). If the correction is not applied the OLS estimates become inefficient, as the calculation of income causes a high association between the non-observable characteristics affecting income and those that simultaneously determine the sector in which the individuals are working (HUESCA AND CAMBEROS, 2010). For instance, the farmers choosing the crop wheat are not a random sample of all farmers that select wheat. Farmers self-select their primary crop. Not all of the reasons for this self-selection are known to the researcher. The researcher does not observe the profits that the farmer would have made had he chosen wheat. Under certain circumstances the omission of potential members of a sample will cause ordinary least squares (OLS) to give biased estimates of the parameters of a model (CURRAN, 2010).
Following BOURGUIGNON (2004) and BOURGUIGNON et al. (2007) DUBIN AND MCFADDEN's Method (1984) is selected, according to which assuming the following linearity condition

EQ (22) $$E\left(\varepsilon_j \middle| \alpha_1, \dots, \alpha_J\right) = \sigma \cdot \sum_{j=1}^{J} r_j \cdot \left(\alpha_j - E(\alpha_j)\right), \text{with } \sum_{j=1}^{J} r_j = 0\,,$$

where ε_j is the error from the profit equation in the second stage of the model, α_j is the error from the first stage's choice equation, σ is the standard error of the net revenue or profit equation, and r_j the correlation between the net revenue equation and choice equations, the selection-bias corrected conditional net revenue function for crop choice *i* can be estimated as follows:

EQ (23) $$\pi_i = X_i \varphi_i + \sigma \cdot \sum_{k \neq 1}^{K} r_k \cdot \left[\frac{P_k \cdot \ln P_k}{1 - P_k} + \ln P_i\right] + \omega_i$$

Where X_i is a set of independent variables that include climate variables, socioeconomic variables, hydrological variables and soils; φ_i represents a vector of parameters to be estimated, and ω_i is the error term. After the correction α in equation 20 and ω in equation 23 are independent.

Equations 21 and 23 form the core of the two-stage analysis. The expected net revenue (*NR*) of the farm is the sum of the probabilities across each choice the farmer faces, multiplied by the conditional net revenue of the selected farm type. Precisely:

EQ (24) $$NR_i(C) = \sum_{k=1}^{K} P_k(C_i, Z_i) \cdot \pi_k(C_i, Z_i)$$

As has been argued in the deliberations on the standard Ricardian method, either land value or net revenue can be used as the measure for conditional income. As the dataset on land values did not include a sub-questionnaire on Agricultural production, a different dataset is used, covering detailed information on farming on the household scale for the year 2007, therefore farm net revenue is the measure for conditional income. Details regarding the dataset, the model specific empirical strategy and the results will be presented in the upcoming considerations in chapter 7.

6 Standard Ricardian Model - Modifications and Application to Pakistan

In this chapter, the final model specification shall be presented. Before estimating the specified model, a detailed account on the study area and the dataset is provided. In this context detailed summary statistics on a variable by variable basis are presented. After presenting the descriptive statistics the econometric estimation is described in detail including relevant misspecification tests. The results are discussed to reveal the sensitivity of agriculture to climatic change and quantify the economic impacts. The Standard Ricardian model is constructed using land values as the left hand side variable.

6.1 Study Area and Data

Pakistan in total has eight administrative units of which four are official provinces and one is a federal capital territory, namely Islamabad. The provinces are Punjab, Sindh, Khyber Pakhtunkhwa (N.W.F.P) and Balochistan. The other units comprise the Federal Administrative Tribal Area (F.A.T.A.), Disputed Area with India and Azad Kashmir (Pakistani part of the disputed territory of Kashmir, the Indian part is known as Jammu & Kashmir). The capital of Islamabad is considered a part of the Punjab province. As second order administrative units the country currently has 114 districts in the above mentioned four principal provinces, with 30 of them in the Balochistan Province, 24 in the Khyber Pakhtunkhwa province, 38 (including Islamabad) in the Punjab province and 23 in the Sindh province. This study covers 96.5% of the country by analyzing 110 districts of Pakistan. The districts of Harnai and Sherani in the Balochistan province, the Chiniot district in the Punjab province and the Umerkot district in the Sindh province are excluded from the analysis due to data unavailability (Figure 6.1).

As the introductory chapters have revealed the country offers a diverse range of climatic regions, which makes a study covering a significant area even more necessary. In Pakistan the climatic and agro ecological setting is very diverse and thus this high variation in terms of precipitation, temperature, elevation, soil types and cropping systems has to be analyzed with all its regional specifics. This study covers all 10 Agro Ecological Zones (AEZ) of the country from the mountains in the north to the Indus delta in the south, from the irrigated plains of the east to the dry mountains of the west.

Figure 6.1 Study Area- Land Value Dataset
Source: own illustration using ArcGIS 10.1 (ESRI, 2011)

For a district level assessment however, in total 24 AEZs using georeferencing in ArcGIS (ESRI-Geographical Information System) have been created. Overlapping zones have been assigned individual AEZ status, more or less as subzones of the 10 major AEZs. As the country as a whole is at study and climatic and agro ecological features have been already explained in the introductory chapters a further repetition of the study area specifics is not deemed necessary, merely the focus will be on the explanation of the dataset with its coverage and summary statistics, this with respect to the chosen four provinces and 24 Agro Ecological Zones. After a general exploration of the dataset a more detailed variable by variable overview will follow.

Generally, cross-sectional analysis using the Ricardian technique requires a very large sample size, covering all climatic regions of the country or region at study. The African study involved surveys from 11000 farms across 11 countries, whereas the South American study collected surveys from 2500 farms across 7 countries (DINAR et al., 2008; SEO AND MENDELSOHN, 2008). The African survey was conducted under the umbrella of a Global Environment Facility (GEF) financed and World Bank managed project "Climate Change and Agriculture in Africa" (CEEPA, 2002), with a total financial support volume of about 1 million US$. To this backdrop, considering the time and financial resources needed to convey a large household survey of the above mentioned extent, the present paper for the purpose of analysis instead, uses a complex household survey (Living Standards Measurement Survey) conducted by the Federal Bureau of Statistics of Pakistan (FBS) for the year 2009 (Annex I). "The Pakistan Social and Living Standards Measurement Survey (PSLM) is one of the main mechanisms for monitoring the implementation of the Pakistan Rural Support Program (PRSP) in the overall context of MDGs" (FBS, 2010). In the survey, based on a unique sampling frame enumeration blocks for every city and town were created, comprising 200 to 250 households each. The sampled households were grouped by income classes. For the purpose of stratification (two-stage stratified sample design) the households were assigned two domains, either urban or rural. In total the survey comprises 75.188 households and 5.298 villages and covers all important aspects of life in rural and urban areas of the country, including estimates of land value. The survey was personally acquired in agreement with the country's data dissemination policy for research purposes in 2010 (FBS, 2009). Based on the possession of agricultural land 22.240 owner operated farms were selected, covering all Agro-Ecological-Zones and major provinces of the country. Figure 6.2 shows the distribution of the farm households by province with the Punjab province contributing significantly to the total sample. As farm data in the survey was not purposely collected for a Ricardian regression, the single farm household datasets were aggregated into 110 districts for the analysis of climatic impacts across different Agro-Ecological settings. In a large cross sectional dataset the issue with outliers is evident. As a remedy a detailed EDA (Explorative Data Analysis) on a single district level preceded the aggregation process to eliminate the most obvious outliers in the dataset, using extreme value statistics and graphical tools such as histograms and box plots in combination. As the EDA is conducted on district level, only the most obvious outliers on a district scale are eliminated not on province scale, this in order to keep the variation which is inherent in the dataset.

In case a value is identified as an outlier or extreme value but the frequency of appearance throughout the district is relatively large (e.g. 3 extreme cases in a total of 20 observations), the values are retained and not eliminated, only if the extreme case is a single appearance in the general distribution exclusion is opted for. In any case exclusion is preferred over other solutions.

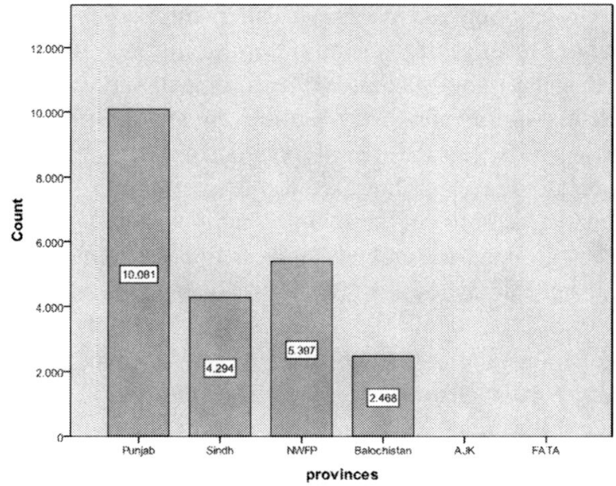

Figure 6.2 Number of Farms by Province – Total Initial Dataset
Source: own illustration using SPSS 17.0 (SPSS, 2009)

Mean, Median and Maximum Likelihood Estimators were compared, before opting for the mean as the best estimator of the aggregated values. Figure 6.3 shows the final working sample size.

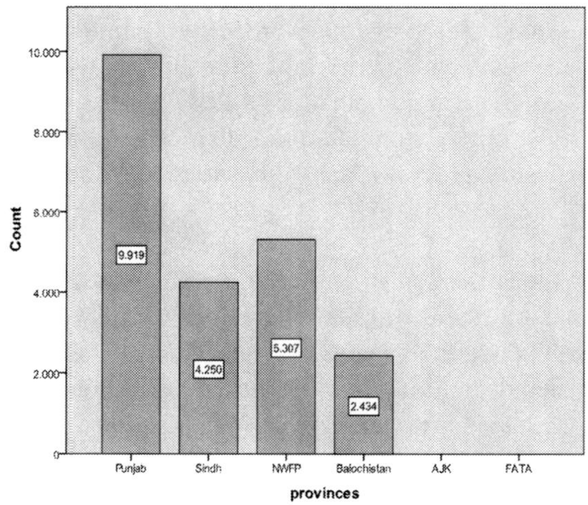

Figure 6.3 Number of Farms by Province – Total cleaned Dataset post- EDA
Source: own illustration using SPSS 17.0 (SPSS, 2009)

As Figure 6.3 and Table 6.1 show, the relative contributions to the sample size do not change after the EDA, the total sample size is reduced to 21.910. The Punjab province contributes 45% to the total sample size, whereas the smallest contribution is made by the Balochistan province with 11%. When weighted, these households represent a great majority of the farmers in the country.

Table 6.1 Sample Size Distribution by farm type, region and scale

Province	Total Farms (N)	Irrigated	Rainfed	Urban Farms	Rural Farms	small scale farms < 5 acres	medium scale farms 5-25 acres	large scale farms >25 acres
Punjab	9919	8531	1388	1420	8499	6060	3506	353
Sindh	4250	3609	641	423	3827	1567	2282	401
NWFP	5307	2569	2738	460	4847	4289	900	118
Balochistan	2434	1364	1070	171	2263	939	1344	151
Total	21910	16073	5837	2474	19436	12855	8032	1023

Source: own calculations (2012)

As the present study aims at answering detailed questions with regard to farming seasons and the behavior of farming to different seasonal climatic patterns, the total dataset comprising 21.910 farms is disaggregated into irrigated, rainfed, small scale and medium scale farms. A regional separation of the households into rural and urban is not undertaken due to twofold reasons. First of all, the sample size for urban farms is relatively small and covers only a small number of districts. On top of small district coverage, comes the fact that the covered districts do not count a sufficient number of households, this would lead to aggregation on the basis of two or three cases, which is deemed insufficient. Especially considering that, the study as a rule of thumb throughout has aggregated single farm information into district values solely when a minimum of ten to twenty farms per district were available. Secondly, additional information regarding the exact location of the farms is missing and thus regional effects which would result from factors such as close proximity of the farms to a city center and road infrastructure cannot be captured, thus rendering a possible regional classification rather superfluous. Large scale farms are not considered due to a much too small sample size, which predominantly only covers the major agricultural provinces of Punjab and Sindh. Farm scales are defined following KIANI (2008), with farms smaller 5 acres assigned the label "small", farms sized 5-25 acres labeled as "medium" and farms owning more than 25 acres defined as "large". A large dataset is constructed by combining the detailed farm-household level information from the Pakistan Living Standards Measurement Survey (PSLM) 2009 (FBS, 2010) with climate and soil data from various sources, including the Food and Agriculture Organization (FAO, 2003, 2006) the Pakistan Meteorological Department and the National Oceanic and Atmospheric Administration (USA) (NCDC, 2011). The household data is combined with data on long-term climate normals, additional sources of water including runoff and surface water and an important range of socio-economic control variables. In following, after deliberations on the quality and reliability of the survey the data, the modeled variables shall be explained in detail. The datasets and surveys used (2008-2009) have been created and collected under the German Federal Ministry for Economic Cooperation and Development's (BMZ) funded project "Support for the Federal Bureau of Statistics (FBS)". The project was carried out by the GIZ (Gesellschaft für Internationale Zusammenarbeit) with the Pakistan Ministry of Economic Affairs and Statistics as the lead executing agency for a period of six years, starting from 2005 (GIZ, 2011). Being the country's major statistical office, the FBS has the function of providing reliable data for e.g. government decision making, research and international community. However, not always the international statistical standards have been met. With the major objective to enable the FBS to release up to date economic statistics of internationally recognized standard, which can be used for purposes of academic research and international decision making, the project execution in 2005 extended the FBS's staff, extended the range of covered subjects and decentralized data collection and most importantly merged the FBS, the Population Census Organization and the Agricultural Census Organization. The GIZ brought- in its expertise for the preparation and analysis and publishing of statistical data. Quality control was introduced, managed and enacted by the GIZ, which also headed the assessment of planned data collection mechanisms and questionnaires. To guarantee a permanent use of international standards and calculation methods such as proposed by the international organizations (UNDP, IMF, ADB etc.) the project introduced a daily application of these methods for gathering, analyzing, calculating and publishing data for third party use. Additionally, the GIZ imparted knowledge on updating old statistical data, filling gaps in the data and on the use of the new classifications. The GIZ experts implemented their strategies through on-the-job training, which was given the highest priority and augmented the transfer of knowledge through workshops, seminars and using illustrative case studies. A significant contribution of the project was the drafting of a National Strategy for the Development of Statistics (NSDS) which fully complies with international guidelines for statistical operations (GIZ, 2011). Additionally, field supervisors were involved in data collection and later in the entry and editing which was carried out at the FBS headquarter in Islamabad. The software used for data entry was equipped with features for checking data consistency. The FBS confirmed that the overall indicators at district scale indeed follow the usual pattern at national and provincial levels (FBS, 2010).

Climate Variables-Temperature, Precipitation, Runoff (Evapotranspiration)

Climatic data for Pakistan was acquired from two different sources. The temperature data was obtained from the NCDC's (National Climatic Data Center) GHCN (Global Historical Climatology Network) of the United States Department of Commerce. The dataset includes a daily global summary of key climatic indicators such as mean temperature, mean wind speed, maximum temperature, minimum temperature and precipitation amount, and is derived from data collected by the USAF (United States Air Force) weather squadron for the United States Department of Defense. Long term summaries of the day are generally available for the time period between 1929 to 2012; however especially in developing countries the range is smaller. For Pakistan a more or less complete long term dataset is only available starting from 1973. In total 130 different meteorological stations are covered throughout the country, representing sufficient variability in climatic observations (Figure 6.4).

Moreover, as a quality criterion a minimum of 4 observations are required to create a daily summary. However, in developing countries especially data on precipitation is less sufficiently reported, thus values for this specific important climate component are missing for most stations (NCDC, 2011). To fill the gaps in the data and generate the long term precipitation averages over Pakistan, an additional Dataset developed by the Environment and Natural Resources Service (SDRN) of the Food and Agriculture Organization (FAO) of the United Nations was utilized. The majority of the datasets cover long term climate observations from 1961-1990.However subject to data availability, 80% of the series exceed 20 years, whereas 45% cover more than 50 years, this significantly varies by continent. Parameters covered, amongst others, can range from temperature and precipitation to water vapor pressure and evapotranspiration (GOMMES et al., 2004). The main rationale for using the dataset developed by the SDRN is to capture real station observations of precipitation.

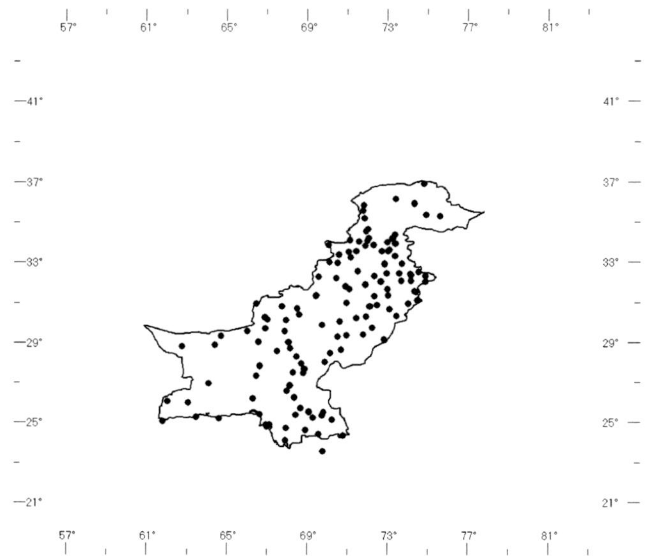

Figure 6.4 Meteorological Station Coverage Pakistan
Source: generated using LocClim 1.10 (FAO, 2005)

MENDELSOHN et al. (2007a) compared satellite soil wetness data (proxy for precipitation) to precipitation records from meteorological stations and based on estimated Ricardian responses came to the conclusion that precipitation data from weather station observations was a more effective measure. Satellite soil wetness data was plagued by distortions from forests, irrigation and water bodies. For a number of districts, precisely 37, that are not directly covered by meteorological stations a spatial interpolation technique was used. Amongst the available techniques for spatial interpolation ranging from nearest neighbor, inverse distance weighted average, modified IDWA, Cressmann's Method, Structure Function, Polynomials, Shepards Method, Kriging and Thin-Plate Spline (FAO, 2005), after weighing the disadvantages and advantages regarding precision of the estimates, Shephards Method was selected as the most accurate interpolation technique. To highlight the precision of Shephard's Method figure 6.5 shows two mean monthly temperature plots for the district Bahawalnagar, one for true station data obtained from the GHCN (Global Historical Climatology Network) and one generated through interpolation based on 11 surrounding stations with a maximum distance of 200 km from the district.

Precisely, 8 stations from Pakistan and 3 from the neighboring country India have been used in the calculations. The nearest station is Montgomery (Pakistan) at an altitude of 170 m, located at 30 km distance from the district Bahawalnagar, whereas the furthest away station is Bahawalpur at an altitude of 113 m and a distance of 201 km.

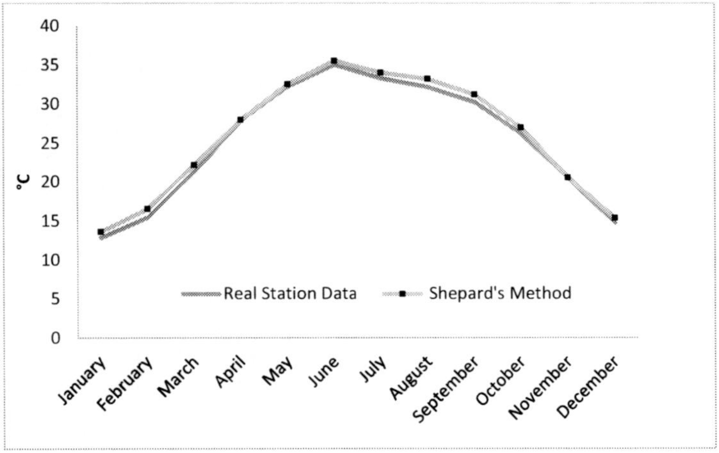

Figure 6.5 Real Station Data vs. Interpolated Climatic Data

Source: own illustration using GHCN Version 3.0 and LocClim 1.10 (NCDC, 2011; FAO, 2005)

After specifying the grid point to be analyzed a circle is drawn containing a minimum of 7 stations. To ensure that the length scale is selected as a constant the least nearest station lies exactly on the frame of the circle. Subject to station density only, the radius is changed. A weight function is applied to the circle, monotonically decreasing with the radius. Directional weights alteration is applied in order to capture the stations' differing location regarding altitude and cardinal direction. This procedure ensures that e.g. two stations that are exactly located in close proximity are individually considered and do not fall into one another's shadow, therefore this procedure is also known as shadow correction. This averaging on the basis of inverse distance weights also ensures that no single point that is closely located to the grid point of interest as compared to the other stations in the circle receives an extraordinary high weight and leads to local maxima and minima that are unreasonable. This situation is avoided by using a small local average-gradient correction where no station is very close to the station at study with respect to grid size. Additionally, grid size and station density is taken into account. A great advantage of Shephard's Method is that its runs automatically, without requiring parameters to be set by the researcher. Station Density and location specific characteristics are taken into account by the calculation. The calculation can be carried out by using LocClim (FAO, 2005).

Table 6.2 Summary Statistics of Climate Variables by Month and Season

Climate Variable/Season	Month	Minimum	Maximum	Mean	Std. Deviation
Average monthly Precipitation (mm)	January	.99	115.00	22.9989	23.93805
	February	2.00	176.70	29.3008	34.60450
	March	1.00	253.70	36.4244	42.97876
	April	1.00	173.00	24.6178	32.78098
	May	.00	88.00	15.6815	18.21297
	June	.00	116.00	19.6453	21.10679
	July	2.00	375.00	80.4246	72.00655
	August	1.00	291.00	74.7447	69.68608
	September	.00	135.00	29.5924	30.34063
	October	.00	51.00	7.4254	10.77305
	November	.00	58.10	6.3794	9.77972
	December	1.00	90.30	14.9242	16.21810
Average monthly Temperature (°C)	January	-.70	18.80	11.3508	4.23237
	February	1.70	20.70	13.9291	4.41887
	March	7.50	26.20	19.0173	4.67693
	April	13.50	31.00	24.5016	4.75728
	May	17.60	35.20	29.4004	4.48073
	June	23.30	38.50	32.4852	3.51123
	July	24.00	37.20	31.5976	2.85650
	August	22.60	35.40	30.3662	2.93054
	September	19.40	33.50	28.5124	3.32424
	October	14.10	29.80	24.0814	4.02579
	November	7.50	24.60	18.0785	4.16089
	December	2.00	20.70	12.9526	4.07867
(summer) Kharif sowing Season	Temperature	18.13	34.47	28.7957	4.16509
	Precipitation	2.00	110.00	19.9815	22.39686
(summer) Kharif growing Season	Temperature	22.57	35.37	30.1588	2.96905
	Precipitation	1.67	267.00	61.5872	56.38403
(winter) Rabi sowing Season	Temperature	8.00	24.60	18.3708	4.04350
	Precipitation	.33	66.23	9.5763	11.81159
(winter) Rabi growing Season	Temperature	2.83	21.60	14.7657	4.42420
	Precipitation	2.00	153.67	29.5747	32.61000

Source: own calculations (2012)

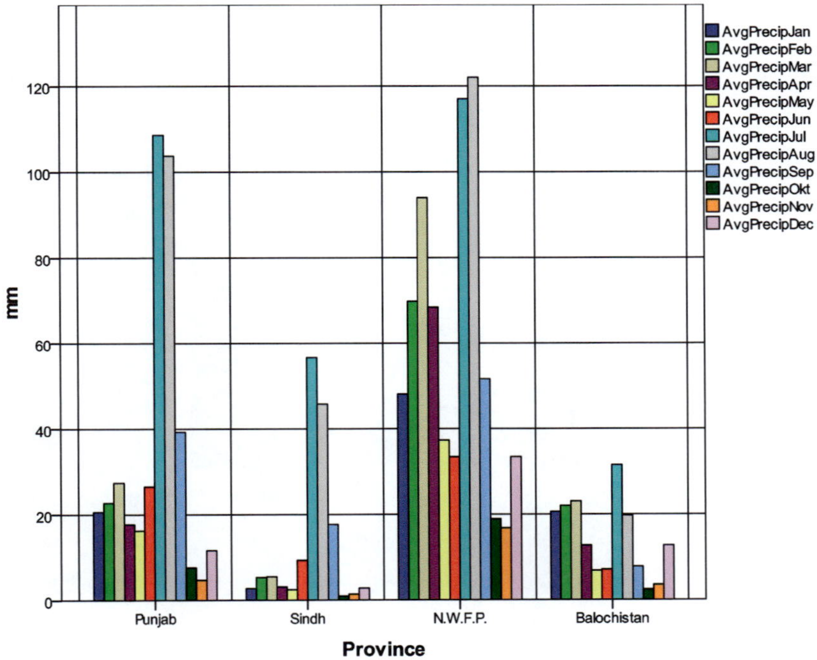

Figure 6.6 Long term Average Precipitation by month and province
Source: own illustration (2012)

As table 6.2 shows the months of July and August are of particular importance as far as precipitation is concerned. In these two months the country receives the highest amount of rainfall through the monsoon. Throughout the country the least amount of rainfall is received in the fall and winter months stretching from October to December, with average monthly precipitation between 7 mm and 11 mm in the Punjab province; 1 mm and 3 mm in the Sindh province; 18 mm and 33 mm in the N.W.F.P. and 3 mm and 13 mm in the Balochistan province (Figure 6.6). The N.W.F.P. is the wettest province of the country with the highest precipitation levels throughout the year. In terms of absolute accumulated precipitation the N.W.F.P. on average receives around 700 mm precipitation annually, whereas the maximum can amount to 1900 mm. The driest province is the Sindh province in the south east of the country with average annual accumulated precipitation to the tune of 150 mm and a maximum of 320 mm. Throughout the country the bulk of the precipitation is received in the summer growing season (Kharif) which also comprises the months of the monsoon period from July to September (Figure 6.6). The summer sowing and growing seasons are characterized by mean temperatures between 28°C and 31.5°C and average monthly precipitation between 20 mm and 60 mm, respectively. Temperatures in the winter sowing season Rabi fluctuate around 18°C, whereas in the growing season they average around 15°C. Winter season precipitation is meager with average monthly precipitation between 9 mm and 30 mm.

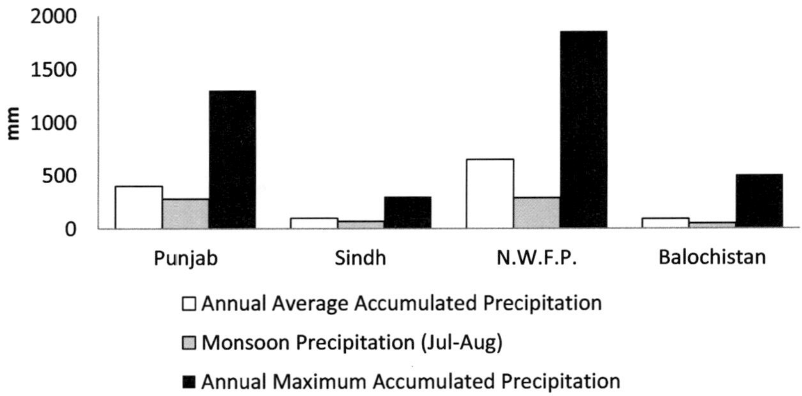

Figure 6.7 Long term Accumulated Precipitation by month and province
Source: own illustration (2012)

As can be seen in figure 6.7 the monsoon period is a pivotal water source for Pakistan. Close to 60 % of the Punjab province's precipitation, 80% of the Sindh province's precipitation, 40% of the N.W.F.P.'s precipitation and 35% of the Balochistan province's precipitation is received during the monsoon months. Although the country is located in the monsoon region, it is generally arid, except for the southern slopes of the Himalayas and the sub-Mountainous tract. Rainfall also occurs from western cyclonic disturbances originating in the Mediterranean (KHAN et al., 2010). A large part of the precipitation in the northern mountain system is in the form of snow which feeds the rivers.

A look at Figure 6.8, which plots the long term average temperature for each month and province, reveals a more or less similar pattern as in the case of precipitation. The country is of arid nature, temperatures are generally high, averaging around 25°C annually. The south east of the country (Sindh province) is the hottest part of Pakistan, with summer season (Kharif) temperatures in the excess of 34°C.

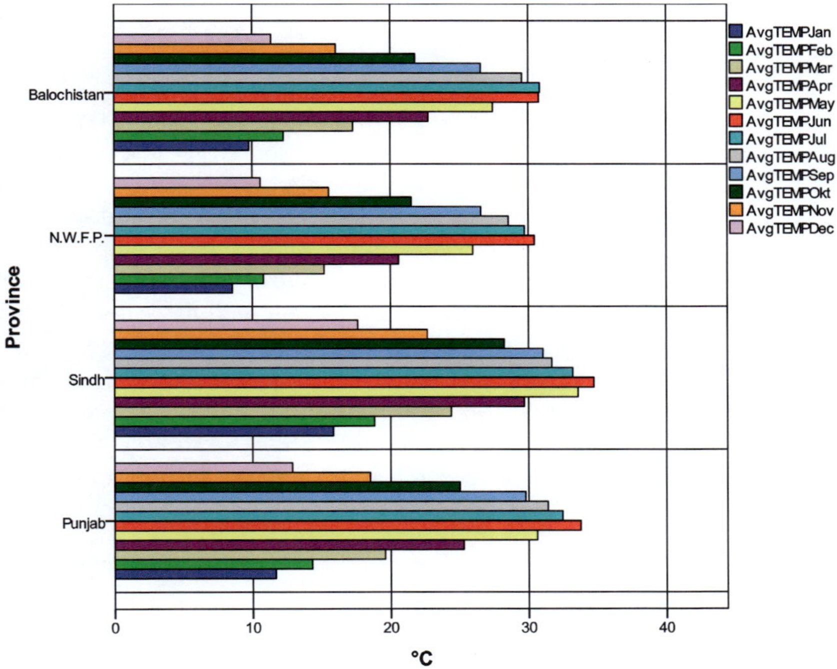

Figure 6.8 Long term Average Temperature by month and province

Source: own illustration (2012)

The Punjab province also experiences summer season temperatures around 32°C, with the exception that it is not as dry as the Sindh province. The other two provinces of Balochistan and N.W.F.P. are relatively cooler, with summer temperatures still around the 30°C mark but winter temperatures around the 10°C mark, whereas in the Punjab and Sindh provinces the long term mean winter season temperatures fluctuate around 15°C and 18°C (December), respectively. Generally, the hottest months stretch from May to September, with June and July being particularly hot. Only in the case of the country's south eastern part the month of May also exceeds 32°C. Figure 6.8 shows long term averages for the provinces (occasional deviations from these normals are not highlighted). However, as the provinces of the country cover a vast amount of land area, fluctuations between the province's different parts are therefore not uncommon and occasionally temperatures close to 0°C are also possible in parts of the Sindh province (e.g. Kirthar mountain range) and the dry western plateau of the Balochistan province.

Runoff Variable:

To capture additional source of water a runoff variable has been incorporated in the model. Runoff in this sense can be defined as "...the amount of water running off, on the average during a year from a unit of land surface in the form of different horizontal flows of water" (BUDYKO, 1974). This process can be understood in three steps. First, when rain falls the location specific vegetation intercepts and stores the water. Second, the rainfall that after being stored by the vegetation reaches the ground infiltrates into the soil until the soil's infiltration capacity is reached. Thereafter in a third step, provided that the precipitation continues even after the soil infiltration capacity is reached, first surface irregularities are filled, then through the remainder of precipitation that comes down runoff is generated (CRITCHLEY AND SIEGERT, 1991) (Figure 6.9). As soon as rainfall drops below the soil's infiltration capacity runoff generation stops.

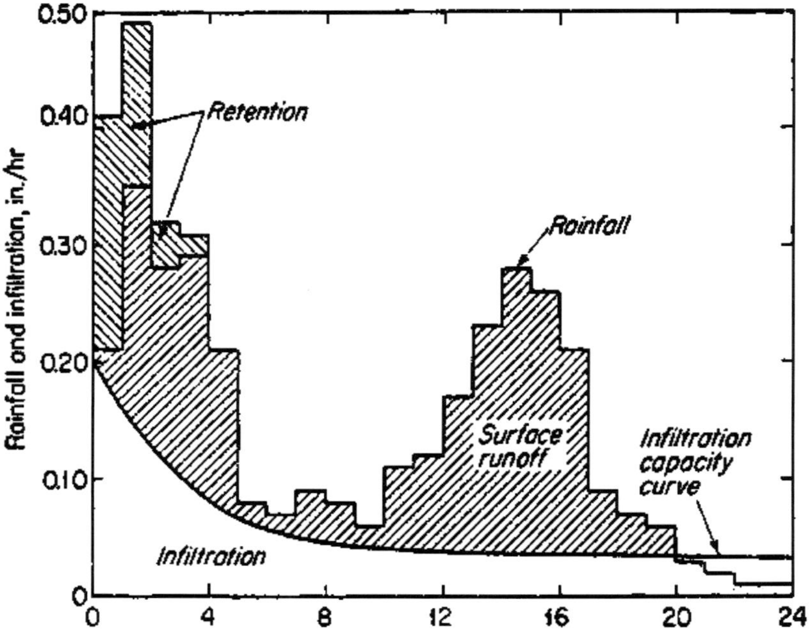

Figure 6.9 Relationship between rainfall, infiltration and runoff
Source: LINSLEY et al., 1958

The SDRN dataset reports evaporation and runoff following BUDYKO (1974). Based on Budyko's Radiational Index of Dryness (RID), which stands for the ratio of the site specific radiation balance and the energy required to evaporate all precipitation (lower index values indicate wet area), evaporation is calculated. Evaporation is generally understood to be the processes whereby water is converted to water vapor (vaporization) and removed from the evaporating surface (vapor removal) (ALLEN et al. 1998). BUDYKO (1974) defines evaporation as being equal to precipitation if the radiational energy (measured by the RID) as compared to precipitation is very large, implying that all precipitation evaporates. If radiation decreases, evaporation regardless of the precipitation amount also decreases. For a detailed discussion of the underlying theory refer to Budyko's seminal work titled "Climate and Life" (1974). The SDRN dataset based on Budyko's climate classes (RID calculated) reports evaporation ratios for the area at study. Based on this data and district level precipitation records, mean annual runoff is calculated simply as precipitation minus evaporation or precisely as average annual precipitation multiplied by the evaporation ratio. Evaporation therefore, is not modeled directly as it reflects temperature and precipitation, which are already in the model.

Surface Water Availability:

In order to capture the effect of easy access to water as an additional factor in climate relevant analysis, following STRZPEK AND MCCLUSKEY'S (2006) approach, for the first time a River Density Index (RDI) is constructed for the districts of Pakistan. In a first step the district areas measured in km² are calculated in ArcMap (ESRI, 2011) using a projected coordinate system. To the district layer a detailed river layer is added and the river layer is converted into a very small grid size (0.001°). In a second step, the individual grids are summed up within a district and the grid sum is converted into a length. By doing this, the calculated river length represents the total amount of all river segments in a district.

$$RDI = \frac{RL_i}{DA_i} \times 1000$$

With *RDI* standing for River Density Index, RL_i representing the total length of the river segments in a district *i* and DA_i denoting the geographic area of district *i*. Higher values for the *RDI* indicate easier access to surface water in a district.

Soil Variables:

The soil dataset was created by using the digital soil map of the world (DSMW) developed by the FAO (2003) in combination with a Geographical Information System (ESRI, 2011). A digitalized image was georeferenced in order to filter the different proportions of soils in a specific district. The initial dataset reports 26 major soil units and defines sub-categories based on texture and land slope. For Pakistan 7 major soil groups are tested, comprising Cambisols, Regosols, Calcisols, Leptosols, Fluvisols and the more marginal categories Sand Dunes and Rock outcrop. Calcisols are most prevalent over the territory.

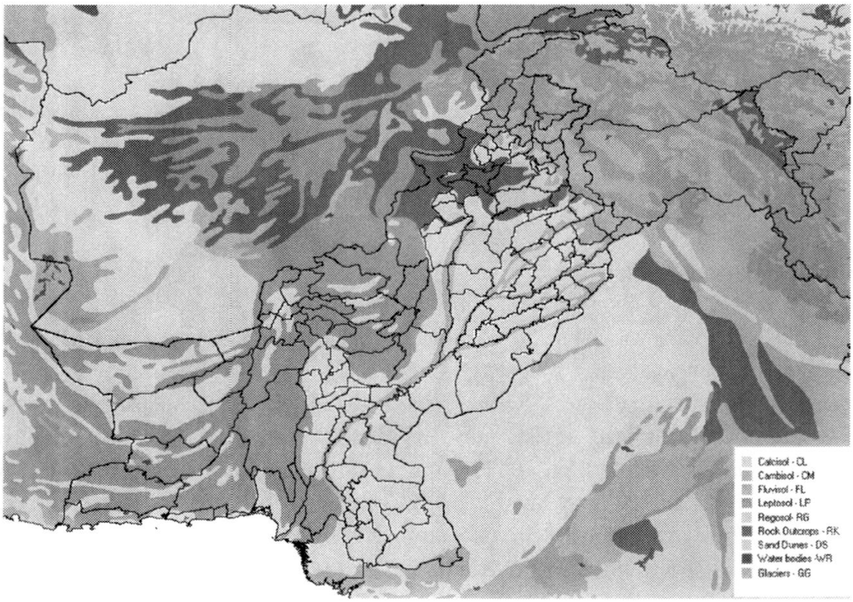

Figure 6.10 Soil Classes of Pakistan
Source: modified after DSMW (FAO, 2003)

Flood Hazard:

The flood hazard was assessed by creating a unique dataset, where those districts most severely affected by the floods in 2010 were identified using observations by the UNO Satellite Observation Program (UNOSAT) and the World Food Program, amongst others (LOGISTICS CLUSTER, 2010). These observations were compared to the long term flood risk assessment of the FAO-

ERCU (Pakistan) (2010) for the period from 1950-2011. By using the maps prepared by the FAO-ERCU (2010) a long term dataset for the regularly flood affected parts of the individual districts was created. By using a dataset prepared by SUPARCO & FAO (2010) which covered all districts of the country with their respective acreage under flooding, the percentage of district area under regular flood was estimated. The calculations made use of total district area and a computed dataset for area under flood.

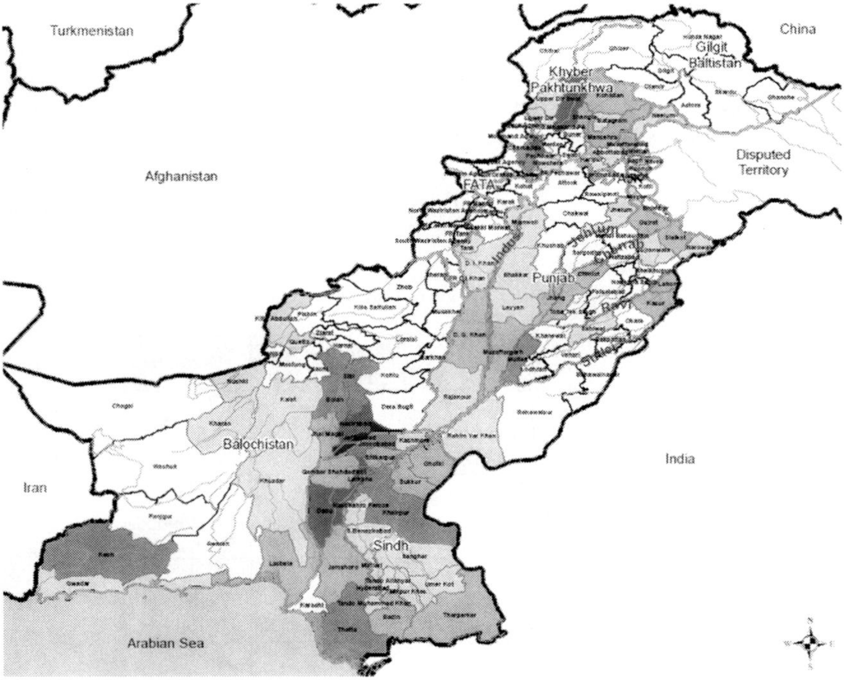

Note: darker colors are associated with a higher frequency of flood occurrences.

Figure 6.11 Regularly flood affected areas 1950-2011

Source: FAO-ERCU (2010)

Socio-Economic Variables:

HDI:

The Human Development Index is obtained from JAMAL AND KHAN (2007). The latest data is available only for the year 2005. A district level dataset is created, covering all modeled districts in the study area. The three different stages of human development can be classified as low (HDI < 0.5), medium (HDI > 0.5 < 0.8) and high (HDI > 0.8). Following these different classifications, in the case of Pakistan only one district can be classified as highly developed, which is the capital territory Islamabad. Similarly, only 11 districts are classified as weakly developed in terms of the HDI, whereas the majority, precisely 102 districts show medium levels of human development.

Population Density:

Population Density per km² has been obtained from the population census organization of the government of Pakistan (PCO, 1998). The average population density is highest in the Punjab province, followed by the Sindh province, the N.W.F.P. and Balochistan, which with an average of 39 people/ km² has the lowest population density in the country. Additional statistics are presented in table 6.3.

Table 6.3 Population Density/km² by province and for selected districts

Province	Statistic	Population Density (km²)	District
Punjab	Mean	524	-
	Maximum	3566	Lahore
	Minimum	90	Rajanpur
Sindh	Mean	346	-
	Maximum	2795	Karachi
	Minimum	39	Jamshoro
N.W.F.P.	Mean	399	-
	Maximum	1606	Peshawar
	Minimum	21	Chitral
Balochistan	Mean	39	-
	Maximum	281	Quetta
	Minimum	4	Washuk

Source: own calculations (2012)

Technological abundance and infrastructure:

As has been discussed in chapter 2.8 on TFP, the introduction of technology in times of the green revolution has played a leading role in increasing farm productivity in Pakistan. Reasoned in the fact that areas with higher abundance of technology have more options to increasing productivity and also to test whether the possession of machinery impacts on land values a proxy variable for technological abundance in form of tractor density per hectare is also introduced. Furthermore, in order to assess the relevance of an improved infrastructure on land values and incidentally capitalized net revenues, a control variable for road density is introduced. Roads attaching the agricultural entities to the general road network are also included. The dataset has been constructed by using the volumes for 2008/2009 of the provincial development statistics for Punjab, Sindh and NWFP (RAFIQUE, 2009; GOS, 2008; GON, 2009) and data obtained from the government of Balochistan (GOB, 2012).

Figure 6.12 Tractor and Road Density by Province 2008-2009

Province	Variable (No. or lenght)/ha	Observations	Mean	Std. Dev.	Min	Max
Punjab	Tractor Density	36	.2208181	.1124301	.0476981	.4218771
	Road Density	36	5.294494	5.867796	1.736037	38.51734
Sindh	Tractor Density	22	.0718673	.13502	.0009841	.6467662
	Road Density	22	1.562359	1.328996	.025878	4.94
N.W.F.P.	Tractor Density	24	.0451648	.0469896	0	.1558176
	Road Density	24	3.051793	2.802539	.5386775	14.60956
Balochistan	Tractor Density	28	.0067694	.0089531	.0004899	.042449
	Road Density	28	.7763959	1.128101	.004	4.532545

Source: own illustration based on various issues of provincial development statistics.

Figure 6.12 reveals the highest tractor density for the Punjab province, followed by Sindh, NWFP and Balochistan. The road density is highest in the Punjab province, followed by the NWFP, Sindh and Balochistan.

Dependent Variable: Land Value (Rupees/acre)

The dependent variable land value is directly obtained from the survey. Precisely, the survey included a detailed sub-questionnaire on asset possession with questions on the ownership of agricultural land, where farmers were asked to value their land. Specifically, farmers replied to the question what they would expect to receive for their total agriculturally operated land if they were to sell it on the market. As mentioned above, an EDA preceded the aggregation process into districts to guarantee the exclusion of the most obvious erroneous outliers or extreme points. Land values are reported in Pakistani Rupees per acre. To show the variability of land values throughout the country figure 6.13 plots the average land value per acre by agro ecological zone.

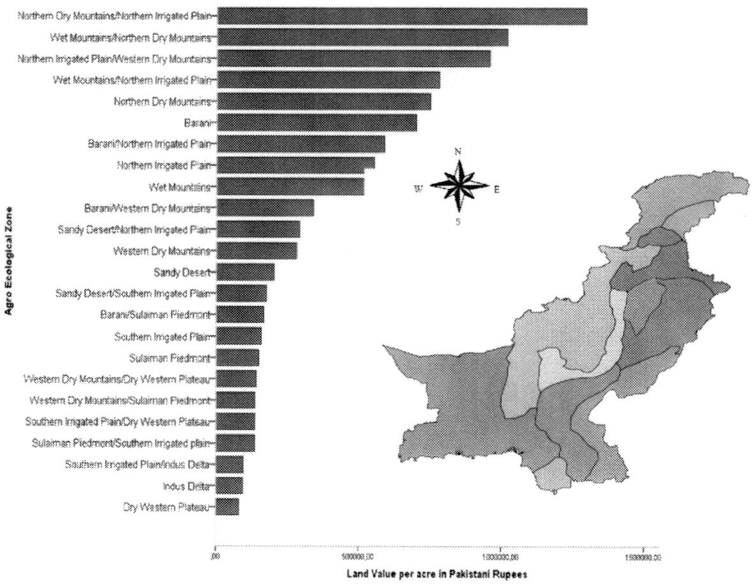

Figure 6.13 Land Value Distribution by Agro Ecological Zone
Source: own calculations (2012)

Consistent with the underlying Ricardian theory that identifies climate as being one factor for determining land values, figure shows that from the wet and relatively cooler northern parts to the very dry and hot southern parts of the country land values steadily decline, confirming the pattern that was expected. The highest land values are recorded in the intersecting Agro Ecological zones of the northern irrigated plains and the northern dry mountains. The highest average land values per acre are recorded for the N.W.F.P., followed by the Punjab, Balochistan and Sindh provinces. The lowest mean land value is aggregated for the district Jhal Magsi in the Balochistan province in the south west, whereas the highest average land value for any individual district is recorded for the capital territory of Islamabad in the north of the country. Looking at the descriptives in table 6.4, similarly to the Agro-Ecological Zone comparison, an expected pattern can be observed, where land values per acre decline consistently, when moving from the northern areas to the southern areas of the country. Climatic differences in a first observation seem to be related to the value of land, which shall be at heart of the upcoming empirical investigation for the above explained dataset. The following chapters will present the empirical modeling strategy and how from an econometric-technical perspective estimates were generated.

Table 6.4 Land Value Rs/acre by province and for selected districts

Province	Statistic	Land Value Rs./acre	District
Punjab	Mean	446768.7	-
	Maximum	2465985	Islamabad
	Minimum	82915.64	Jhelum
Sindh	Mean	144819.6	-
	Maximum	231372	Sanghar
	Minimum	23723.84	Tharparkar
N.W.F.P.	Mean	724322.4	-
	Maximum	1592870	Swat
	Minimum	86597.67	Tank
Balochistan	Mean	152355.2	-
	Maximum	909945.5	Ziarat
	Minimum	16164.89	Jhal Magsi

Source: own calculations (2012)

6.2 Econometric estimation

Multiple regression analysis under the notion of ceteris paribus can be a powerful tool for researchers that are working with non-experimental data and are interested in the analysis of causal effects. In this way a set of explanatory variables are tested for their influence on a dependent variable. To verify the results of a regression analysis and to ensure that the findings are not misleading certain assumptions dependent on the choice of the estimation technique have to be met. In a first step this study applies linear regression to a cross sectional dataset to estimate the parameters of the multiple regression model. The linear model or OLS (Ordinary Least Squares) regression has certain specific underlying assumptions which are: linearity, normality and homoscedasticity. Additionally, other important issues that are of particular importance regarding the model quality are collinearity and influence, which will be elaborated on further in the upcoming deliberations. To sum up, the major issues regarding standard linear regression include *multicollinearity*, *heteroskedasticity* and significant *impact of outliers* alongside *leverage*. However, as the present study deals with districts as measurement units, which have spatial features additional issues related to *spatial dependence* might arise. To this effect, detailed considerations regarding this particular issue are presented. The following considerations intend to explain the econometric estimation strategy regarding the diagnosis and solutions for the above mentioned issues. After explaining diagnosis and solutions for multicollinearity, the impact of outliers and leverage is assessed. Subsequent deliberations will present a detailed account of spatial dependence and heteroskedasticity. Based on the diagnostics for the standard OLS regression (Annex), the final econometric specification is selected. All estimates and diagnostics are produced by using a STATA econometric package (Version 11.2) (STATACORP, 2009) and GeoDA 1.0.1 (ANSELIN et al., 2006).

6.2.1 Structural multicollinearity, outliers, leverage and influence

Multicollinearity is generally understood to be a high but not perfect correlation between two explanatory variables. Without going against the assumption of multiple regression analysis of no perfectly collinear predictor variables in a model, multicollinearity still causes issues (WOOLDRIDGE, 2001). If multicollinearity is present, the individual parameter estimates are rendered less reliable as high correlation between predictors tends to inflate the variances of the estimated regression coefficients, thus lowering t-values and increasing the confidence intervals. However, multicollinearity can have different origins. Either it can be inherent in some of the modeled predictors that are chosen, such as for instance any two variables that are measuring the same thing, when similar trends are introduced in a time series setting or when for instance relational lags are incorporated into the model. On the other hand a different type of multicollinearity which is more or less to some extent introduced by the researcher himself is referred to as *structural multicollinearity*. Structural multicollinearity in this sense is a mathematical artifact caused by generating polynomials from one single variable. The present study uses the Ricardian climate valuation approach which is based on a quadratic formulation of climate variables, thus multicollinearity is inherent in the model. But as GUJARATI (1995) has noted, the extent of multicollinearity is detrimental not its general presence, for it cannot be completely eliminated.

Multicollinearity in the model is assessed by exploring the Variance Inflation Factor (VIF) that indicates the extent by which the estimated variance of the ith regression coefficient is increased above what it would be if the ith variable is orthogonal to the other predictor variables, thereby depicting the effects of multicollinearity on the respective coefficient's variance (O'BRIEN, 2007). The VIF is defined as $1/(1 - R^2_i)$, with R^2_i as the coefficient of determination of the ith VIF regression, in which each of the independent variables X_i once is a function of all other predictor variables. The thresholds for deciding whether a VIF value is indicating a serious amount of multicollinearity have been controversially discussed in literature, these are mere arbitrary cut-off points ranging from 2 to 10 in the bulk of the studies. A detailed discussion about VIF scores can be found in O'BRIEN (2007). In this context for instance HAIR et al. (1995) suggest that VIF values smaller 10 are inconsequential for multicollinearity. Similarly, KENNEDY (1992) and KUTNER (2004) also propose a VIF threshold value of 10, above which collinearity becomes harmful. CHATERJEE AND PRICE (1991) provide a collection of different informal rules of thumb for the VIF scores. The present study based on recommendations by several researchers uses a threshold value of VIF > 10 for defining serious multicollinearity. In a first step all climatic variables and their respective squared values are tested for multicollinearity. The collinearity diagnostics are presented in Table 6.5.

Table 6.5 Structural Collinearity Diagnostics – Climate Variables

Variable	VIF	SQRT VIF	Tolerance	R^2
Kharif sowing season Temp. and Temp.²	182.36	13.50	0.0055	0.9945
Kharif growing season Temp. and Temp.²	341.66	18.48	0.0029	0.9971
Rabi sowing season Temp. and Temp.²	56.47	7.51	0.0177	0.9823
Rabi growing season Temp. and Temp.²	32.00	5.66	0.0312	0.9688
Kharif sowing season Precip. and Precip.²	8.44	2.90	0.1185	0.8815
Kharif growing season Precip. and Precip.²	8.82	2.97	0.1134	0.8866
Rabi sowing season Precip. and Precip.²	7.83	2.80	0.1277	0.8723
Rabi growing season Precip. and Precip.²	8.15	2.85	0.1227	0.8773

Source: own computation using STATA 11.1 (STATACORP, 2009)

As the collinearity diagnostics in table 6.5 reveal, especially the temperature variables are plagued by a high extent of multicollinearity. The VIF values for the temperature terms are considerably higher than the arbitrary threshold of 10, for the summer (Kharif) growing season the VIF even surpasses 340. Where temperature variables show a high extent of multicollinearity the precipitation terms remain below the threshold. Nevertheless, both will be further dealt with in the multicollinearity context. As by the nature of the Ricardian approach and on the basis of agronomic research generally a hill shaped relationship for land values (capitalized net revenues) and climate variables is modeled, the multicollinearity is self-induced. A first approach would be to neglect the presence of collinearity between climate coefficients as key assumptions of the OLS model are not hurt. However, as parameter estimates are not efficient and reliable due to variance inflation and consequently high standard errors, a solution is most essential. Additionally, KAUFMANN (1998) noted that without accounting for multicollinearity in Ricardian Models, parameter estimates would be too unstable and cause unrealistic marginal effects. Following KIM (1999) the present study opts for standardizing the climate variables. Standardization is achieved by a simple linear transformation of the following form:

EQ (25) Standardized Climate Variable: $(C) = (X - \bar{X})/S_X$

EQ (26) Standardized Squared Variable: $(C^2) = (\frac{X-\bar{X}}{S_X})^2$

With X as the predictor value, $\bar{X}$ as the mean of the predictor values and S_X as the standard deviation. The rationale is that, no information is lost as a result of the standardization and the original values can always be recovered using the untransformed variables mean and standard deviation. One of the major advantages of this multicollinearity correction is that the other included predictors are not affected and generally calculations become more accurate (KIM, 1999). The following table presents the collinearity diagnostics after standardizing the temperature variables for the different farming periods.

Table 6.6 Structural Collinearity Diagnostics – Standardized Temperature Variables

Variable	VIF	SQRT VIF	Tolerance	R^2
Kharif sowing season Temp. and Temp.²	1.81	1.34	0.5540	0.4460
Kharif growing season Temp. and Temp.²	2.12	1.46	0.4723	0.5277
Rabi sowing season Temp. and Temp.²	1.21	1.10	0.8242	0.1758
Rabi growing season Temp. and Temp.²	1.18	1.09	0.8489	0.1511

Source: own computation using STATA 11.2 (STATACORP, 2009)

Table 6.6 clearly reveals the advantages of standardization. The VIF scores are significantly reduced and multicollinearity is largely eliminated. The Stata econometric package contains a variety of graphical and numerical tests for regression diagnostics. Using these, the following considerations aim at detecting potential problems in the regression. First of all the diagnostics explore the possible presence of unusual and influential data by checking for outliers (observations with large residuals), leverage (extent of an observation's deviation from the mean) and influence (caused by outliers and leverage).

In order to detect outliers in the data, in a first step the studentized residuals are analyzed using a stem and leaf plot in combination with a table depicting the observations that particularly stick out (Annex A). The residual -3.8 for the district Jhelum seems to be one needing attention. A closer look at data however, does not indicate any data entry error; therefore a deletion of the observation is not considered. In an additional step the leverage is explored to identify observations that might substantially influence the regression estimates. The highest leverage as shown in Annex B is exerted by the district Lower Dir, followed by Islamabad, Abbottabad, Kohistan and Karak. As a rule of thumb, leverage points higher than (2*k+2/n) should be carefully examined (STATACORP, 2009). With 28 predictor variables (k) and 110 observations (n) the threshold value is given by 0.53. Theory suggests that the points (observations) with both large residual and large leverage exert the highest overall influence on the regression coefficients. Using a residual versus leverage plot the idea is to depict those observations that are jointly high in both measures (Annex B). The districts Lower Dir and Jhelum are observations that are instantly identified as being of potential concern, with the former having the highest leverage statistic and the latter having the highest residual squared. As the foregone single indicator diagnostics could not reveal a joint influence of any observation, in following overall influence measures such as COOK's D and DFITS are utilized (Annex C). Yet again the district Lower Dir sticks out being influential. Consequently to assess how each coefficient changes by deleting the problematic observations DFBETA statistics are calculated (Annex D). By setting a threshold value of (2/square root (n)) the results are evaluated. Values higher than the threshold value are found for the predictor variables summer sowing temperature, winter sowing temperature, Human Development index, Monsoon precipitation, Population density, Elevation, road density and soil type Calcisols in the districts of Jhelum, Karachi, Upper Dir and Bolan. These potential problematic DFBETA scores are plotted in a graph which includes the threshold interval (Annex D). Nothing irregular as far as data entry is concerned is found. Influential points are Islamabad with its road density and Karachi with its population density, both of which can be expected due to the fact that Karachi is the capital of the Sindh province and Islamabad is the Punjab province's capital. Karachi with a population close to 20 million is one of the globally most populous places. Although some observations have appeared as potential outliers statistically and others have been identified exerting leverage on the coefficients of the regression model, a joint influence was not revealed. Therefore, before any final model specification which controls for the influence in one or the other case can be introduced, additional reflections regarding the other properties of the OLS regression model are needed. Normality of residuals, which is a requirement for valid hypotheses testing is also assessed using a combination of graphical and numerical tools, namely Kernel-Density plot, standardized normal probability plot (P-P), quantile- quantile plot (Q-Q) and Hamilton's interquartile range test (iqr-test) (HAMILTON, 1992). The iqr-test defines severe outliers as points that are three inter-quartile-ranges below the first quartile or three inter-quartile-ranges above the third quartile. If severe outliers are present normality is rejected at the 5% significance level. The test confirms that the normality of residuals assumption is not violated in the present study's case, thus the p-value for the *t* and *F* tests are be valid. The above normality plots and the iqr test output are presented in the Annex E. Additionally, as the model includes quite a number of variables added variable plots are utilized to depict possible problematic outliers in the aggregate sample, the output is presented in Annex F. In following, a more serious concern regarding econometric models that have spatial features will be studied in detail. Alongside spatial autocorrelation diagnostics the focus will also be on other post-estimation tests for issues of multicollinearity and foremost heteroskedasticity. After these tests have been conducted a final econometric approach will be finalized.

6.2.2 Climate Sensitivity and Spatial Autocorrelation

Earlier studies such as LIPPERT et al. (2009) have raised concerns regarding the issue of spatial autocorrelation in the Ricardian model. The spatial autocorrelation causes the dependent variable to be influenced by the neighboring observations. To the background of a theoretical and a data-based argument the introduction of spatial features in the Ricardian estimation is called for. The theoretical argument stems from the observation that farmers or farms in close proximity to one another tend to interact and therefore stick to similar practices (ANSELIN, 2006; POLSKY, 2004). This phenomenon of interaction is also known as "Tacit Knowledge" (BRETSCHGER, 1999). From the data-based perspective spatial effects are important to consider for achieving econometric soundness, foremost efficiency (SCHLENKER et al.; 2006; DESCHENES AND GREENSTONE, 2007). The present study has opted for the approach used by POLSKY AND EASTERLING (2001) who considered a larger spatial scale for their analysis. Thus, instead of using single farms in close proximity, the study uses districts, with the expectation that thereby spatial influences will be controlled for. In order to verify the efficiency of the proposed method, in following both graphical and analytical tools are utilized.

Two types of spatial dependence have been identified in literature, namely spatial error and spatial lag. In case of the spatial error in OLS regression the error terms across different spatial units are correlated, whereas in case of the spatial lag the independent variable y in place *i* is influenced by the independent variables in location *i* and *j*.

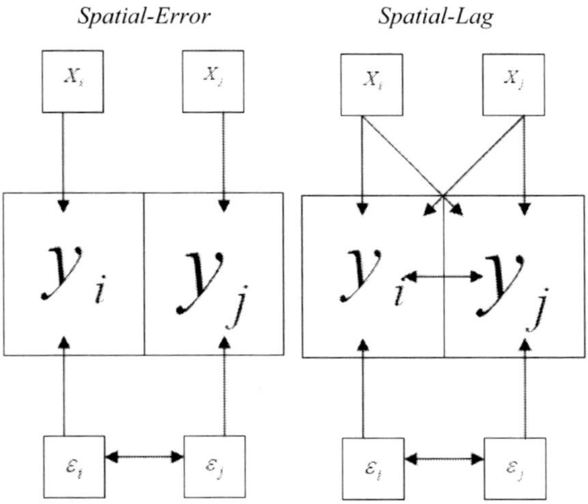

Figure 6.14 Types of spatial dependence
Source: S4 GIS INSTITUTE (2006)

A logical expectation is that close observations will be more similar than far apart ones. This is quantified by assigning weights to each pair of observations. Before spatial autocorrelation can be measured a structure has to be applied to the assumed spatial relationship (KUMAR et al., 2009). This structure is delivered by a spatial weighting matrix. The matrix defines a local neighborhood around each geographic unit and the weights file identifies the neighbors. This is achieved by taking the weighted average of neighboring polygons; the result is the spatial lag. The options for constructing the weights are either contiguity based or distance based. Either contiguity to the polygon (e.g. district) is used as reference or distance between single points. Moran's I, which is an extension of the Pearson correlation coefficient, is one prominent measure of spatial autocorrelation, which is present when spatial randomness (location's values are independent of neighbor's values, the spatial pattern is random, if the location is altered the values do not change) is violated (S4 GIS INSTITUTE, 2006; PARADIS, 2012).

Moran's I:

EQ (27)
$$I = \frac{n}{S_0} \frac{\sum_{i=1}^{n} \sum_{j=1}^{n} w_{ij}(x_i - \bar{x})(x_j - \bar{x})}{\sum_{i=1}^{n} (x_i - \bar{x})^2}$$

With w_{ij} as the weight between observation *i* and *j*, and S_0 representing the sum of all w_{ij} 's.

EQ (28)
$$S_0 = \sum_{i=1}^{n} \sum_{j=1}^{n} w_{ij} \cdot$$

The expected value of *I* under the null hypothesis of no autocorrelation is given by $I_0 = -1/(n-1)$. Knowing the expected variance of I_0 the null hypothesis can be tested. "If the observed value of *I* (denoted $\hat{I}$) is significantly greater than I_0, then values of *x* are positively autocorrelated, whereas if $\hat{I} < I_0$, this will indicate negative autocorrelation" (PARADIS, 2012). In other word +1 indicates strong positive autocorrelation, 0 indicates random spatial ordering and -1 indicates strong negative spatial autocorrelation. In this way several one- or two- tailed tests can be conducted to test the significance of the spatial dependence. The present study works with district level data, therefore the weights are calculated using contiguity. Figure 6.15 presents a map for local spatial autocorrelation. For each district a tendency to have an attribute that is correlated with neighboring observation's values is presented. The map reveals high spatial dependence in six districts. The chosen significance level is 0.01.

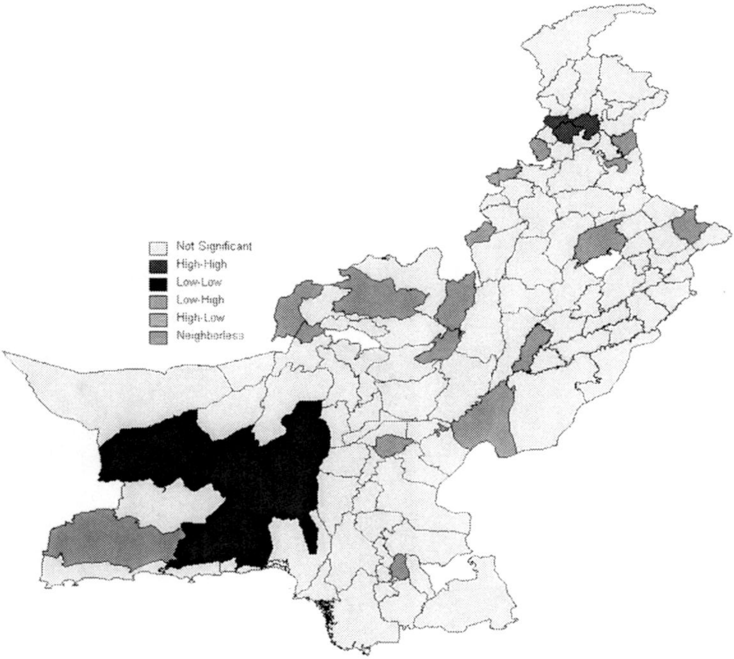

Figure 6.15 Local Indicators of Spatial Autocorrelation - Clusters for Land Value /ha
Source: created using GeoDA 1.0.1 (ANSELIN, 2006)

The scatterplot in Figure 6.16 shows the value of the original variable (land value per hectare in the district) on the horizontal axis and the spatial lag of the variable (average land value in the neighboring districts) on the vertical axis. The variables are standardized and the graph is divided into four equal quadrants, namely high-high (upper right) and low-low (lower left) indicating positive spatial autocorrelation, whereas the quadrants low-high (upper left) and high-low (lower right) indicate negative spatial autocorrelation. The slope of the regression line is Moran's I.

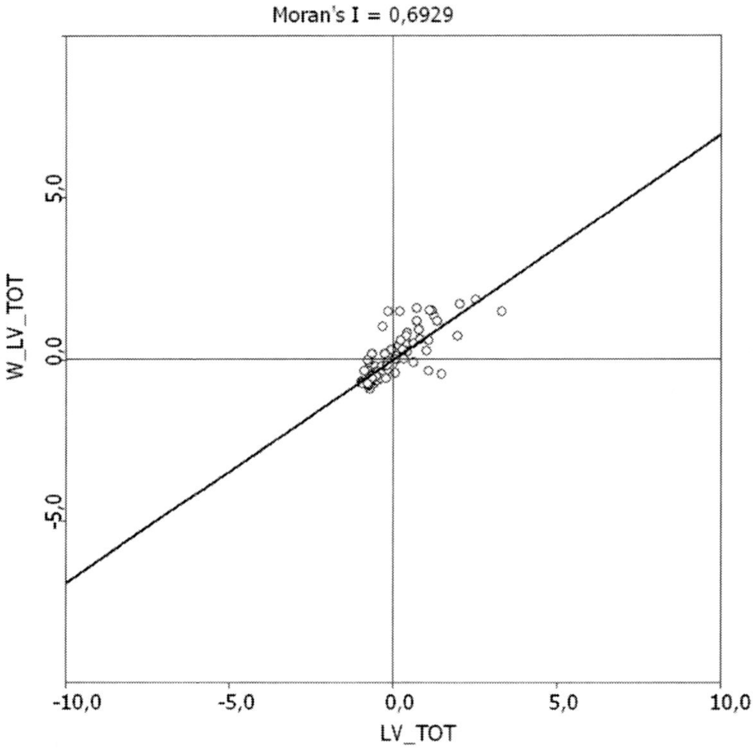

Figure 6.16 Moran's I Univariate Scatterplot for Land Value / ha
Source: own calculation using GeoDA 1.0.1 (ANSELIN, 2006)

Moran's univariate scatterplot reveals a slope of 0.69 for the Moran spatial autocorrelation measure, indicating that "Tacit Knowledge" has an effect on land value. However, before any final decision regarding the severity and impact of spatial dependence in the model can be drawn, additional tests are necessary. Thus, as with spatial error and spatial-lag in OLS regression the assumption of uncorrelated error terms is violated, rendering the results inefficient and affecting inference, the following plot analyses spatial dependence between the error terms.

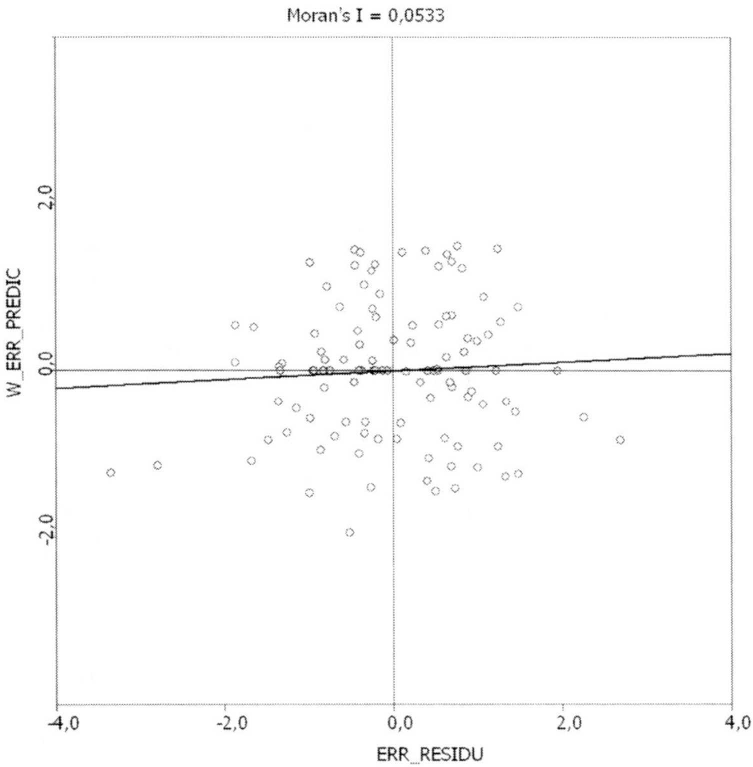

Figure 6.17 Moran's I univariate Scatterplot for Error terms from OLS regression
Source: own calculation using GeoDA 1.0.1 (ANSELIN, 2006)

Moran's I in Figure 6.17 is close to zero, indicative of no issues related to spatial error. However, as the autocorrelation indicator value is relatively high for the dependent variable, some additional statistical tests have to be undertaken. For this, the open source geographical information system GeoDa by Luc ANSELIN (2006) provides a wide range of diagnostic tests. Furthermore, Maximum likelihood approaches for spatial lag or spatial error models are also provided. In a first step a contiguity based weights file is created. Using this weights file a standard OLS regression is performed, predicting land value per hectare with several indicators including climatic, socio-economic and hydrological variables. The weights file is used to compute spatial dependence diagnostics for the model. The regression output is presented in the Annex G. Here the idea is to specifically focus on the spatial dependence and other regression diagnostics.

As far as multicollinearity is concerned, the Jarque-Bera test on normality of errors is carried out. The test score is not significant, indicating the normal distribution of the error term, which is essential for the tests of spatial dependence as they are based on the assumption of normally distributed error terms. To test for the homogeneity of variance of the residuals, a key assumption of the OLS model, graphical and numerical tools are utilized. The graphical tool is the residual versus fitted value plot (rvf-plot). If the homogeneity of variance of the residuals is confirmed, the rvf-plot should not reveal any pattern. The rvf-plot presented in Annex G does not show a particular pattern however. For a final assessment of heteroskedasticity appropriate numerical tests have to be additionally consulted. To numerically test for a possible presence of heteroskedasticity the Breusch-Pagan and Koenker Bassett tests are computed, with mixed results. The Breusch-Pagan test indicates the presence of heteroskedasticity, whereas the Koenker-Bassett test confirms the homogeneity of variance of the residuals. The Breusch-Pagan test is sensitive to the assumption of normality, therefore KOENKER (1981) and KOENKER AND BASSETT (1982) suggest the computation of the Lagrange Multiplier (LM) test on a more robust estimator of the variance of the squared errors (GREENE, 2008). Moreover, WALDMAN (1983) has found that if the vector of independent variables contains the same independent variables for both tests, then both are algebraically the same. In other words, the Koenker-Bassett test is essentially the same as the Breusch-Pagan test, except that it is carried out on studentized residuals, which are robust to non-normality. To this background the Koenker-Bassett test's results are chosen as more reliable. Moreover, as already mentioned, the rvf-plot did not reveal any heteroskedastic pattern. Furthermore, the question that arises is, whether the error variance could be affected by the possible presence of spatial dependence in the regression model. In order to test for spatial effects six different tests are performed (Table 6.7). Table 6.7 reports six diagnostic tests for spatial dependence. The first measure that is used is the Moran statistic for spatial autocorrelation of the residuals. In order to test for any missing spatially lagged variables, the Lagrange Multiplier (lag) test is performed. In addition spatial error is tested for by computing the Lagrange Multiplier (error) test. These three tests are used to generally test for the presence of spatial dependence. To test error dependence in the presence of an omitted lagged dependent variable, the Robust LM (error) test is performed. The other robust LM (lag) test, tests for spatial lag dependence in the possible presence of spatial error dependence. The robust tests help to understand the nature of the spatial dependence, e.g. if the presence of a lagged dependent variable compensates the effects of spatial error. The Langrage Multiplier (SARMA) test is a combination of the Langrage Multiplier (error) and Robust

LM (lag) test (S4 GIS INSTITUTE, 2006). None of the tests confirm the presence of spatial dependence.

Table 6.7 Spatial Dependence Diagnostics – standard OLS regression

TEST	MI/DF	VALUE	PROB
Moran's I (error)	0.014866	0.8997765	0.3682392
Lagrange Multiplier (lag)	1	0.1685753	0.6813811
Robust LM (lag)	1	0.1545554	0.6942196
Lagrange Multiplier (error)	1	0.0366258	0.8482289
Robust LM (error)	1	0.0226059	0.8804865
Lagrange Multiplier (SARMA)	2	0.1911812	0.908836

MI=Moran's I
DF=Degrees of Freedom

Source: computed in GeoDA 1.0.1 (ANSELIN, 2006)

Although the spatial autocorrelation is not found to be of concern, the model is re-estimated using the maximum likelihood procedure and controlling for spatial dependence. First, a spatial lag model is estimated by incorporating a spatial lag term of the dependent variable. The coefficient on the lagged variable will provide insights to the severity of a spatial dependence originating from a spatial lag. In as second step, an additional model including a control for spatial error (LAMBDA) is estimated. Results of the spatial lag regression and detailed diagnostics can be found in the Annex (G). In both models, the indicator coefficients (lagged dependent variable; LAMBDA) are not significant; moreover the values on the coefficients are very small. Neither does the overall model-fit improve, as indicated by almost constant R^2 and Log-Likelihood values over the three models (OLS; spatial lag; spatial error), nor do the coefficients on the independent variables change or are rendered significant as a result of controlling for spatial dependence. A wide range of tools, graphical and statistical, have been used to assess the possible presence of spatial dependence in the spatial units of observation, namely districts. As the graphical tools did not provide a conclusive picture, the combination with statistical diagnostic tests was deemed necessary. The test statistics for spatial dependence in the standard model did not raise any concerns regarding spatial lag or spatial error. In addition, multicollinearity was not found to be an issue either, as it initially controlled for by demeaning or standardizing the problematic variables. Heteroskedasticity did not appear to be of major concern. Furthermore, to test whether the overall model fit improves by controlling for spatial lag and spatial error, two additional models incorporating control variables for both types of spatial dependence were estimated. The controls did not improve the model.

To sum up, a variety of assumptions of the OLS regression model were tested in this chapter for their validity in the present study's case. Issues regarding structural multicollinearity were resolved using standardization. Issues were particularly depicted regarding the impact of outliers and leverage on the regression coefficients. Specific observations were found to exert influence on the regression coefficients. The residuals were found to be normally distributed. A detailed analysis of any spatial dependence in the model alongside tests for the homogeneity of variance of the residuals was carried out. The presence of heteroskedasticity was tested and disproved. The most striking issue that prevails from the diagnostics is the presence of outliers, leverage and consequently influence, which require further attention. As deviant observations may carry vital information instead of discarding them or fixing the underlying data without justification, the idea is to apply statistical procedures that manage to accommodate them in the model (BARNETT AND LEWIS, 1978). In this sense, robust regression is considered as a solution regarding the issue of influential observations. Robust regression (*iteratively reweighted least squares regression*) is proposed as an alternative to OLS regression in cases where the data is contaminated with influential observations. Especially, when after fitting a least squares regression post-estimation diagnostics do not reveal any data entry errors and the exclusion of the problematic observations is not justified, robust regression is a sound alternative to OLS. Robust regression can be used as an alternative to OLS in any case, as it weighs the observations based on their behavior. Thus, if OLS and robust regression are used on well-behaved observations without serious issues of influence, the results should not be very different (FOX AND LONG, 1990). To sum up, in robust regression most influential observations are dropped, before in a second step the remaining observations with large absolute residuals are down-weighted. Basically, the robust regression procedure in the STATA econometric package first calculates COOK's D (distance) (COOK AND WEISBERG, 1982) for each observation and based on the distance (>1) drops problematic observations. An iteration process ensues, calculating the weights on the basis of absolute residuals. This iteration comes to an end, when the maximum difference between the weights from one iteration to the next falls below tolerance. The weighting scheme consists of two different types of weights, namely Huber weights and bi-weights, both belonging to the group of M-estimators. In case of the Huber weighting, observations with small residuals are assigned a weight of 1, thus the larger the residual is the smaller will be the weight that is assigned to it. Bi-weighting simply down-weighs observations with a non-zero residual. Two types of weights are used due to the sensitivity of Huber weights to severe outliers and the weakness of bi-weights to offer multiple solutions, when applied without correction. The correction is provided by applying the Huber weights in the first iteration (VERARDI AND CRUX, 2009).

Several models are estimated, with the following criteria being of paramount importance for selecting the working models:

- Goodness of Fit: R-square, F-value, root mean squared error
- Significance: selection on the basis of significance levels 10% ,5%, 1%
- Plausibility: Direction of impact as postulated by theory
- t-values: higher t-values preferred; variables with extremely low values are generally dropped
- Strategy for variable retainment and exclusion (GELMAN AND HILL, 2007):
    - Insignificant Predictor, expected sign: generally kept if sufficient t-value
    - Insignificant predictor, wrong sign: dropped
    - Significant predictor, expected sign: retained
    - Significant predictor, wrong sign: review

6.3 Interpretation of the parameter estimates

The previous chapters presented the methodology that is at heart of the present study with its conceptual understanding and revisions for an application to a developing country like Pakistan. Following the revised unique methodology developed for application to the Pakistan case the upcoming deliberations present a set of Ricardian models for different farm types.

The analysis was conducted using a step-wise approach, where modeling started with only climate variables and was augmented with other relevant control variables. In total three separate models were estimated for the available dataset. First, a model was estimated for the entire sample, covering 109 districts. To test whether the impacts of climate differ for rainfed and irrigated farming systems, two additional models were estimated, covering 77 and 109 districts, respectively. The individual farm households that were used to generate the farm type samples (rainfed and irrigated) sum up to the entire sample, this however on the single farm household level not the district level. Several different econometric specifications were examined before settling on the specification in table 6.8.

The second and third columns of the table show separate regressions of rainfed and irrigated farms. For the different tested specifications, the same control variables were used and the same non-linear relationship between land value per acre and climate was assumed. The different specifications included a two-seasonal, a three-seasonal and a four-seasonal model. The first major specification included the agricultural growing seasons of the country, namely Kharif (summer) and Rabi (winter) only. In the second specification an additional season was added with spring, to test whether pre-sowing conditions are important for explaining the variation in land values. In the final specification a fourth season was added. The final specification thus was made up of a summer sowing (compares to spring), a summer growing (compares to summer), a winter sowing (compares to fall) and a winter growing season (compares to winter). In this way it was ensured, that all important annual climatic peculiarities were included in the model and the agricultural growing seasons which can be altered by the farmer were not the sole representatives of annual climate. Moreover, by disaggregating the season into sowing and growing periods the model was prepared to test for specific features of climatic timing.

When comparing the two-season and the three-season models to the four-season model, the improvements in the four-seasonal model proved substantial, with clearly more significant climate coefficients. Ultimately, the four-seasonal model with disaggregated sowing and growing periods seemed to capture the individual inter-seasonal effects best. In the next step the idea was to test the assumption whether season specific interactions between precipitation and temperature are important. The results proved not very promising and particularly led to a significant amount of instability in the model, mainly through substantially raising variance inflation (multicollinearity). Furthermore, the season-specific interaction terms were neither significant nor did they have the expected signs and thus were excluded. Instead, to test whether surface water availability and temperature interact, an interaction term between surface water (runoff and river density) was introduced. Additionally, the irrigated model added an interaction term between irrigated acreage and annual temperature.

Adding these annual interaction terms instead of seasonal ones did not raise issues with multicollinearity and the results remained stable. The result indicated that these interactions are significant and important. Agronomic research suggests a hill-shaped relationship between temperature and crop growth (see chapter 3.1), however several other functional forms are possible (KAUFMANN, 1998). Positive linear and negative quadratic terms translate into a hill-shaped functional relationship, whereas vice versa the relationship is U-shaped. A hill-shaped relationship indicates that up to a certain level of increase, climate will have a beneficial effect on land values (or capitalized net revenues) but after having reached a maximum, land values will start to decline with an increase in the respective climate variable. In contrast to the hill-shaped relationship the U-shaped relationship between climate variables and land values indicates an initial decrease in land values with an increment in the climate variable, which however after reaching a minimum, again starts increasing.
As Table 6.8 shows, most climate variables in all three models are significant. In general all three models confirm the high sensitivity of farming in Pakistan to climate. Detailed summary statistics are presented in the Annex H.

Table 6.8 Ricardian Climate Regressions- Estimation Results

Variables		Total Farms	Rainfed Farms	Irrigated Farms
Land Value/acre (in Rs.)			***Climate Variables***	
Summer Sowing Temperature	(°C)	0.0192	-0.0630	0.0404
		(0.0344)	(0.0637)	(0.0386)
Summer Sowing Temperature²	(°C)	0.0195***	0.0189***	-0.00728***
		(0.00419)	(0.00677)	(0.00276)
Summer Growing Temperature	(°C)	-0.110***	-0.0769**	-0.168***
		(0.0217)	(0.0346)	(0.0506)
Summer Growing Temperature²	(°C)	-0.00474*	-0.00425	0.0108**
		(0.00258)	(0.00408)	(0.00429)
Winter Sowing Temperature	(°C)	0.241***	0.300***	0.158**
		(0.0662)	(0.112)	(0.0687)
Winter Sowing Temperature²	(°C)	-0.00506	-0.00330	-0.0181**
		(0.00744)	(0.0147)	(0.00697)
Winter Growing Temeprature	(°C)	-0.216***	-0.239*	-0.103
		(0.0684)	(0.120)	(0.0655)
Winter Growing Temperature²	(°C)	-0.0172***	-0.0171	0.0113*
		(0.00597)	(0.0111)	(0.00610)
Spring Precipitation	(mm)	-0.000393	0.00710*	0.000805
		(0.00232)	(0.00388)	(0.00243)
Spring Precipitation²	(mm)	-2.62e-05***	-5.43e-05***	-9.42e-06
		(9.48e-06)	(1.58e-05)	(5.77e-06)
Monsoon Precipitation	(mm)	-0.0146***	-0.0177***	-0.0103***
		(0.00172)	(0.00309)	(0.00173)
Monsoon Precipitation²	(mm)	3.55e-05***	1.91e-05	1.31e-05
		(1.24e-05)	(2.08e-05)	(1.25e-05)
Winter Growing Precipitation	(mm)	0.000447***	-0.00263	0.0177***
		(0.000166)	(0.00895)	(0.00612)
Winter Growing Precipitation²	(mm)	1.30e-07**	0.000360***	5.88e-05
		(6.47e-08)	(0.000110)	(5.21e-05)
Latitude	Decimal Degrees	0.0481	-0.0819	-0.0468
		(0.0338)	(0.0628)	(0.101)
			General Control Variables	
Human Development Index	HDI	1.696***	1.551**	2.943***
		(0.364)	(0.583)	(0.410)
Population Density	Population/km²	0.0323	0.00100***	0.270***
		(0.0342)	(0.000262)	(0.0354)
Elevation	m	-0.189***	-0.150**	0.197***
		(0.0388)	(0.0649)	(0.0434)
Tractor Density	tractors/ha	0.116***	0.108***	0.0488**
		(0.0182)	(0.0306)	(0.0194)
Road Density	road km/area	0.0495***	0.0359***	0.0311***
		(0.00828)	(0.0133)	(0.00901)

continued………		Hydrology and Interactions		
Runoff	mm	0.199***	0.213***	0.0195*
		(0.0349)	(0.0593)	(0.0113)
River Density Index	River lenght km/area	0.0296**	0.0498*	0.0135
		(0.0129)	(0.0249)	(0.0297)
Flood Risk	% of total area affected	-0.496***	-0.697*	-0.131
		(0.141)	(0.375)	(0.148)
Surface*Temperature	mm*(°C)	-0.000303***	-0.000273**	-6.96e-05
		(7.36e-05)	(0.000123)	(8.99e-05)
Irrigation*Temperature	ha*(°C)	-	-	3.73e-06*
				(2.17e-06)
			Soil Type	
Soil Calcisols	% in total area	0.337***	-0.346	0.237**
		(0.115)	(0.256)	(0.117)
Soil Fluvisols	% in total area	0.588*	0.946	0.652**
		(0.325)	(0.736)	(0.324)
Soil Rock Outcrop	% in total area	0.316*	0.541*	0.494**
		(0.183)	(0.294)	(0.196)
Soil Regosols	% in total area	-0.163	1.017**	-0.865***
		(0.299)	(0.483)	(0.321)
Constant		12.12***	11.65***	9.791***
		(0.311)	(0.483)	(0.335)
		Robust regression measures of fit		
Observations		109	77	109
R-square		0.69882549	0.70607471	0.7039479
AICR		224.48602	153.31708	219.75103
BICR		318.68092	235.55413	315.25755
deviance		5.2503218	5.9265026	5.4961853

Standard errors in parentheses *** $p<0.01$, ** $p<0.05$, * $p<0.1$

Dependent Variables is logarithm of Land Value/acre in Rupess (80 Rupees=1$)

Note: the respective season's sowing period represents the opposite season's harvesting period!

In the regression of total farms, six out of eight temperature coefficients are significant, five of which at the 1% significance level. Only the linear coefficient for the summer sowing temperature and the squared term on the winter sowing temperature are not significant. The results confirm the agronomic and agro-climatic wisdom of a non-linear relationship between climate and crop growth and ultimately farm incomes (land values), indicated by significant linear and squared coefficients. In case of precipitation all variables except for the linear coefficient on accumulated spring precipitation are highly significant, four coefficients on the 1% significance level and one at the 5% level. Summer sowing temperature shows an upward trend in relation with land values in the models for the total and rainfed farms, however a hill-shaped relationship is found for the irrigated farms model. The summer growing season on the other hand has a hill-shaped relationship with land values for both the total and rainfed farms' models, whereas irrigated farmland values show a U-shaped response to a temperature increase in the later stages of the growing season. In the sowing period of the winter season all three models reveal a hill-shaped relationship between farmland values and temperature, with the same relationship remaining for the growing period months in case of the models for total farms and rainfed farms; for irrigated farms in the later stages of the winter season a positive trend is observed, as revealed by a U-shaped relationship between temperature and dependent variable. The precipitation effects remain consistent over all three models, with spring precipitation being beneficial up to certain point after which a decline in land values entails. The important months of the summer growing season, characterized by the monsoon rainy period, clearly benefit from additional precipitation. A similar U-shaped relationship is found for the winter precipitation. As was expected, according to agronomic-wisdom, the majority of the temperature land value relationships are hill-shaped, seven out of twelve to be precise.

By distinguishing between sowing and growing months the idea was to determine the importance of the timing of climatic changes and weather occurrences. However, the season cannot be disaggregated in a definite way, thus the entire farming season has to be considered when assessing the impact of climate change on agriculture. To this background, a marginal impact analysis which is to follow in the upcoming deliberations, will determine the true nature of the relationship between climate variables and land values, by providing a farming season (sowing plus growing months) examination. Nevertheless, in cases were an upward trend was observed for temperatures in relation to land values inter alia for summer sowing temperatures, the results indicate that higher temperatures in the early part of the respective farming season may be beneficial. Likewise, a hill-shaped trend for the later stages of the farming season (i.e. summer growing) may indicate a negative impact for farming. The proxy for solar radiation (Latitude) is not significant in the full model. General control variables such Human Development Index, population density, elevation etc. and more specific control variables such as soils and hydrology are mostly significant and have the expected signs, clearly indicating the importance of the inclusion of controls when assessing the impact of climate change in a hedonic setting. In the regression for rainfed farms, four out of eight temperature coefficients and four out of six precipitation coefficients are significant. In case of irrigated farms, six out of eight temperature coefficients and two out of six precipitation coefficients are significant. Both disaggregated models have the expected sign for the solar radiation proxy (latitude). A detailed interpretation of the control variables will be presented following the chapter on the Ricardian climate sensitivities (Marginal Climate Impacts).

The interpretation of the climate coefficients in raw format is rather difficult as linear and quadratic terms are included. These terms can however indicate totally different functional relationships between climate and agricultural land values. As mentioned earlier, based on the signs of the linear and quadratic coefficients, the relationship can be either hill or U-shaped, asking for a further interpretive appraisal. To this background, in order to get an unmitigated understanding of the results regarding climate, in following the marginal impacts or so-called Ricardian climate sensitivities (POLSKY, 2004) of temperature and precipitation are examined.

6.3.1 Ricardian Climate Sensitivities

The Ricardian climate sensitivities are evaluated for each seasonal and annual variable and are used to assess the extent of the climate effects on land values. In the standard Ricardian model climate is modeled using a quadratic formulation, whereas the other variables are linear in the function. Thus recalling equation 11 from chapter 5.1 and differentiating it with respect to a climate component f_i (e.g. summer sowing temperature) yields the following marginal value or climate sensitivity for each climate component:

EQ (29) $$MCI = \frac{dLV}{df_i} = \beta 1 + 2 * \beta 2 * f_i$$

With MCI denoting Marginal Climate Impact, $\beta 1$ representing the linear coefficient of the climate component and $\beta 2$ denoting the squared coefficient and f_i representing the level of the climate components' long term normal value. These marginal values of each climate variable depend on the linear and quadratic coefficients. The respective climate component's marginal impact is computed by replacing E [f_i] for f_i and estimating the mean of the climate component; in this regard the climate variable's level of measurement also matters. When seasonal marginal values are obtained, summing them up over the entire year yields the annual marginal value for all included climate components. The marginal assessment reveals how a change of 1°C in temperature or 1 mm in rainfall translates into a change in land values per acre or hectare, not only annually but also seasonally and for different farm types. The evaluation is achieved by estimating equation 29 (32) at the mean level of the respective climate component. Additionally, to get an even more comprehensive interpretation of the regression output, Ricardian elasticities are calculated for the climate components, this is achieved by using the following equation (BENHIN, 2008):

EQ (30) $$\varepsilon = MCI * \frac{\bar{f_i}}{\overline{LV}}$$

With ε standing for the respective climate variable's elasticity, MCI representing the Marginal Climate Impact, $\bar{f_i}$ and $\overline{LV}$ denoting the mean of the climate component and the mean of the land value variable, respectively. The elasticity indicates i.e. how a land value responds to a 1 % increase in temperature or precipitation. As the model has been revised and reasoned in the right skewed nature of the land value dataset a semi-logarithmic formulation has been selected instead of the aforementioned standard Ricardian function, a series of additional considerations is required to correctly estimate the marginal impacts in the present particular case. One great advantage of the semi-log or log-linear form is that the marginal value is not anymore independent of all other variables in the model as is the case when opting for the standard Ricardian formulation with un-transformed land value data. The control variables that also determine land values are thus also captured when calculating the marginal impacts. This relationship is shown in the following equations:

EQ (31) $$Ln\, V = \beta_0 + \beta_1 F + \beta_2 F^2 + \beta_3 Z + \beta_4 G + u$$

EQ (32) $$MCI = \frac{dLV}{df_i} = V * (\beta 1 + 2 * \beta 2 * f_i)$$

Moreover, for reasons of econometric soundness regarding the presence of significant multicollinearity when including linear and their associated squared terms in the econometric model, the variables have been demeaned. However, as the marginal impact of climate also depends on the level of the climate parameter $f_{i,}$ the regression coefficients that have been calculated based on centered or demeaned data have to be undemeaned in a first step for the marginal climate assessment. Consequently, the squared terms originated from the demeaned linear variable. Regarding the estimation and interpretation of the marginal impacts this implies, that the linear coefficients of the climate variables consequently denote the marginal impacts at the mean climate. From this it follows that the demeaned variables can be undemeaned in the following way:

EQ (33) $$\beta 1 = \frac{MCI}{V} - (2 * \beta 2 * f_i)$$

The coefficient of the uncentered linear variable is found by dividing the demeaned variable's linear term, which equals the Marginal Climate Impact, by the average land value and subtract from it two times the demeaned quadratic variable's coefficient times the centered mean of the climate variable. Undemeaning the variables is important for two reasons. First of all, the interpretation of the results becomes more logical and most importantly although overall country estimates of climate impacts are accurate, on a district level the question, whether a district is below or above a certain average is of great importance, which indeed requires undemeaned coefficients (compare SCHLENKER, 2000). Ultimately, as the chosen functional form is semi-logarithmic, the coefficients represent the percentage change of land value. To verify the significance level of the Marginal Climate Impacts, an F-Test is run (see AMIRASLANY, 2010) using the following procedure:

First the Standard Deviation (SD) is calculated using:

EQ (34) $$SD(\frac{dLV}{df_i}) = 2 * \beta 2 * SD(f_i)$$

After calculating the SD, the F-statistic is obtained by dividing the MCI by the SD and taking the square of the result. Precisely:

EQ (35) $$F = \{(MCI)/(2 * \beta 2 * SD(f_i))\}^2$$

Table 6.9 presents the results.

Table 6.9 Ricardian Climate Sensitivities (Marginal Climate Change Impacts)

	Temperature				Precipitation		
	Total Farms	Rainfed Farms	Irrigated Farms		Total Farms	Rainfed Farms	Irrigated Farms
Summer Sowing	1.92%	-6.30%	4.04%	Spring Precipitation	-0.04%	0.71%	0.08%
Summer Growing	-11.00%	-7.69%	-16.80%	Monsoon Precipitation	-1.46%	-1.77%	-1.03%
Summer Farming Season (Kharif)	-9.08%	-13.99%	-12.76%	Summer Farming Season (Kharif)	-1.50%	-1.06%	-0.95%
Winter Sowing	24.10%	30.00%	15.80%				
Winter Growing	-21.60%	-23.90%	-10.30%	Winter Farming Season (Rabi)	1.46%	-0.26%	1.77%
Winter Farming Season (Rabi)	2.50%	6.10%	5.50%				
Annual	-6.58%	-7.89%	-7.26%	Annual	-0.04%	-1.32%	0.82%
Annual ($)	-$123.87	-$115.28	-$167.43	Annual ($)	-$0.81	-$19.33	$18.92
Annual Elasticity	-1.51	-1.79	-1.67	Annual Elasticity	-0.01	-0.82	0.19

Source: own calculations based on regression output (2012)

Table 6.10 Significance Level of estimated Marginal Climate Impacts

	Total Farms			Rainfed Farms			Irrigated Farms		
	SD	F-statistic	MCI	SD	F-statistic	MCI	SD	F-statistic	MCI
Summer Sowing	0,1700	0,012756	1,92%	0,1740	0,131094	-6,30%	-0,064	0,398477	4,04%
Summer Growing	-0,0270	16,59808	-11,00% ***	-0,0262	8,628065	-7,69% ***	0,0626	7,202278	-16,80% ***
Winter Sowing	-0,0410	34,55146	24,10% ***	-0,0270	123,4568	30,00% ***	-0,14589	1,17297	15,80%
Winter Growing	-0,1500	2,0736	-21,60% ***	-0,1570	2,317376	-23,90% ***	0,0996	1,069438	-10,30%
Spring Precipitation	-0,0049	0,006566	-0,04%	-0,0114	0,388638	0,71%	-0,00175	0,2116	0,08%
Monsoon Precipitation	0,0033	19,57392	-1,46% ***	0,0020	79,11164	-1,77% ***	0,0012	73,67361	-1,03% ***
Winter Precipitation	0,0090	2,618068	1,46% ***	0,0266	0,009767	-0,26%	0,00385	21,13611	1,77% ***

**** denotes significant at 1% level*

Source: own calculations (2012)

As the significance test show, the MCI is highly significant for the summer growing, winter sowing and winter growing season temperatures for the country as a whole and rainfed farms. In case of the irrigated farms, the summer growing temperature is found to be the only highly significant temperature related MCI. Generally, the monsoon season and winter season precipitation MCIs are found to be highly significant. The MCI analysis reveals some interesting and important results. The results for the modeled climate variables have the expected signs and the importance of timing regarding the influence of climatic impacts is also revealed. Moreover, from a technical point of view the significance of estimating separate models for rainfed and irrigated farms is clearly affirmed. For the regression of all farms, an increase in temperature by 1°C is expected to have a positive impact on land values in the summer sowing period, which compares to the spring season. The same relationship is found for the winter sowing period and land values, indicating that higher temperatures in the initial growth period of crops may be beneficial. However, the results for the respective farming season's growing period are consistently negative throughout all three models, showing that although the initial crop growth reacts positively to higher temperature, in later stages however, especially for the summer season where temperatures are at their annual peak, temperature increases would be clearly damaging to agriculture. Furthermore, higher temperatures in the growing period can be expected to increase the water demand. The overall annual marginal climate impact for the full sample reveals an annual loss in agricultural land values to the tune of -6.58% from a 1°C increase in temperature. Interestingly, when comparing rainfed and irrigated farms the results as expected reveal contrasting results for the different farm types. Although the model for all farms revealed a positive impact from a 1°C increase in temperature for the summer sowing period, the effects on rainfed or in other words dryland farms are different from the impacts on irrigated farms, a result also found by earlier studies by AHMED AND SCHMITZ (2011b). Rainfed farms are expected to suffer from any increase in temperature in the summer farming season, be it sowing or growing period. The results for the irrigated farms are consistent with the general results obtained for the mixed farm sample (total farms). While different impacts are observed for the summer season, for the winter farming season however, they are consistent over all models, indicating a positive impact from a temperature increase for the early sowing period and negative effects from a marginal warming in the growing period of the winter farming season.

For a more intuitive understanding of the MCIs and given that the two separately modeled periods (sowing and growing) are part of the summer and winter farming seasons, both results have to be seen in aggregate. As the individual sowing and growing periods present fractions of the full farming seasons, the respective agricultural farming season's Ricardian climate sensitivity is obtained by totaling the sowing and growing period's values. For the full farms sample this implies, that a marginal increase in temperature in the summer farming season is expected to reduce land values (or capitalized crop net revenues) by -9.08%. For dryland farms the damages from a 1°C in summer farming season temperature is estimated to incur losses to farmland values to the tune of -13.99%, whereas the damages for irrigated farms amount to -12.76% in the assumed marginal summer season temperature change. Thus the results clearly reveal the sensitivity of agriculture to additional warming in semi-arid Pakistan, with average summer season temperatures around 31°C. When comparing irrigated and rainfed farming, rainfed farming in particular is found to be highly vulnerable to additional warming. Although irrigated farms too show a very high vulnerability to warming, they seem to be slightly less worse off due to the fact that additional irrigation provides an option to farmers. Nevertheless, rainfed and irrigated farms likewise, both are expected to suffer great losses incurred by higher temperatures in the summer farming season. Warming in the winter growing season is found to have a positive impact on land values in Pakistan, with a 1°C temperature increase translating into gains of 6.1% and 5.5%, for rainfed farmland values and irrigated farmland values, respectively. From a farming season point of view the results confirm the hill-shaped trend for temperature. That is, for the farming season up to a certain point increases in temperature will be beneficial, however after which they will start harming agriculture. The sowing period marginals are generally positive and significant, indicating that a longer season is beneficial. However, the winter growing and summer growing temperature marginals are negative. As SCHLENKER et al., (2006) have shown, hot summers put additional stress on crops, whereas cold winters help to keep back pests. By aggregating the seasonal impacts, the annual impact of higher temperatures is disclosed. Clearly, the annual impact of a marginal temperature increase is found to be negative over all farm types, with dryland farms being hit most severely. For the regressions of the mixed sample, a temperature increase by 1°C is found to decrease land values on average by -6.58% per hectare. Over all seasons, rainfed farms are to suffer losses in land values to the tune of -7.89% and irrigated farms to the tune of -7.26%.

The precipitation impacts are clearly more consistent over all types, suggesting that the country's arid nature is very much dependent on precipitation throughout different farming systems. As the parameter estimates in the foregone deliberations have revealed, the majority of the relationships between precipitation and land values are U-shaped. This finding is essential to understand the discussion of the marginal impacts regarding a 1 mm increase in precipitation. Both farm types, rainfed and irrigated, show a positive response to an addition in accumulated spring precipitation. This shows that sufficient precipitation ahead of the actual sowing period is of paramount importance to agriculture, although the impact is relatively small with 0.71% increase in land values for rainfed farms and 0.0805% for irrigated farms. The marginal precipitation effect in the summer growing period which is the period where the bulk of the country's annual precipitation is received, is found to be negative with -1.77% loss in rainfed farmland value and -1.03% in irrigated farmland value. Over the entire summer farming season this implies that losses in land values should be expected from a 1 mm in precipitation in the summer months, with rainfed farmland values depleting by -1.06% and irrigated farmland values by -0.95%. However, it has to be kept in mind that the relationship between monsoon season precipitation and land values, as reported by the parameter estimates of the Ricardian regressions, is U-shaped. The negative values thus do not indicate a negative impact of precipitation on land values as such, they explain that a 1 mm increase is simply not sufficient and thus is to be understood as an indicator for a precipitation deficiency in the hot summer months. The u-shape indicates that with every millimeter of additional rain the losses will become smaller up until the minimum extreme point, after which land values will start actually increasing. For the winter season the results are similar, however the irrigated farms show a positive response to a 1 mm increase with an increment in land values of 1.77%, whereas rainfed farms disclose a negative impact with -0.263%. With a U-shaped relationship calculated for the winter season as well, this goes to show that the rainfed farms do not receive the minimum amount of precipitation that they require, whereas irrigated farms have already surpassed this minimum and hence are benefitting from any further increase. Over the farming seasons the precipitation deficiency is evident in both farming systems; however more so in rainfed systems, which solely depend on rainfall for farming. Mixed results are revealed for the winter farming season, with irrigated farms being positively impacted by a marginal precipitation increase and rainfed farms being negatively hit, however again with the U-shaped relationship in mind, the negative impact revealing a precipitation deficiency. Annually, the rainfed systems show a negative impact, driven by the significant precipitation deficiency in the summer growing season. Irrigated systems show a positive trend. Therefore, rainfed farms require additional precipitation to reach a point after which land values start increasing. Irrigated farms rely heavily on groundwater and precipitation is essential in this regard, therefore considering the paramount importance of irrigated farming to the country, additional water availability will be beneficial.

The estimated annual elasticity indicates that a 1% increase in temperature will lead to 1.51% decrease in land values for the whole of Pakistan, 1.79% for rainfed farms and 1.67% for irrigated farms. As the majority of the seasonal temperature coefficients show a hill-shaped trend, the results have the expected signs. Another striking finding is that the rainfed farms are particularly vulnerable to increases in temperature, with a marginal temperature increase causing the highest losses to this particular farming system. Higher temperatures are expected to increase the demand for water. Moreover, Pakistan has particularly high Evapo-Transpiration rates, thus a considerable fraction of the precipitation that is received evaporates again. In a lot of areas evaporation rates are as high as 99%, showing the additional pressure on groundwater resources. In the rainfed areas i.e. of the Punjab, the amount of annual rain missing to fully compensate Potential-Evapo-Transpiration (PET) can reach dimensions close to 1400mm/year (e.g. Rabwah, Rawalpindi, Sargodha) (FAO, 2004a; 2005). The weather conditions in barani (rainfed) areas are fluctuating severely, thus in some years precipitation can be very sparse and plentiful in others. Irrigated systems seem to be better equipped for mitigating the effects of additional warming, as the estimated elasticity shows. Although the impact of a marginal temperature increase is still negative, compared to rainfed systems it is much smaller, indicating that irrigation buffers the impacts from an increase in temperature, however provided that water for irrigation is available.

The elasticities calculated for precipitation indicate that for the country as whole and for rainfed farms, annual net losses to the tune of -0.01% and -0.82% are expected with a 1% increase in precipitation, whereas for irrigated farms annual gains of 0.19% are revealed. Departing from a predominantly U-shaped relationship between precipitation and land values, the elasticities confirm the finding that the country and more specifically rainfed farms currently suffer from a precipitation deficit. The initial negative impact for the mixed and rainfed farms' samples is also reasoned in the fact that the positive effects in the spring and winter season are outweighed by the negative effects in the summer growing season (Monsoon-period), in which due to high temperatures water demand irrespective of the monsoon rains nevertheless is relatively high, also driven by the earlier mentioned high rates of water losses due to Evapo-Transpiration. For the irrigated farms the results confirm that the minimum level of precipitation that is required is given, thus any further increase in precipitation will be beneficial for irrigated farming. In sum, the precipitation results indicate that any further increase in precipitation will put all farming systems marginally better off than in the current situation. Overall, the temperature effects are (negatively) elastic, whereas as expected the precipitation effects are particularly (negatively) elastic for rainfed farming. In the irrigated and total farms' case the seasonal effects offset each other; therefore the annual elasticities are more or less close to zero.

In following, the goal is to assess the marginal impacts of climate on the different Agro-Ecological Zones and provinces of Pakistan. For this purpose, the respective climatic normals are used.

Table 6.11 Ricardian Climate Sensitivities on the Agro-Ecological-Zone level

	Temperature								
AEZ	**Land Value ($/ha)**	**Summer Sowing**	**Summer Growing**	**Winter Sowing**	**Winter Growing**	**Summer Farming Season (Kharif)**	**Winter Farming Season (Rabi)**	**Annual**	**Annual ($)**
Barani	3543	1.1%	-10.2%	25.1%	-18.0%	-9.1%	7.1%	-2.0%	-$72
Barani/Northern Irrigated Plain	2971	-4.6%	-8.2%	25.5%	-14.5%	-12.8%	11.0%	-1.8%	-$54
Barani/Sulaiman Piedmont	844	7.0%	-12.5%	23.1%	-24.4%	-5.5%	-1.2%	-6.7%	-$57
Barani/Western Dry Mountains	1720	-0.6%	-10.6%	24.7%	-18.5%	-11.2%	6.2%	-5.0%	-$85
Dry Western Plateau	410	5.0%	-11.0%	22.0%	-31.0%	-6.0%	-9.1%	-15.1%	-$62
Indus Delta	476	6.6%	-10.4%	18.7%	-42.7%	-3.8%	-24.0%	-27.8%	-$132
Northern Dry Mountains	3792	-30.7%	-5.5%	30.6%	6.3%	-36.2%	36.9%	0.7%	$27
Northern Dry Mountains/Northern Irrigated Plain	6556	-10.2%	-9.3%	26.8%	-12.5%	-19.4%	14.3%	-5.1%	-$337
Northern Irrigated Plain	2796	5.2%	-11.9%	23.7%	-22.2%	-6.7%	1.5%	-5.1%	-$144
Northern Irrigated Plain/Western Dry Mountains	4842	-8.9%	-10.8%	25.8%	-13.0%	-19.7%	12.7%	-7.0%	-$339
Sandy Desert	1024	7.0%	-12.5%	23.1%	-24.4%	-5.5%	-1.2%	-6.7%	-$69
Sandy Desert/Northern Irrigated Plain	1474	10.9%	-12.7%	22.6%	-26.4%	-1.8%	-3.8%	-5.5%	-$82
Sandy Desert/Southern Irrigated Plain	896	19.5%	-12.8%	20.6%	-35.6%	6.7%	-15.0%	-8.3%	-$75
Southern Irrigated Plain	804	19.5%	-13.4%	19.7%	-37.7%	6.2%	-18.0%	-11.8%	-$95
Southern Irrigated Plain/Dry Western Plateau	692	20.2%	-13.1%	19.6%	-39.0%	7.1%	-19.4%	-12.2%	-$85
Southern Irrigated Plain/Indus Delta	490	17.8%	-11.9%	18.1%	-44.4%	6.0%	-26.3%	-20.3%	-$99
Sulaiman Piedmont	756	22.6%	-14.0%	17.9%	-43.6%	8.6%	-25.7%	-17.1%	-$129
Sulaiman Piedmont/Southern Irrigated plain	691	21.8%	-14.0%	20.4%	-35.7%	7.8%	-15.3%	-7.5%	-$52
Western Dry Mountains	1424	-18.5%	-8.3%	29.1%	-4.1%	-26.7%	24.9%	-1.8%	-$26
Western Dry Mountains/Dry Western Plateau	712	-23.7%	-10.7%	30.8%	-1.4%	-34.4%	29.4%	-5.0%	-$35
Western Dry Mountains/Sulaiman	693	14.9%	-13.8%	22.6%	-28.7%	1.1%	-6.1%	-5.0%	-$35
Wet Mountains	2646	-32.4%	-6.8%	30.2%	2.1%	-39.2%	32.4%	-6.8%	-$180
Wet Mountains/Northern Dry Mountains	5156	-22.7%	-5.8%	29.5%	0.1%	-28.4%	29.6%	1.2%	$61
Wet Mountains/Northern Irrigated Plain	3946	-21.3%	-6.8%	28.5%	-5.2%	-28.1%	23.3%	-4.7%	-$186

continued…….	Precipitation							
AEZ	**Spring Precipitation**	**Monsoon Precipitation**	**Winter Precipitation**	**Summer Farming Season (Kharif)**	**Annual**	**Annual ($)**	***Total MCI (%)***	***Total MCI ($)***
Barani	-0.2%	-1.0%	1.9%	-1.2%	0.6%	$22.7	*-1.4%*	***-$49***
Barani/Northern Irrigated Plain	-0.2%	-1.1%	1.8%	-1.3%	0.5%	$15.7	*-1.3%*	***-$39***
Barani/Sulaiman Piedmont	0.0%	-1.6%	1.2%	-1.6%	-0.4%	-$3.2	*-7.1%*	***-$60***
Barani/Western Dry Mountains	-0.3%	-1.3%	1.8%	-1.6%	0.3%	$4.9	*-4.7%*	***-$81***
Dry Western Plateau	0.2%	-1.7%	1.1%	-1.5%	-0.4%	-$1.7	*-15.5%*	***-$64***
Indus Delta	0.3%	-1.5%	0.8%	-1.2%	-0.4%	-$1.9	*-28.2%*	***-$134***
Northern Dry Mountains	-1.7%	-1.3%	3.9%	-3.0%	0.9%	$35.6	*1.6%*	***$62***
Northern Dry Mountains/Northern Irrigated Plain	-0.8%	-1.1%	3.4%	-1.9%	1.5%	$95.8	*-3.7%*	***-$241***
Northern Irrigated Plain	0.0%	-1.4%	1.3%	-1.3%	0.0%	-$.1	*-5.1%*	***-$144***
Northern Irrigated Plain/Western Dry Mountains	-0.4%	-1.4%	2.1%	-1.8%	0.3%	$15.1	*-6.7%*	***-$324***
Sandy Desert	0.2%	-1.6%	1.0%	-1.4%	-0.4%	-$4.4	*-7.2%*	***-$73***
Sandy Desert/Northern Irrigated Plain	0.1%	-1.5%	1.0%	-1.4%	-0.4%	-$5.5	*-5.9%*	***-$87***
Sandy Desert/Southern Irrigated Plain	0.3%	-1.6%	0.8%	-1.3%	-0.5%	-$4.9	*-8.9%*	***-$79***
Southern Irrigated Plain	0.3%	-1.6%	0.8%	-1.3%	-0.6%	-$4.5	*-12.4%*	***-$100***
Southern Irrigated Plain/Dry Western Plateau	0.3%	-1.6%	0.7%	-1.3%	-0.5%	-$3.7	*-12.8%*	***-$88***
Southern Irrigated Plain/Indus Delta	0.3%	-1.5%	0.7%	-1.2%	-0.5%	-$2.4	*-20.8%*	***-$102***
Sulaiman Piedmont	0.2%	-1.8%	1.1%	-1.6%	-0.5%	-$3.5	*-17.6%*	***-$133***
Sulaiman Piedmont/Southern Irrigated plain	0.3%	-1.7%	0.8%	-1.4%	-0.6%	-$4.0	*-8.1%*	***-$56***
Western Dry Mountains	-0.1%	-1.6%	1.6%	-1.7%	-0.1%	-$1.4	*-1.9%*	***-$27***
Western Dry Mountains/Dry Western Plateau	0.2%	-1.8%	1.4%	-1.6%	-0.2%	-$1.6	*-5.2%*	***-$37***
Western Dry Mountains/Sulaiman	0.3%	-1.7%	0.9%	-1.4%	-0.5%	-$3.8	*-5.6%*	***-$38***
Wet Mountains	-1.2%	-0.6%	3.4%	-1.9%	1.5%	$40.8	*-5.3%*	***-$139***
Wet Mountains/Northern Dry Mountains	-1.1%	-0.9%	3.4%	-2.0%	1.5%	$75.7	*2.7%*	***$137***
Wet Mountains/Northern Irrigated Plain	-1.2%	-0.8%	3.9%	-2.0%	1.9%	$74.8	*-2.8%*	***-$112***

Source: own calculations, 2012

Table 6.12 F-Test of joint significance (two-tailed)

AEZ	Temperature				Precipitation		
	Summer Sowing	Summer Growing	Winter Sowing	Winter Growing	Spring	Monsoon	Winter
Barani	0.1	59.2 ***	577.2 ***	31.8 ***	1530.8 ***	12.5 ***	1.7 *
Barani/Northern Irrigated Plain	0.1	3.0 ***	122.1 ***	1.9 **	247.7 ***	16.5 ***	0.9
Barani/Sulaiman Piedmont	0.4	19.7 ***	114.8 ***	8.3 ***	2.5 **	17.4 ***	0.6
Barani/Western Dry Mountains	0.0	19.4 ***	62.5 ***	2.5 **	199.2 ***	5.1 ***	1.7
Dry Western Plateau	0.4	144.1 ***	31.8 ***	5.8 ***	405.0 ***	727.5 ***	7.1 ***
Indus Delta	0.7	128.5 ***	23.1 ***	11.0 ***	690.6 ***	548.4 ***	4.0 ***
Northern Dry Mountains	54.2 ***	24.0 ***	1006.0 ***	19.6 ***	67750.1 ***	1.3	5.4 ***
Northern Dry Mountains/Northern Irrigated Plain	5.9 ***	69.3 ***	775.3 ***	77.5 ***	14360.3 ***	1.0	4.0 ***
Northern Irrigated Plain	0.2	55.3 ***	114.8 ***	5.8 ***	4.0 ***	13.2 ***	2.4 **
Northern Irrigated Plain/Western Dry Mountains	0.8	81.1 ***	149.0 ***	2.0 **	716.2 ***	91.3 ***	16.2 ***
Sandy Desert	9.0 ***	160.5 ***	518.2 ***	55.8 ***	3509.3 ***	187.4 ***	4.3 ***
Sandy Desert/Northern Irrigated Plain	8.4 ***	211.5 ***	1156.6 ***	117.3 ***	1237.5 ***	64.0 ***	4.6 ***
Sandy Desert/Southern Irrigated Plain	480.4 ***	134.9 ***	227.7 ***	76.0 ***	20544.1 ***	1621.8 ***	2.6 **
Southern Irrigated Plain	83.2 ***	646.2 ***	144.4 ***	34.0 ***	4792.9 ***	7718.3 ***	6.2 ***
Southern Irrigated Plain/Dry Western Plateau	1347.8 ***	317.3 ***	62.5 ***	20.2 ***	3041.8 ***	24555.4 ***	53.2 ***
Southern Irrigated Plain/Indus Delta	72.4 ***	127.8 ***	2937.6 ***	3864.0 ***	13986.0 ***	1474.8 ***	9.6 ***
Sulaiman Piedmont	9.0 ***	42.8 ***	134.6 ***	68.6 ***	382.7 ***	1095.8 ***	56.6 ***
Sulaiman Piedmont/Southern Irrigated plain	8.4 ***	42.8 ***	174.3 ***	46.0 ***	1277.6 ***	998.8 ***	29.2 ***
Western Dry Mountains	2.7 ***	19.2 ***	104.7 ***	0.2	29.7 ***	33.6 ***	4.1 ***
Western Dry Mountains/Dry Western Plateau	1.2	56.7 ***	41.8 ***	0.0	40.6 ***	562.0 ***	16.8 ***
Western Dry Mountains/Sulaiman Piedmont	3.9 ***	41.6 ***	215.0 ***	29.7 ***	962.8 ***	995.1 ***	33.7 ***
Wet Mountains	2.0 *	23.9 ***	39.6 ***	0.0	2076.3 ***	0.4	0.9
Wet Mountains/Northern Dry Mountains	0.6	17.5 ***	16.8 ***	0.0	900.3 ***	0.6	0.8
Wet Mountains/Northern Irrigated Plain	1.8 *	23.9 ***	140.3 ***	0.3	5229.5 ***	0.9	1.5

Table reports F-Values for degrees of freedom 24; * denotes significant at 10% level;** denotes significant at 5% level;*** denotes significant at 1%

Source: own calculations, 2012

The marginal impacts or so-called Ricardian climate sensitivities are the slopes of the climate variables modeled in the Ricardian framework. As has been mentioned before, the presentation or summary of the climatic relationships of this non-linear specification is not straight forwardly achieved by presenting the simple slopes as constants (POLSKY, 2004). Furthermore, KAUFMANN (1998) explains that the slope of a nonlinear specification is not a constant value but a function of where the variable is evaluated. To this effect, the climate sensitivities vary across the region (Agro-Ecological Zone, province, district etc). Accordingly, in order to calculate these spatially variable climate sensitivities, the parameter estimates from the regression are jointly evaluated across long term normal climatic means for the respective region. Moreover, to test whether the marginal impact of climate (temperature, precipitation) is significant, a joint hypothesis test is calculated.

The results of the Agro-Ecological Zone analysis are in line with expectations and the climatic-region specific marginal climate impacts have the anticipated signs. Using the parameter estimates of the total farms model, table 6.11 presents the marginal impacts of climate at the Agro-Ecological Zone level and table 6.13 at the provincial level. First, for each seasonal sub-period (sowing, growing) an individual marginal impact is calculated, secondly the table provides aggregated values for the entire farming season (summer, winter). Additional columns provide annual marginal impacts in percent and dollars per hectare. In a first step, the temperature sensitivities will be presented, before moving on to precipitation sensitivities and finally reporting the aggregate impacts for the Agro-Ecological Zones. As far as the individual sub-periods are concerned, the calculations show that an increase in temperature by 1°C in the summer sowing period, which begins around the month of May, will benefit most of the country's agro-ecological regions. Particularly, the southern parts of the country, with fruits and rice production as their mainstay, are expected to benefit from a longer sowing period. The mountainous regions reveal a negative trend for the summer sowing season. The Ricardian temperature sensitivities for the growing period of the Kharif farming season show consistently negative impacts for all zones throughout the country. On the other hand, the winter sowing period is characterized by consistently positive marginal impacts for all Agro-Ecological-Zones. However, in the growing period of the Rabi season, the impacts from a 1°C increase in temperature on land values are predominantly negative, with the exceptions of the Northern Dry Mountain zone and the Wet Mountain zone, for which marginally positive sensitivities are calculated. The marginal impact analysis indicates that in general, all Agro-Ecological-Zones except for the Northern Dry Mountains Zone and the shared Wet Mountain/

Northern Dry Mountains Zone will experience negative mean annual impacts. The Northern Dry Mountains Zone is mainly used for grazing. Mild summers and particularly cold winters are characteristic for this region, therefore the positive impact from a marginal warming in winter over-compensates the negative impact from a 1°C warming in the Kharif season. As precipitation levels in this zone are generally low, the positive impact from a 1mm increase in precipitation is no surprise. The overall impact from a joint marginal increase in temperature and precipitation is associated with an increase in land values per hectare by 1.65% or 62.4 $/ha in monetary terms. The land values in this zone are generally high with 3792 $/ha. The other beneficiary from a joint increase in temperature and precipitation is the intersecting zone between Wet Mountains and Northern Dry Mountains. Especially warming in the winter months positively impacts this zone, for in these agro-ecological settings winter temperatures as low as 0°C are not uncommon. The summers are mild in general, however temperature records have also shown temperatures to peak around 35°C, therefore the negative impact for the summer growing period is in line with climatic trends. The winter gains clearly over-compensate the negative impact for the summer season, leaving a net mean annual gain of 1.188% in land values per hectare from a 1°C increase in temperature, or 61.24 $/ha in monetary terms. A humid classification of the Wet Mountains Zone does not apply to the intersecting zone Wet Mountains Zone/ Northern Dry Mountains alike, therefore another interesting finding for the latter is the beneficial impact from an increase in precipitation. Particular gains are reported for the winter months, namely + 3.4%. The precipitation deficit in the summer months is not as grave as in the other zones, which is indicated by the marginal precipitation impact for the monsoon period of -0.88%. For the spring precipitation a deficit of considerable amount is computed leading to losses in land value per hectare to the tune of -1.076%. Recalling the U-shaped relationship between precipitation and land values found through the parameter estimates, which indicates that having achieved a minimum level of rainfall any further increases will be beneficial, the negative impacts from a mere 1 mm increase in precipitation go to show that these minimum levels of precipitation that are required have not been achieved yet, therefore losses are a logical consequence.

For a more complete understanding of the marginal impacts from a change in climate the entire farming seasons have to be considered, especially to the background of individual periods' effects offsetting one another. The majority of Agro-Ecological Zones suffer from additional warming in the Kharif season, foremost the northern dry mountains, the wet mountains and western dry mountains zones. In contrast, marginal benefits are found for southern irrigated plain and the Sulaiman Piedmont. The winter farming season Rabi benefits from an additional warming in the zones Barani, northern dry mountains, northern irrigated Plain, western dry mountains and the wet mountains zone. An annual analysis only, reveals the true nature of the marginal impacts for the different Agro-Ecological settings of the country, as individual seasonal effects are aggregated over the year. As expected the zones most severely hit also have the lowest land values per hectare, precisely under 500$ per hectare. The dry and hot regions of Sindh and Balochistan comprising the zones Dry Western Plateau, Indus Delta and the Sulaiman Piedmont will suffer losses in land values from a marginal increase in normal temperature to the tune of -15%, -27.76% and -17.1%, respectively. The shared zone Southern irrigated Plains/ Indus Delta will likewise suffer a loss as high as -20.29%. The Northern Mountains zone is expected to slightly benefit from an additional warming, with benefits ranging between 0.7% and 1.18%.

The importance of a seasonal analysis is revealed as the annual estimates hide the interseasonal differences; these however could be crucial for policy. As the marginal analysis shows, the growing periods of the respective farming seasons seem to be of particular relevance as they do not intersect with the opposite season's harvesting period. While in the summer farming season almost all zones with the exception of three, experience negative impacts on land values, in the winter season half of the zones benefit, whereas the other half suffers losses. These differences are crucial as the aggregate impacts are dependent on the relative impacts of the individual seasons. This relation is particularly evident in case of the Northern Dry Mountains and Wet Mountains/ Northern Dry Mountains zone, where although the trends in the Kharif and Rabi farming seasons more or less follow the general trend in the other Agro-Ecological Zones, the positive impact from the winter season is sufficient to offset the negative impact from warming in the summer farming season. In this case the advantages from the beneficial effects should be taken, whereas the negative impacts should be limited (compare BENHIN, 2009). The test of joint significance of the linear and quadratic terms in table 6.11 reveals that the winter farming season impacts from a 1°C warming are not significant for the shared zone Wet Mountains/Northern Dry Mountains zone. Summer Sowing temperature impacts are solely significant for the Southern irrigated Plains and their intersecting zones. The other periods' impacts with respect to temperature are generally significant.

Table 6.13 Climate Sensitivities on the Province level and joint significance tests

		Temperature							Precipitation						
Province	**Land Value ($/ha)**	**Summer Sowing**	**Summer Growing**	**Winter Sowing**	**Winter Growing**	**Summer Farming Season**	**Winter Farming Season**	**Annual**	**Spring**	**Monsoon**	**Winter**	**Summer Farming Season**	**Annual**	***Total MCI (%)***	***Total MCI ($)***
Punjab	2261	7%	-12%	24%	-23%	-5%	1%	-4%	0.04%	-1.33%	1.29%	-1.29%	0.00%	*-4%*	*-$90*
Sindh	733	18%	-13%	20%	-39%	5%	-19%	-14%	0.31%	-1.59%	0.76%	-1.29%	-0.53%	*-15%*	*-$106*
N.W.F.P	3666	-11%	-8%	27%	-10%	-19%	17%	-2%	-0.68%	-1.25%	2.59%	-1.93%	0.66%	*-1%*	*-$49*
Balochistan	771	-6%	-11%	26%	-16%	-17%	10%	-7%	0.14%	-1.71%	1.24%	-1.57%	-0.32%	*-7%*	*-$56*

	Temperature				Precipitation		
Province	**Summer Sowing**	**Summer Growing**	**Winter Sowing**	**Winter Growing**	**Spring**	**Monsoon**	**Winter**
Punjab	0.42	72.27 ***	139.12 ***	7.65 *	0.03	14.61 **	5.59
Sindh	18.85 **	127.22 ***	196.47 ***	46.17 **	357.34 ***	318.89 ***	214.79 ***
N.W.F.P	0.49	6.01 *	70.46 ***	0.64	1.16	9.47 **	4.89
Balochistan	0.09	14.60 **	26.94 **	0.84	1.12	261.00 ***	14.25 **

Table reports F-values for degress of freedom 4; * denotes significant at 10% level;** denotes significant at 5% level;*** denotes significant at 1%

Source: own calculations, 2012

On the province level some interesting results analogous to the previous findings on the AEZ level are revealed. Thus, the results for the Sindh province, which has the lowest average land value per hectare of farmland in the country, indicate the highest losses from a marginal change in climate, namely -15%. The Balochistan province with the second lowest average land value per hectare is also expected to suffer significant losses, precisely to the tune of -8%. The northern parts of Pakistan with the Punjab province in particular, will also experience shrinkage in average land value per hectare. Losses in the N.W.F.P. are the lowest with -1% (Land Value/ha), whereas the Punjab will record losses to the tune of -4%. In absolute terms as well, the highest monetary reduction is encountered by the Sindh province with -110$ per hectare with an increase in temperature and precipitation by one unit each, the first being a hill shaped relationship and the second being a U-shaped relationship. In the Punjab province from an absolute point of view, decreases around -90$/ha are found, followed by -62$ for Balochistan and -37 $/ha for the N.W.F.P. province. From a farming season perspective the three provinces Punjab, Balochistan and N.W.F.P. follow a similar trend, especially for temperature. Thus, a 1°C increase in average summer farming season temperature results in negative impacts for the three mentioned provinces, whereas the impact for the Sindh province is positive. In the winter farming season except for the Sindh province all other provinces experience positive impacts from warming. Moreover, except for the Sindh province in all other provinces the summer farming season impact is stronger and clearly over-compensates the positive impact in the winter farming months. In Sindh the negative effect in winter nullifies the benefits from a longer sowing period in the summer farming season. Both the provinces Punjab in the North East and Sindh in the South East benefit from a longer sowing period in summer, revealed by increases in land value per hectare as high as 7% and 18%, respectively. In the growing period of the summer farming season (Kharif) the impacts from a marginal warming are throughout negative, especially in Sindh with losses in land value per hectare around -13%. However, as the sowing period benefits in Sindh are higher than the losses in the growing period, the overall farming season impact is positive. Although, the sowing period impacts are important, only the sowing period effect for Sindh is significant, as calculated by a two-tailed F-Test of joint significance including both, linear and quadratic parameters. The winter sowing period is positively affected by a temperature increase, revealed by particularly high increments in average land values, precisely between 20% and 27%. All marginal impacts for the winter sowing period are significant, stressing the importance of this particular period for agriculture. On the other hand, the growing period of the winter farming season reveals significant losses for all provinces, ranging between -10% in the N.W.F.P. to -39% in Sindh, however with statistically significant impacts in the Punjab and Sindh provinces only. The overall marginal impact in the winter farming season from a 1°C temperature increase is positive for all provinces, except for the Sindh province, where the losses in the growing period (-39%) clearly and significantly from a statistical point of view over-compensate the gains in the sowing period (20%). As far as temperature is concerned annually, all provinces experience losses in land value per acre when annual normal temperatures increase by 1°C. The losses range from -2% in N.W.F.P. and -4% in the Punjab province to -7% in Balochistan and -14% in the Sindh province.

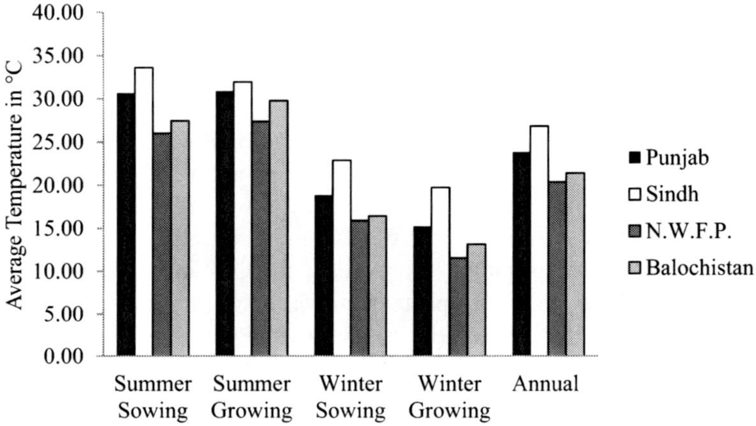

Figure 6.18 Province wise long-term temperature averages by season

Source: own illustration, 2012

Figure 6.18 shows the seasonal temperature averages for all provinces in the country. The highest temperature averages are found for the Sindh province, this throughout the year. Rice, cotton and sugarcane alongside fruits represent the Sindh province's major crops. A longer sowing period seems to be beneficial. However, it remains to be tested what impacts occur from an additional warming, in excess of the above tested 1°C increase in temperature. This will be subject of the climate simulation chapter. The summer farming season's average temperature in Sindh is estimated at 32.7°C. According to agronomic research the optimum temperature for the normal development of rice ranges from 27 to 32 °C (YIN et al., 1996). Research has also come to find that cotton emerges well from warm and moist soil, therefore low temperatures in combination with soil moisture may hinder germination by slowing metabolic processes (RITCHIE et al., 2007). OOSTERHUIS AND SNIDER (2011) come to find that although an optimum temperature range of 20°C to 30°C has been reported for cotton (REDDY et al., 1991), cotton is successfully grown at temperatures in excess of 40°C in India and Pakistan for example. There is no clear consensus about the optimum temperature for cotton as plant response varies with plant developmental stage and plant organ (BURKE AND WANJURA, 2009).

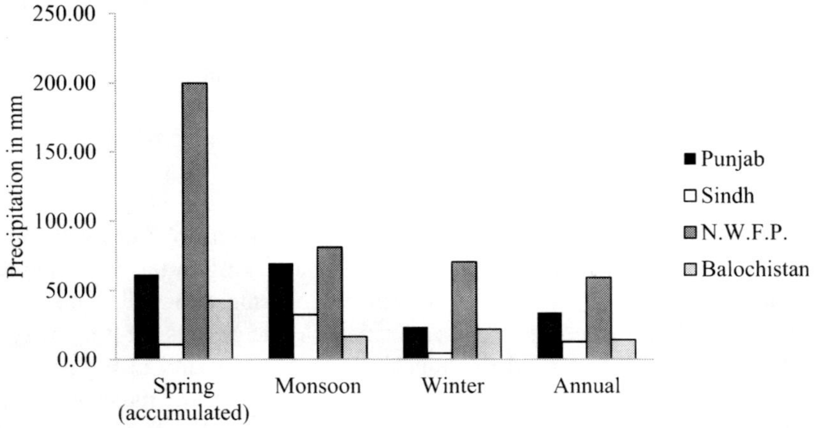

Figure 6.19 Province wise long-term precipitation averages by season
Source: own illustration, 2012

As far as precipitation is concerned a remarkable picture is revealed. As figure 6.19 shows the country's Southern parts comprising the Sindh and Balochistan provinces are particularly dry. Precipitation levels are generally very low, however even more so in the Southern provinces. The spring period is particularly dry in the Sindh province, with average normal precipitation below 15 mm. Thus consistent with the climatic records, the results of the marginal analysis for precipitation reveal that the Sindh province is expected to benefit from any increase in precipitation in the spring months; the joint hypothesis test is clearly significant at the 99% level. An increase in precipitation by 1 mm is expected to raise land values per hectare by 0.305% in the Sindh province for the sowing period. Also consistent with the climatic records shown in figure 6.18 the results show positive impacts from additional precipitation in the provinces Punjab and Balochistan. The highest average levels of accumulated precipitation in spring are received in the N.W.F.P.; this is also reflected in the results, as further precipitation in the spring season seems to be harmful for agriculture in the north western part of the country, also including the mountainous regions with sufficient rainfall. In line with the averages presented in figure 6.19, the results reveal the highest spring season benefits from additional accumulated rainfall for the driest province Sindh, followed by the second driest province of Balochistan and ultimately the Punjab province. Benefits range from an additional 0.305% in land value per hectare in the Sindh province and 0.139% in the Balochistan province to merely 0.041% in the Punjab province. The wettest spring is found in the N.W.F.P.; accordingly the results reveal a negative impact from additional precipitation in this period to the tune of -0.684%. These results are based on a hill-shaped relationship between spring precipitation and land values, clearly revealing that the optimal levels of rainfall have not yet been achieved in three out of four provinces, only in the N.W.F.P. the extreme point has already been achieved, after which any further increases are found to be harmful.

The hot summer months are highly dependent on the monsoon rains, which play an integral part in any successful agricultural undertaking in the Sub-continent. Although this particular period is considered to bring about the bulk of the rainfall for all provinces (except Balochistan), the results of the marginal analysis based on a U-shaped relationship, which was found in the parameter estimates (negative linear and positive quadratic term), indicate a clear precipitation deficiency for all provinces. The minimum levels of precipitation required have not been achieved as yet. Again, very interestingly in line with the long term climatic observations, the driest part of the country in this particular season is the Balochistan province, thus the high losses in land values per hectare from a mere one unit increase in rainfall is the highest in this province amongst all Pakistani provinces, precisely -2.591%. What has to be remembered is that due to the found U-shape relationship between monsoon precipitation and land values per hectare only the marginal impact is negative not the absolute impact. This is important to keep in mind as otherwise the wrong inference could be drawn about additional precipitation in a dry country like Pakistan. Every increase in precipitation although being absolutely beneficial, will be marginally negative up and until the minimum required levels of precipitation have been achieved, after which even the marginal impacts will be throughout positive. The analysis finds the second highest precipitation deficiency for the monsoon period in the Sindh province, followed by Punjab and relatively wet N.W.F.P, where although having higher levels of precipitation on average, the optimum has not been achieved as yet. For the monsoon period all marginal climate impacts or Ricardian sensitivities are highly significant for the driest provinces Sindh and Balochistan at the 99% level and strongly significant at the 95% level for the provinces N.W.F.P. and Punjab. Although all four provinces seem to benefit from additional precipitation in the winter period, the marginal climate impacts are only significant for the two driest provinces Sindh and Balochistan, at the 99% and 95% significance level, respectively. On an annual basis as well, Sindh and Balochistan are the two provinces where a precipitation deficiency is found and where the benefits from a 1mm increase in precipitation for the spring and winter periods are not sufficient to offset the negative impacts from a precipitation deficiency in the monsoon or hot summer period in

Pakistan. In the Punjab province the marginal impacts annul each other, whereas in the N.W.F.P. the benefits over-compensate the losses and lead to an overall positive impact from a mere 1 mm increase in annual average precipitation.

6.3.2 Control Variables

The importance of including relevant control variables is clearly revealed by the results. It is important to account for the non-climate features that determine the value of land in particular geographic regions, for this different controls are necessary as no Ricardian model automatically adapts to a country's socio-economic, hydrological and pedological conditions. Although earlier research related to the Ricardian model of climate change valuation has also made extensive use of different control variables, the present work includes some new variables. Another reason for using control variables is to make sure that econometric soundness is achieved regarding the issue of omitted variables. The new variables in the present study include the Human Development Index, Tractor Density, Road Density, Flood Risk, Solar Radiation proxy (latitude) and interaction terms between surface water and temperature. Other, more established control variables are population density, elevation and soil types.

Theory suggests that Pakistan is one of the most urbanized countries in South-Asia, e.g. urbanization rates as mentioned in the earlier chapters can be as high as 36% (i.e. Lahore). To capture the urbanization alongside the effects of land use competition (demand side of hedonic model) the population density per km^2 is used as a control variable in the regression model. In line with expectations, the population density variable is positive and highly significant for the two farm types rainfed and irrigated. This shows that an increase in population by one unit positively impacts on land values. As population in the different areas of the country increases, the higher demand for land for competing purposes substantially rises and ultimately leads to an increase in the land values. Thus population pressure is found to be a key determinant of land values in Pakistan, especially for the irrigated farms the impact from a unit increase in population ceteris paribus leads to a stronger increase in land values as compared to rainfed farms. Other Ricardian studies have used per capita income as a control variable for the demand side of the hedonic pricing model. The Human Development Index (HDI) offers a much more sophisticated way of measuring the level of human development in a country or in different regions, as it uses a multidimensional approach for the assessment, not only a one-dimensional approach comprising income only. The HDI is a composite index that summarizes the level of human development using the three dimensions health, knowledge and income. Thus, the HDI serves as an indicator of both social and economic development. The HDI ranges from 0 to 1, with one being the highest level of development (HUMAN DEVELOPMENT REPORT, 2011). The inclusion of the HDI in the model is reasoned in the fact that higher levels of social and economic development will increase the demand for land and thus values will rise. The interesting additional advantage of the HDI over simple income as a demand side indicator is that the HDI also indicates the policy choices of the district (in the modeled case) to some extent. Thus, for instance an account of the emphasis that is put on the three components is presented, whereas simple income data would not reveal specific spending choices e.g. when a rich country has a high per capita income this does not imply that the average education levels also have to be high, different priorities can be prevalent (e.g. Kuwait), whereas a country with low per capita income can have higher educational levels than richer countries. In line with the proposed theory and hypothesis, the impact of human development on land values per hectare in Pakistan is significantly positive over all farm types; significant at the 99% percent level for irrigated and mixed farms and at the 95% for rainfed farms. The results underline the importance of including social and economic development in combination when estimating the impacts of climate change on agriculture in a hedonic setting.

Higher elevations in general for the whole country are found to be negatively associated with land values. A similar result is found for the rainfed farms. The irrigated farms on the other hand seem to benefit from higher elevations, also reasoned in the fact that the country's northern parts (particularly comprising Punjab and N.W.F.P.) have excellent agro-climatic foundations for successful agricultural production, these benefits translate into higher net revenues that are capitalized in land values. The infrastructure in the northern provinces of the Punjab and N.W.F.P. is generally well enough developed to provide a functioning logistical pathway for agriculture. In this sense the elevational disadvantage in the other regions is owing to the negative side-effects of higher elevations which more or less are not only environment related but more infrastructurally related to land values. It is clear that in the northern areas the beneficial growing conditions as opposed to the dry and harsh growing conditions in the southern parts of the country make the difference, not elevation alone, thus in the north these conditions are clearly offsetting the negative effects from higher elevation, which as one would intuitively assume be related to higher transportation costs and difficult use of agricultural machinery and access to markets. The positive parameter for the rainfed farms clearly reveals that the better technically and environmentally equipped irrigated farms have a rather positive impact from elevation. Where rainfed farms have to rely on beneficial climatic conditions and depend on the rainfall for cultivation, the irrigated farms can always, provided water is available, compensate for climatic deficiencies, particularly for rainfall.

A proxy variable for technological abundance is also introduced. This is reasoned in the fact that in areas where the tractor density is higher the farmers can be expected to have higher financial resources, and more importantly can be more productive in agricultural cultivation. Higher productivity thus when associated with economies of scale will lead to higher net revenues which will be capitalized in land values, according to the Ricardian theory. Consistent with the deliberated reasoning, the sign on the tractor density per hectare of farmland coefficient is positive and the regressor is highly significant on all three levels, namely mixed farming, rainfed farming and irrigated farming. However, especially in rainfed areas the impact from a one unit increase in tractor density/ha, other things equal, seems to be relatively more beneficial for land values (capitalized net revenues) than in other farming systems. Recent research by MALIK (2010) has come to find that apart from its usual physical functions, farm technology particularly in resource scarce rainfed (barani) areas, performs a multiplicity of socio-cultural and aesthetic functions. The tractor is the most expensive machine used at present in barani areas and besides its original devotion is also used for other ecological, infra-structural, cultural and transportation purposes. Moreover, according to MALIK (2010) in rainfed (barani) areas technology is valued as a symbol of high economic status. This explains the significant positive impact of tractor density per hectare on land values. Farmers have to transport their produce to the markets, which is linked to transportation costs and time, not to forget the maintenance of the machinery or vehicle when the road network is in a weak shape. To assess the relevance of a better road network on land values, which can be expected to be positively correlated, a control variable for road density was introduced. The dataset for the road-density also included the roads which attach the agricultural entities to the general road network. In line with the underlying hypothesis, a higher road density per km^2 is found to have a significant positive impact on land values per hectare for all farming systems. This is reasoned in the fact that as distances are travelled quicker and predominantly on paved roads, the transportation costs will become smaller and the income of the farmer which is capitalized in land values will increase. The model also includes a set of additional hydrological control variables, simply to the background that additional water availability will positively impact land values. Two variables have been employed to control for the effect of surface water availability on land values, namely annual mean runoff and river density km^2. In line with the theoretical expectations both variables show a positive and significant impact for agriculture, foremost annual mean runoff. The runoff variable is positive and highly significant for the total farm sample and the rainfed farming systems and significant for irrigated farms. River density plays positive significant role in determining land values in Pakistan as a whole and rainfed systems. Although also positive for irrigated farms, the effect is not statistically significant. Access to surface water is thus found to give more options to farmers as they gain moderate independence from the temporal distribution and sequence of rainfall, particularly in rainfed areas.

As Pakistan is a country which is frequently struck by extreme events, particularly floods, the model includes a control for flood risk, measured as the long term (1951-2010) percentage area affected by flood in each district. The sensitivity of the response variable to climatic change related extreme events shall be explored. A negative impact would underpin the underlying theory, that from a long term perspective farmers and in the proposed theoretical case land values react to climatic changes; in a broader sense land markets react to climatic extremes and environmental changes. As the results clearly indicate, the impact of flood on land values is clearly negative over all farming systems, however significant only for the total farms' and rainfed farms' samples. To see whether interactions between the climatic and hydrological components of the model exist two types of interaction terms were introduced, one between annual mean runoff and annual temperature and one between irrigated percentage in district and temperature (only for irrigated farms). Interestingly the results reveal a contrasting picture; the runoff-temperature interaction is consistently negative over all models and significant for the mixed and rainfed sample. Contrary to the expectations in Pakistan more surface water availability does not seem to compensate for higher temperatures. The annual mean runoff does not seem to be sufficient to offset the impacts from a temperature increase, especially to the background of notably high PET rates which are close to 100%. For the irrigated farming sample an interaction term between irrigated acreage and temperature was introduced, which is found to be positive and significant. Irrigated farms are worth more than rainfed farms, simply because from a supply side viewpoint irrigation systems are costly, and from a demand side point of view irrigated farms are more productive. The results indicate that additional irrigation, provided water supplies are available, can compensate for higher temperatures. As shown in table 6.8, the climate sensitivity of irrigated farms is different to that of rainfed farms; irrigation seems to slightly compensate for the harmful effects of warming. The marginal losses in irrigated farming systems are smaller than in rainfed systems, although only marginally. Precipitation as shown in table 6.8 proved to be beneficial for irrigated farm types, as explained by the U-shaped relationship, revealing that these farming systems do not suffer from a precipitation deficiency, whereas rainfed systems do. The control variable 'latitude' was introduced to account for solar radiation. Interestingly the results have the expected signs for both rainfed and irrigated farming systems, however they are not significant. The negative sign of the coefficient can be explained by relating latitude and solar radiation scientifically. Owing to the near spherical shape of the earth the parallel beams of solar radiation that reach the earth have a more direct impact on surfaces in lower latitudes than at higher latitudes, because at higher latitudes radiation spreads over a wider area and thus is less intense per unit surface area than at lower latitudes (NSIDC, 2012).

Another important feature of Ricardian analysis is application of a control for soil differences. Work that does not include soil control variables and is restricted to one province only for studying climate change impacts (such as HANIF et. al, 2010), suffers from overestimation bias in favor of the climate variables. By doing this, the researcher may be tempted to draw wrong inferences from the results of the Ricardian estimation, as possible soil impacts on land values are wrongly accounted to the climatic impacts. Moreover restricting the analysis to a small geographical area, without climatic variation can cause misleading inferences. For these reasons, soil control variables covering the country's major soil types were introduced. The soil classes considered comprise: Calcisols, Fluvisols, Rock Outcrop and Regosols. Some fascinating results are revealed by the estimations. The impact of soil type Calcisol, ceteris paribus, is positive for irrigated systems and negative for rainfed farms. Moreover, for irrigated farms the impact is significant. According to the international soil reference system, Calcisols are soils with substantial secondary accumulation of lime. These soils are made up of "alluvial, colluvial and aeolian deposits of base-rich weathering material". Calcisols are mineral soils found in arid and semi-arid environments like Pakistan. In rainfed systems the use is limited due to dryness and stoniness; however Calcisols, when irrigated, can be highly productive under different crops (FAO, 2001). Another soil type included in the model is Fluvisol. The World Reference Base for Soil Resources (WRB) describes Fluvisols as very young soils with weak horizon differentiation and a texture that depending on the site can range from coarse sand to heavy clay. From a hydrological point of view, Fluvisols are wet soils, originating from marine deposits. As far as the chemical characteristics are concerned, the majority of Fluvisols possess neutral or near-neutral pH values and the nutrient availability is generally good, making Fluvisols naturally fertile soils. Paddy rice cultivation is practiced in areas where Fluvisols are dominant, provided irrigation is sufficient. Furthermore, occasional dry periods, as are common for arid/ semi-arid Pakistan, can stimulate the microbial activity and promote mineralization of organic matter (FAO, 2001). Throughout all three models the coefficient for Fluvisols is positive and significantly positive for irrigated and mixed systems. This is a very interesting result, as in line with theory and pedological research, Fluvisols require additional management for rainfed systems but are rather well suited for irrigated farming (e.g. rice in the Punjab province). To this background the significant positive effect for irrigated farms seems plausible. The results for the soil class Rock Outcrop are positive and significant over all models. According to the WRB the soil class Rock outcrop is made up of 33% Leptosols, 33% Gypsisols and 34% rock outcrop. Leptosols are shallow soils with minute amounts of fine earth and are predominantly encountered in areas of high or medium altitude, however foremost in eroding areas. From a hydrological point of view Leptosols are characterized as free-draining soils, with stoniness and shallowness being additional limitations. On hill slopes Leptosols are fertile and can be used for crop cultivation, however with a high risk of erosion. One way to make Leptosols cultivable is through terracing, which is widely practiced in the medium to high mountainous areas of the country. As the country's mountainous parts generally receive sufficient amounts of rainfall, weak water holding capacity can be compensated. The other component of the Rock outcrop soil classification is Gypsisol, soils with substantial secondary accumulation of calcium sulphate, more than 40% clay and a decent water holding capacity. Gypsisols are found in level to hilly areas in arid regions and when located near water sources can be cultivated with a diverse range of crops such as wheat, apricots, dates, maize and grapes. However, one crucial criterion for cultivation of this soil type is that the gypsum content is not very high. Gypsisols are also used for cultivation in rainfed farming; however measures such as water harvesting are necessary. Another soil type that has been included in the model is Regosol. The results indicate a positive impact for land values in the rainfed areas and negative impacts for irrigated farms. Regosols consist of finely grained weathering material and are common in arid and mountainous regions. In Pakistan, Regosols are found in the wet sub-mountainous northern regions. Furthermore, medium texture and good drainage are characteristic for this soil class. The negative impact for irrigated farms could stem from the fact, that irrigated farming on Regosols is capital-intensive. Therefore, the FAO (2001) suggests to improve rainfed (barani or dry) farming practices in regions with sufficient precipitation levels, rather than to stick with expensive irrigation farming practices. As Regosols in Pakistan are located in areas with sufficient rainfall, the parameter estimates of the regression seem to be consistent with the theoretical expectations.

7 Structural Ricardian Model for the economic impact assessment of Climate Change

The upcoming deliberations present the Structural Ricardian Model which serves as a remedy for the black box of adaptations implicit in the standard Ricardian approach. Instead of land value, the Structural model uses net revenues on the farm scale as the dependent variable, in this way an observed preference model is constructed, unveiling specific adaptations that farmer's make to adapt to climatic change. Another difference is that the Structural model uses farm level data as opposed to aggregated data. The results of both models can be compared and supplemented to understand the economic impacts of climate change on agriculture in Pakistan. Multinomial logit models are nonlinear and include many parameters, which makes the interpretation of the output difficult. In following the empirical strategy is explained and results are presented alongside deliberations on model fit and testing. This is followed by the interpretation of the results. The theoretical considerations are based on the extensive review on models for categorical outcomes by LONG AND FREESE (2006).

7.1 Dataset and Study Area

A large dataset is constructed by combining the detailed farm-household level information from the Pakistan Living Standards Measurement Survey (PSLM) 2007-2008 (FBS, 2009) with climate and soil data which was used for the land value model (Annex I). The PSLM survey covered 106 administrative districts (4 provinces) in the country between 2007 and 2008. Farming households in the sample were administered a detailed farm production (input and output) questionnaire. Based on land ownership and data completeness 2369 farm households were selected covering 79 districts and all Agro-Ecological Zones in the country. When weighted, these household represent a great majority of the crop farmers in the country. The dataset covers the seven most important crops of the country, namely wheat (975 farms), cotton (408 farms), sugarcane (178 farms), rice (507 farms), maize (108 farms), fruits (94 farms) and vegetables (99 farms). Table 7.1 presents the sample properties by district. The dependent variable for the second stage income regression net crop revenue per hectare is constructed by valuing all variable inputs (except land) and outputs at the market price, as provided by the Federal Bureau of Statistics Pakistan. Total annual farm income in this paper includes annual earnings from crop farming only.

Table 7.1 Sample size properties by district

No.	District	Frequency	%	Cumulative %	No.	District	Frequency	%	Cumulative %
1	Abbotabad	16	1	0.7	41	Lower Dir	10	0.4	53.8
2	Attock	45	2	2.6	42	Malakand	10	0.4	54.2
3	Badin	46	2	4.5	43	Mandi Bahauddin	17	0.7	54.9
4	Bahawal Nagar	29	1	5.7	44	Mansehra	23	1.0	55.9
5	Bahawalpur	40	2	7.4	45	Mardan	39	1.7	57.5
6	Bannu	26	1	8.5	46	Mianwali	10	0.4	58.0
7	Batagram	17	1	9.2	47	Mirpur Khas	40	1.7	59.7
8	Bhakar	44	2	11.1	48	Multan	45	1.9	61.5
9	Bonair	10	0	11.5	49	Muzaffar Garh	32	1.4	62.9
10	Chakwal	20	1	12.4	50	Narowal	23	1.0	63.9
11	Charsada	7	0	12.7	51	Naseerabad	66	2.8	66.7
12	Chitral	15	1	13.3	52	Nawabshah	36	1.5	68.2
13	D.G.Khan	44	2	15.2	53	Nowshera	7	0.3	68.5
14	D.I.Khan	22	1	16.1	54	Nowshero Feroze	8	0.3	68.8
15	Dadu	19	1	16.9	55	Okara	22	0.9	69.7
16	Faisalabad	32	1	18.2	56	Pak Pattain	17	0.7	70.5
17	Ghotki	76	3	21.4	57	Peshawar	21	0.9	71.3
18	Gujranwala	29	1	22.7	58	Quetta	32	1.4	72.7
19	Gujrat	29	1	23.9	59	R.Y.Khan	33	1.4	74.1
20	Hafizabad	16	1	24.6	60	Raajanpur	14	0.6	74.7
21	Hangu	12	1	25.1	61	Rawalpindi	29	1.2	75.9
22	Haripur	17	1	25.8	62	Sahiwal	24	1.0	76.9
23	Hyderabad	23	1	26.8	63	Sanghar	24	1.0	77.9
24	Islamabad	34	1	28.2	64	Sargodha	51	2.2	80.1
25	Jacobabad	129	5	33.6	65	Shangla	13	0.6	80.6
26	Jhang	43	2	35.5	66	Sheikhupura	48	2.0	82.7
27	Jhelum	14	1	36.1	67	Shikarpur	82	3.5	86.1
28	Kalat	78	3	39.3	68	Sialkot	24	1.0	87.1
29	Karak	16	1	40.0	69	Sibbi	42	1.8	88.9
30	Kasur	25	1	41.1	70	Sukkar	52	2.2	91.1
31	Khairpur	88	4	44.8	71	Swabi	12	0.5	91.6
32	Khanewal	10	0	45.2	72	Swat	23	1.0	92.6
33	Khusab	12	1	45.7	73	T.T.Singh	19	0.8	93.4
34	Kohat	11	0	46.2	74	Tank	12	0.5	93.9
35	Kohistan	29	1	47.4	75	Tharparkar	8	0.3	94.2
36	Lahore	15	1	48.0	76	Thatta	32	1.4	95.6
37	Lakki Marwat	17	1	48.8	77	Upper Dir	8	0.3	95.9
38	Larkana	69	3	51.7	78	Vehari	21	0.9	96.8
39	Layya	27	1	52.8	79	Zohb	76	3.2	100.0
40	Lodhran	13	1	53.4		**Total**	2,369		100

Source: calculated using FBS (2009) dataset

The livestock sector could not be included due to data constraints. Additional earnings by hiring out labor and equipment were only partly reported and therefore are excluded. The total costs are estimated by considering variable costs associated with hired labor, seeds, fertilizer and chemical sprays, tractor charges, irrigation costs, transport/packaging costs, and other incidental farm expenses. Net revenues per hectare are defined as the difference between revenues and costs divided by the number of hectares farmed. The household data is combined with data on long-term climate normals, soil and hydrological variables alongside socio-economic characteristics.

7.2 Empirical strategy

The term choice in this specific setting refers to the alternative that is actually revealed in the survey, which can be thought of as the „most preferred" alternative. Thus if the dependent variable in the first stage model is the crop chosen in the last growing season, the alternatives might be cotton, wheat, vegetables, fruits, rice, sugarcane and maize. For nominal outcomes like the choice of a certain crop the multinomial logit model can be used. As seven choices are possible, the multinomial model, which is an advanced version of the binary logit models as it simultaneously estimates all comparisons among the different alternatives, is the preferred econometric model. These models are not only restricted to be used for building models for outcomes occurring from a choice, as the name might indicate (LONG AND FREESE, 2006). One great advantage of the multinomial model over the binary model is its simultaneous estimation, whereas the binary model compares two choices, leaving out a possible third alternative. For the single comparisons thus different samples have to be used every time. For example when having three crops at choice e.g. cotton, wheat, maize; the aforesaid would imply that when comparing cotton and maize, the third crop wheat would be dropped.

As indicated in the above deliberations, although there are many crops that farmers in Pakistan grow, this study focuses on seven primary crops that generated the highest crop income for the farming year 2007-2008: wheat, maize, sugarcane, rice, cotton, fruits and vegetables. Previous studies have found all agricultural seasons to be important for determining crop choice (MENDELSOHN et al., 1994). Therefore, the probability of choosing each crop type as a function of summer, fall and winter temperature, spring, and monsoon precipitation is modeled. The winter season in Pakistan spans from January to March and thus intersects with the spring season. The majority of the precipitation is received during the monsoon season in the country, to this background the emphasis is laid on this particular season for modeling the precipitation impact. However, as most of the crops of the summer season are sown in the months starting from May/June, the accumulated precipitation in the soil is of paramount importance for agriculture. Therefore, the accumulated spring precipitation is separately modeled. Fall and winter precipitation are scarce, agriculture mainly relies on the precipitation that is received during the summer monsoon period, incidentally a separate inclusion of the fall and winter precipitation does not provide sufficient variation and is therefore left out. Various other explanatory variables are also included such as dominant soil types and socioeconomic household characteristics. Furthermore, own prices are used to identify the multinomial choice model. The choice of wheat has been left out of the regression as the base case. Following previous studies, this model also identifies the choice equation using cross prices, annual runoff (proxy for surface water flow) and soil variables (EKELAND et al., 2004). For testing the effects of the included independent variables or more precisely to test whether the variable has no effect requires a test that J-1 coefficients are simultaneously equal to zero. Formally the hypothesis that the independent variable x_k does not affect the dependent variable can be stated as follows:

EQ (36) $$H_0: \beta_{k,1|b} = \cdots = \beta_{k,J|b} = 0$$

with b as the base category (wheat). As the statement $\beta_{k,b|b}$ is 0, the above hypothesis imposes constraints on J-1 parameters, which are not imposed by estimations relying on binary logits. In this way the remaining comparison, is the difference between the sets of the estimated coefficients. The hypothesis is tested using the Wald and Likelihood-Ratio (LR) tests. The LR test fits a full model and retrieves a first LR statistic, in a second step the model is estimated without the variable at study and a second LR statistic is obtained. The difference between both statistics reveals whether the variable has a significant effect on the independent variable or not. An alternative test, which is computationally less demanding and is interpreted in a similar way, is the Wald test. Both these test statistics for the selected empirical model are presented in Table 7.2.

The results of both tests show that the effects of the majority of the variables are significant at the 1% level, thus the hypothesis that all coefficients with given variables are simultaneously equal to 0 can be rejected. Monsoon precipitation and household size are both not significant in both tests.

Table 7.2 Wald and LR tests for independent variables (N=2369)

Independent Variable	Wald Test			Likelihood Ratio Test		
	chi2	df	P>chi2	chi2	df	P>chi2
Summer Temperature	38.125	6	0	43.491	7	0
Summer Temperature²	26.234	6	0	27.019	6	0
Fall Temperature²	35.079	6	0	34.793	6	0
Winter/Spring Temperature²	47.554	6	0	47.638	6	0
Spring Precipitation	38.395	6	0	42.959	6	0
Spring Precipitation²	16.913	6	0.01	20.915	6	0.002
Monsoon Precipitation	9.221	6	0.162	10.212	6	0.116
Monsoon Precipitation²	14.212	6	0.027	18.648	6	0.005
Irrigated Land	29.171	6	0	73.198	6	0
Wheat value	164.517	6	0	231.383	6	0
Cotton value	220.392	6	0	715.594	6	0
Sugarcane value	119.744	6	0	179.798	6	0
Rice value	232.440	6	0	410.248	7	0
Maize value	53.073	6	0	72.319	6	0
Elevation	65.778	6	0	76.487	6	0
Household size	7.609	6	0.268	7.627	6	0.267
Education	18.267	6	0.006	18.188	6	0.006
Runoff	21.399	6	0.002	13.691	6	0.033
Soil Cambisol	30.081	6	0	38.597	6	0
Marginal Soil	28.343	6	0	34.765	6	0
Farmland Area	26.099	6	0	77.831	6	0

Ho: All coefficients associated with given variable(s) are 0.

Source: own calculations using STATA 11.2 (STATACORP, 2009)

Moreover, the Wald test finds the squared term of Monsoon Precipitation² to be significant at the 5% level and the LR test suggests runoff to be significant at the 5% significance level. Both insignificant variables are nevertheless retained as although overall the significance is not confirmed by the tests, individually they are significant in the regression for certain crops, i.e. Monsoon precipitation is significant for sugarcane and household size is found to be significant for cotton. This approach is in line with the proposed strategy by LONG AND FREESE (2006), who state that before models are revised the individual coefficients involved in each test should be examined carefully as although the statistical significance as a group might be not be given, individual statistical significance can still be present. Next, the idea is to test whether the alternatives are mutually exclusive or indistinguishable. To this effect, a Wald test is run in order examine whether two alternatives can be combined for all pairs of alternatives. More precisely the hypothesis that all coefficients except the intercepts associated with a given pair of alternatives are 0 is tested. Table 7.3 presents the test output.

Table 7.3 Wald tests for combining alternatives (N=2369)

Alternatives tested	chi2	df	P>chi2
Cotton-Sugarcane	216.246	21	0
Cotton-Rice	223.003	21	0
Cotton-Maize	88.366	21	0
Cotton-Fruits	98.597	21	0
Cotton-Vegetables	121.932	21	0
Cotton-Wheat	235.975	21	0
Sugarcane-Rice	258.328	21	0
Sugarcane-Maize	108.043	21	0
Sugarcane-Fruits	106.546	21	0
Sugarcane-Vegetables	92.645	21	0
Sugarcane-Wheat	200.852	21	0
Rice-Maize	146.655	21	0
Rice-Fruits	203.645	21	0
Rice-Vegetables	179.887	21	0
Rice-Wheat	331.784	21	0
Maize-Fruits	81.977	21	0
Maize-Vegetables	101.628	21	0
Maize-Wheat	118.688	21	0
Fruits-Vegetables	66.651	21	0
Fruits-Wheat	134.722	21	0
Vegetables-Wheat	174.386	21	0

Ho: All coefficients except intercepts associated with a given pair of alternatives are 0 (i.e., alternatives can be combined).

Source: own calculations using STATA 11.2 (STATACORP, 2009)

The test output shows that the choices or alternatives are mutually exclusive and distinguishable, thus the null hypothesis can be rejected at the 1% level and alternatives do not need to be combined. The validity of multinomial logit models is conditional on the Independence of Irrelevant Alternatives (IIA) assumption that states that the deletion or inclusion of alternatives does not affect the odds amongst the remaining choices. For example, if IIA holds, a person's choice between going to a concert or watching a movie does not depend on a certain football match taking place that particular day (VIJVERBERG, 2011). In the proposed model, this implies that for instance if the choice is between cotton and wheat and the farmer prefers cotton, the availability of fruits would not suddenly make wheat the preferred choice over cotton. However, for modeling crop choices the restrictive assumption, that if two crops have similar shares, substitution with a third crop will be the same for the two original crops, is applied (ARRIAGADA, 2005). Crop choices are modeled as being distinct. LONG AND FREESE (2006) state, that the advice to use choice models with dissimilar alternatives (AMEMIYA, 1981), although reasonable, seems to be an ambiguous one. Substitutive relations between crops can be expected to exist, however even if the IIA assumption would not hold, in the full set of choices, the violation will be a problem only in the residuals of the model. Moreover, problems will be caused only if the appropriate variables to explain the similarities between the choices are not included (KROPKO, 2010). The proposed model includes price terms to identify crop choices.

Table 7.4 Hausman tests of IIA (N=2369)

Omitted	chi2	df	P>chi2	evidence
Cotton	0	1	1.000	for Ho
Sugarcan	0	1	1.000	for Ho
Rice	0	1	1.000	for Ho
Maize	0	1	1.000	for Ho
Fruits	14.485	96	1.000	for Ho
Vegetables	0	1	1.000	for Ho
Wheat	46.968	92	1.000	for Ho

Ho: Odds(Outcome-J vs Outcome-K) are independent of other alternatives.

Source: own calculations using STATA 11.2 (STATACORP, 2009)

The Hausman McFadden (HM) test output shows that the IIA has not been violated. For the HM test, HAUSMAN AND MCFADDEN's (1984) suggestion that negative chi-squared values be recorded as 0 with the corresponding p value of 1 was followed. Negative test statistics for the chi^2 are found to be common, HAUSMAN AND MCFADDEN (1984) have suggested that this is evidence that the IIA has not been violated. For assessing the overall goodness of fit of the model, different tests were conducted. The general STATA output reports the log-likelihood chi^2 and pseudo R^2 for the estimated model, where the log-likelihood chi^2 is two times the difference between the log-likelihood of the estimated model and the log likelihood of the constant only model. Pseudo R^2

although a different measure, is interpreted in the same manner as the well know R^2 statistic, the higher the better. The pseudo R^2 captures the proportion of change in terms of likelihood and not the proportion of variance that is explained by the model as in the case of R^2, reasoned in the fact that in logistic regression the variance is fixed as the variance of the standard logistic distribution (LONG AND FREESE, 2006; MENARD, S., 1995). Another test that is commonly used in binary logit regression is the Hosmer-Lemeshow test for comparing predicted probabilities with the observed data (HOSMER AND LEMESHOW, 1980). After computing the predicted probabilities, they are grouped into 10 groups (G), with the first group containing the smallest values of the predicted probabilities. Within each group the means of predicted and observed probabilities are calculated and compared, with the outcome statistic being a Pearson chi^2 statistic with G-2 degrees of freedom. To assess the fit of the multinomial model, few methods exist, therefore FAGERLAND et al., (2008) have developed a similar test for models with multiple choices. A significant value of the test statistic indicates a weak model fit, thus the model fails to accurately predict the probabilities of being in a certain group. Table 7.5 summarizes the three tests for the overall model fit.

Table 7.5 Measures of Fit for Multinomial Logit Model

Log-Likelihood Chi^2 Test:		**Fagerland et al. Statistic:**	
Log-Lik Intercept Only:	-3776		
D(2237):	2984	Number of observations	2369
Log-Lik Full Model:	-1492	Number of outcome values	7
LR(126):	4.569	Base outcome value	1
Prob > LR:	*0*	Number of groups	10
Pseudo-R^2:		chi-squared statistic	72.051
McFadden's R2:	0.605	degrees of freedom	60
McFadden's Adj R2:	*0.57*	*Prob > chi-squared =*	*0.137*

Source: own calculations using STATA 11.2 (STATCORP, 2009)

The three tests suggest that the model fits the data well. In following the results of the choice model are presented.

7.3 First stage crop choice model- Interpretation of results

Although the tests for model specification have shown model to fit the data well, it is important to register that "there is no convincing evidence that selecting a model that maximizes the value of a given measure of fit results in a model that is optimal in any sense other than the model's having a larger value of that measure" (LONG AND FREESE, 2006). Keeping in mind the previous statement, the following deliberations are devoted to discuss the results of the parameter estimates. Table 7.6 shows the summary statistics of the modeled variables.

Table 7.6 Summary Statistics for the Multinomial Crop Choice Model (N=2369)

Variable	Unit	Mean	Std. Dev.	Min	Max
Net_Crop_Revenue	$/ha	233.8699	133.6996	17.08333	1085.625
Summer Temperature	°C	30.60787	2.649737	18.23	34.67
Fall Temperature	°C	19.27459	3.305058	9.63	24.57
Winter/Spring Temperature	°C	15.68348	3.675906	5.23	21.73
Accumulated Spring Precipitation	mm	67.35965	85.0064	5	506.1
Monsoon Precipitation	mm	31.34104	3.213903	23.1	36.15
Wheat Value	$/kg	10.52638	4.912004	0	50
Cotton Value	$/kg	6.79274	12.9194	0	87
Sugarcane Value	$/kg	.209371	.8567478	0	25
Rice Value	$/kg	4.863233	8.393704	0	62
Maize Value	$/kg	1.891938	4.815706	0	30
Elevation	m	420.1144	635.3687	5	3314
Household Size	Number	7.896581	4.001565	1	61
Education	Years	3.68721	4.628752	0	23
Runoff	mm/year	38.49388	103.3493	1	627
Soil Cambisol	%	.0889194	.2142947	0	1
Soil Calcisol	%	.5056535	.3674237	0	1
Soil Fluvisol	%	.0834549	.1161126	0	.4273678
Soil Rock Outcrop	%	.0347731	.1460655	0	1
Soil Sand	%	.1348121	.2717466	0	.8937818
Soil Regosol	%	.0313969	.1297228	0	.738439
Soil Leptosol	%	.1189034	.2713905	0	.9844586
Weak Soil	%	.169371	.2944473	0	1
Total Cultivated Land	acres	7.499367	13.34708	1	450
Irrigated Land	%	.8111785	.3835235	0	1

Source: own calculations using FBS (2009) dataset

Table 7.7 presents the results of the multinomial logit model. However, Table 7.8 provides a more intuitive interpretation of the parameter estimates in the language of odds or so-called risk ratios which define the probability of choosing one outcome over the probability of choosing the base-case (wheat).

Table 7.7 Multinomial logit crop selection model for the farming season 2007/08

Independent Variables	Cotton	Sugarcane	Rice	Maize	Fruits	Vegetables
	Coefficients	Coefficients	Coefficients	Coefficients	Coefficients	Coefficients
Summer Temperature	0.120	0.239	0.675***	0.185	0.402***	0.0855
	(0.204)	(0.165)	(0.128)	(0.217)	(0.120)	(0.0921)
Summer Temperature²	0.0395	-0.160*	0.0224	0.0635***	0.0485***	0.00131
	(0.0744)	(0.0888)	(0.0201)	(0.0175)	(0.0144)	(0.0134)
Fall Temperature²	-0.225	-0.411**	-0.434***	-0.00278	0.0634**	-0.0398
	(0.250)	(0.183)	(0.102)	(0.0738)	(0.0260)	(0.0247)
Winter/Spring Temperature²	0.187	0.377***	0.401***	-0.0304	-0.0558*	0.0689***
	(0.186)	(0.141)	(0.0778)	(0.0656)	(0.0287)	(0.0249)
Spring Precipitation	-0.0292**	0.00364	-0.0126*	0.0267***	0.0269***	-0.00828
	(0.0117)	(0.00674)	(0.00659)	(0.00794)	(0.00715)	(0.00547)
Spring Precipitation²	7.89e-05	1.55e-05	-2.90e-06	-4.10e-05**	-0.000197***	7.23e-06
	(6.49e-05)	(3.84e-05)	(5.54e-05)	(1.71e-05)	(6.66e-05)	(1.41e-05)
Monsoon Precipitation	-0.000156	0.0294**	0.00956	0.00344	0.00401	0.0102
	(0.0140)	(0.0126)	(0.00583)	(0.0116)	(0.00891)	(0.00762)
Monsoon Precipitation²	-7.05e-05	-0.000584***	-0.000156**	-4.52e-05	-0.000349*	-8.73e-06
	(0.000245)	(0.000221)	(7.19e-05)	(8.11e-05)	(0.000203)	(6.56e-05)
Irrigated Land	0.0852	0.544***	0.263**	0.216**	0.247***	0.0492
	(0.0640)	(0.188)	(0.111)	(0.107)	(0.0777)	(0.0370)
Wheat value	-0.268***	-0.167***	-0.132***	-0.302***	-0.168***	-0.251***
	(0.0412)	(0.0285)	(0.0272)	(0.0450)	(0.0251)	(0.0251)
Cotton value	0.189***	-0.0107	-0.130***	0.0224	-0.568	0.0110
	(0.0139)	(0.0119)	(0.0267)	(0.0386)	(21.11)	(0.0171)
Sugarcane value	-1.435***	1.138***	-1.333***	-0.545	0.358	-2.594**
	(0.327)	(0.156)	(0.301)	(0.466)	(0.251)	(1.043)
Rice value	-0.0758***	-0.00963	0.207***	-0.0314	-0.00460	-0.0585*
	(0.0248)	(0.0176)	(0.0145)	(0.0377)	(0.0280)	(0.0331)

continued on next page...

Maize value	-0.180	0.0112	-0.0325	0.250***	-0.0555	-0.0944*
	(0.112)	(0.0363)	(0.0612)	(0.0388)	(0.0440)	(0.0487)
Elevation	-0.00456**	-0.00225***	-0.00470***	0.000248	0.00196***	0.000496*
	(0.00197)	(0.000849)	(0.000886)	(0.000744)	(0.000406)	(0.000285)
Household size	0.573**	0.209	0.116	-0.164	-0.145	0.361
	(0.254)	(0.241)	(0.206)	(0.341)	(0.276)	(0.281)
Education	-0.00467	-0.0167	0.00233	0.0581*	0.0949***	-0.0142
	(0.0274)	(0.0256)	(0.0210)	(0.0337)	(0.0262)	(0.0299)
Runoff	0.0147	-0.0277	0.0159***	0.00111	-0.00152	0.00323
	(0.00914)	(0.0251)	(0.00408)	(0.00201)	(0.00291)	(0.00220)
Soil Cambisol	-0.00766	-0.00978	0.0128***	-0.00640**	-0.00751	-0.00394
	(0.0357)	(0.00791)	(0.00272)	(0.00310)	(0.00653)	(0.00400)
Marginal Soil	-0.318	0.109	-1.878***	-5.599**	0.0541	-0.424
	(0.499)	(0.458)	(0.413)	(2.180)	(0.630)	(0.505)
Farmland Area	-0.0850	-0.575***	-0.224**	-0.182*	-0.241***	-0.0295
	(0.0631)	(0.189)	(0.112)	(0.106)	(0.0794)	(0.0362)
Constant	-3.143***	0.713	-1.974***	-1.753**	-0.934	-1.116*
	(0.932)	(0.608)	(0.579)	(0.886)	(0.642)	(0.655)
Observations	2,369	2,369	2,369	2,369	2,369	2,369
Log Likelihood	-1492	-1492	-1492	-1492	-1492	-1492
pseudo R^2	0.604	0.604	0.604	0.604	0.604	0.604
chi^2	4570	4570	4570	4570	4570	4570

Standard errors in parentheses.*** p<0.01, ** p<0.05, * p<0.1

Note: Wheat is the omitted choice.

Source: own calculations in STATA 11.2 (STATACORP, 2009)

Table 7.8 Multinomial logit crop selection model for the farming season 2007/08 - Odds

Independent Variables	Cotton	Sugarcane	Rice	Maize	Fruits	Vegetables
	Odds	Odds	Odds	Odds	Odds	Odds
Summer Temperature	1.127	1.270	1.964***	1.204	1.495***	1.089
	(0.230)	(0.210)	(0.251)	(0.262)	(0.179)	(0.100)
Summer Temperature²	1.040	0.852*	1.023	1.066***	1.050***	1.001
	(0.0774)	(0.0757)	(0.0205)	(0.0187)	(0.0151)	(0.0134)
Fall Temperature²	0.799	0.663**	0.648***	0.997	1.065**	0.961
	(0.199)	(0.121)	(0.0660)	(0.0736)	(0.0277)	(0.0238)
Winter/Spring Temperature²	1.206	1.458***	1.494***	0.970	0.946*	1.071***
	(0.224)	(0.205)	(0.116)	(0.0637)	(0.0272)	(0.0266)
Spring Precipitation	0.971**	1.004	0.988*	1.027***	1.027***	0.992
	(0.0114)	(0.00676)	(0.00650)	(0.00816)	(0.00734)	(0.00542)
Spring Precipitation²	1.000	1.000	1.000	1.000**	1.000***	1.000
	(6.49e-05)	(3.84e-05)	(5.54e-05)	(1.71e-05)	(6.66e-05)	(1.41e-05)
Monsoon Precipitation	1.000	1.030**	1.010	1.003	1.004	1.010
	(0.0140)	(0.0130)	(0.00589)	(0.0117)	(0.00895)	(0.00770)
Monsoon Precipitation²	1.000	0.999***	1.000**	1.000	1.000*	1.000
	(0.000245)	(0.000221)	(7.19e-05)	(8.10e-05)	(0.000203)	(6.56e-05)

continued on next page...

Irrigated Land	1.089	1.722***	1.301**	1.241**	1.280***	1.050
	(0.0697)	(0.323)	(0.145)	(0.132)	(0.0994)	(0.0389)
Wheat value	0.765***	0.846***	0.876***	0.739***	0.845***	0.778***
	(0.0315)	(0.0242)	(0.0238)	(0.0332)	(0.0212)	(0.0196)
Cotton value	1.208***	0.989	0.878***	1.023	0.567	1.011
	(0.0168)	(0.0118)	(0.0234)	(0.0394)	(11.97)	(0.0173)
Sugarcane value	0.238***	3.121***	0.264***	0.580	1.430	0.0748**
	(0.0777)	(0.487)	(0.0793)	(0.270)	(0.359)	(0.0779)
Rice value	0.927***	0.990	1.230***	0.969	0.995	0.943*
	(0.0230)	(0.0175)	(0.0179)	(0.0365)	(0.0278)	(0.0312)
Maize value	0.835	1.011	0.968	1.285***	0.946	0.910*
	(0.0934)	(0.0368)	(0.0592)	(0.0498)	(0.0416)	(0.0443)
Elevation	0.995**	0.998***	0.995***	1.000	1.002***	1.000*
	(0.00196)	(0.000847)	(0.000882)	(0.000744)	(0.000407)	(0.000285)
Household size	1.774**	1.233	1.123	0.849	0.865	1.435
	(0.450)	(0.297)	(0.232)	(0.290)	(0.239)	(0.403)
Education	0.995	0.983	1.002	1.060*	1.100***	0.986
	(0.0273)	(0.0252)	(0.0211)	(0.0357)	(0.0288)	(0.0295)
Runoff	1.015	0.973	1.016***	1.001	0.998	1.003
	(0.00928)	(0.0244)	(0.00414)	(0.00201)	(0.00290)	(0.00220)
Soil Cambisol	0.992	0.990	1.013***	0.994**	0.993	0.996
	(0.0355)	(0.00783)	(0.00275)	(0.00308)	(0.00648)	(0.00399)
Marginal Soil	0.728	1.116	0.153***	0.00370**	1.056	0.654
	(0.363)	(0.511)	(0.0632)	(0.00807)	(0.664)	(0.330)
Farmland Area	0.919	0.563***	0.799**	0.833*	0.786***	0.971
	(0.0580)	(0.106)	(0.0894)	(0.0884)	(0.0623)	(0.0352)
Constant	0.0432***	2.039	0.139***	0.173**	0.393	0.327*
	(0.0402)	(1.241)	(0.0805)	(0.154)	(0.252)	(0.215)
Observations	2,369	2,369	2,369	2,369	2,369	2,369
Log Likelihood	-1492	-1492	-1492	-1492	-1492	-1492
pseudo R^2	0.604	0.604	0.604	0.604	0.604	0.604
chi^2	4570	4570	4570	4570	4570	4570

Standard errors in parentheses. *** $p<0.01$, ** $p<0.05$, * $p<0.1$

Note: Wheat is the omitted choice.

Source: own calculations in STATA 11.2 (STATACORP, 2009)

The results indicate that crop choice is sensitive to climate. Furthermore, the importance of identifying the model using a multi-seasonal approach is revealed. Many of the seasonal climate variables are significant at the 1% level. Temperature coefficients are not significant for the cotton crop. Most of the individual coefficients in the model are significant. A positive coefficient implies that the probability of choosing the respective crop increases with an increment in the corresponding variable, whereas the opposite is indicated when the coefficient is negative. In relative risk terminology for a one unit change in a variable, the relative risk or the odds of choosing a crop over the base-case (wheat) increase (positive coefficient) or decrease (negative coefficient) by exp(+-coefficient).

Some interesting findings are revealed by the estimates. The coefficient on education is negative for cotton, sugarcane and vegetables, whereas it is positive for rice and positive significant for maize and fruits. This result implies that educated farmers tend to choose maize, fruits and rice (not significant) over wheat. In Pakistan a certain and effective market demand for maize is yet not sufficient, whereas triggered by government intervention wheat and rice are quite effectively procured. Maize is one of the most profitable crops in Pakistan, however access to the main outlets for the crop which are the livestock and feed industry, requires initiative by the farmer himself. Educated farmers thus seem to be better informed about these marketing channels (TARIQ AND IQBAL, 2010). Similarly, fruit production in Pakistan is a complicated undertaking as marketing channels are not well developed in rural areas. Marketing is dominated by the private sector and farmers often lack the necessary information about marketing systems for their produce, which leads to inefficient production structures. Especially the mango is a fruit crop with international marketing potential. However, to successfully market their crops farmers have to put in extra effort themselves to gather information on how to successfully produce and market according to export standards. Complicated and opaque market structures hinder wider access of farmers to the lucrative fruits export sector in Pakistan. Foremost, the Middle East and Central Asia are major export markets for Pakistani fruits, while a growing scope, although gradually, is also observed for the developed countries of Europe and America. The terminal markets with well established trade infrastructure are mostly concentrated in or around large urban centers, whereas smaller markets such as assembly markets (shopkeepers and retailers mostly) are generally located in small towns and in close proximity to farms. Farmers tend to bring the bulk of their fruit produce to these markets, thereby missing out on higher prices and export possibilities. Access to the far-away terminal markets and better information about marketing channels could bring the farmers valuable revenue from growing fruits. Education it seems, as in the case of maize and fruits plays an important role when the marketing channels are rather opaque and export potential is high. Educated farmers seem to be better informed about the value of their produce and quality standards to meet export quality requirements (AUJLA et al., 2007; KHUSHK et al., 2006).

The odds for selecting cotton over wheat are significantly higher for larger households. Precisely, farms with a larger family size are 1.774 times (odds ratio) more likely to grow cotton than farms with smaller household sizes. Cotton picking is a highly labor-intensive activity and also an important source of employment for the women in rural Pakistan (ASHFAQ et al., 2012).

Although not significant, larger households also tend to opt for sugarcane, rice and vegetables cultivation, whereas they are less likely to choose maize and fruits over wheat. Sugarcane cultivation in Pakistan has been found to be labor-intensive undertaking, especially in the planting phase (FAROOQ et al., 1992). Maize on the other hand, with the increasing cultivation of spring varieties has experienced mechanization, thereby reducing the manual labor input (MUHAMMAD, 2005). The wheat crop since the green revolution has experienced a widespread adoption of tractors and related tillage equipment in Pakistan. Mechanized threshing of wheat has also been widely adopted. Similar results have also been obtained by SEO AND MENDELSOHN (2008) for South America, where the findings showed that larger farm families are less likely to choose maize and wheat, because of the relative ease they are mechanized with. Same is inferred for the case of Pakistan on the basis of the choice model. As far as the different soil types are concerned, when the share of marginal soils (calcic, rock, sand) is higher, the farmers are significantly less likely to choose rice and maize over wheat. Rice in Pakistan is grown on the better soils of the country (cambisols). The odds of picking rice over wheat on cambisols are clearly positive significant. Cambisols (or Ochrepts) occur in a wide range of moisture regimes and physiographic positions, particularly in Asia they are important rice soils with a relatively good drainage (MOORMANN AND VAN BREEMEN, 1978). Thus wheat, fruits and sugarcane seem to be the dominant choice for farmers, where the marginal soil share is greater. Surface water availability on the district level as measured by the annual mean runoff is highly significant and positive for rice. This implies that with a higher abundance of surface water farmers are likely to foremost opt for rice cultivation. All of the own prices are significant. Farmers are more likely to choose these crops over wheat when their prices are higher. The highest odds ratios are obtained for maize, sugarcane, rice and cotton. The three crops rice, sugarcane and cotton in particular are all are important cash crops of the country, with cotton being the most important one, followed by sugarcane and then rice. For all these crops (including maize) the farmers can achieve high producer prices. Especially sugarcane with an odds ratio of 3.121 for its own price is a very lucrative source of income. Farmers with higher sugarcane prices are 3.121 times more likely to stick with sugarcane, whereas for cotton the odds ratio is 1.208, 1.230 for rice and 1.285 for maize. Many of the cross price terms are also significant. When the wheat price is higher, the odds of choosing wheat as the primary crop increase. According to the odds ratios, farmers are least likely to choose maize over wheat with higher wheat prices, followed by cotton, vegetables, fruits, sugarcane and rice. When the cotton price is higher, farmers are less likely to pick rice with an odds ratio of 0.878. However, when the sugarcane price is higher the odds for picking cotton, rice and vegetables over wheat significantly decline in favor of sugarcane; with sugarcane least likely to be chosen with an odds ratio of 0.0748 followed by cotton with an odds ratio of 0.238 and rice with 0.264. When the rice price is higher, farmers are significantly less likely to pick cotton and vegetables over wheat in favor of rice. When the maize price is higher the odds for selecting vegetables over wheat become significantly smaller with an odds ratio of 0.910. To sum up, farmers are significantly least likely to pick rice, cotton and vegetables when the sugarcane price is higher. Negative cross price terms from the viewpoint of cross price elasticity of supply, in the eyes of the farmers imply a substitute relationship between the two commodities in question. On the other hand, positive cross price terms indicate complementarities between the crops considered (GRECO, 2005). Interestingly, when the own price effects are compared to the cross price effect of wheat, which is significant for all crops, the results reveal that the reaction of the rice, sugarcane and maize farmers to a change in the respective crop's own price (ceteris paribus) is more elastic than their reaction to a cross price change for wheat. Cotton farmers however, react more elastic to a change in the wheat prices as compared to a change in the own price of cotton, thus when both cotton and wheat prices increase, they can be expected to choose wheat over cotton.

The coefficients for elevation are found to be significantly affecting the crop choice for all crops except maize. The odds for selecting cotton, sugarcane and rice over wheat with higher elevation are only marginally negative significant, whereas almost no effects are observed for fruits and vegetables, with positive significant odds. Thus, only very cautiously it can be inferred, that with increasing elevation wheat is chosen over cotton, sugarcane and rice, whereas fruits and vegetables are as well as likely to be chosen. Higher altitude areas in Pakistan are mainly rainfed, approximately 55% of the cultivated area is cultivated under rainfed conditions; wheat is the primary crop for rainfed areas of the country (MAJEED et al., 2010). Not many new studies related to highland agriculture in Pakistan exist, an older study however found that most of the inhabitants of the mountain regions prefer wheat which consequently has the largest market and a reasonable price which can be obtained by the farmers, provided they have production to sell (STALEY, 1966). Many of the modeled crops have several highly significant temperature coefficients. Especially for sugarcane, rice, maize and fruits, temperature sensitivities are confirmed. Sugarcane choice is sensitive to all seasonal temperature variables. Figures 7.1 and 7.2 reveal that the crop choice in Pakistan in general is temperature and precipitation sensitive. The graphs plot the estimated probability of choosing a crop on the y-axis and annual mean temperature in (°C) or annual mean precipitation in (mm) on the x-axis. The annual mean temperature in Pakistan is 21.76°C (chosen sample).

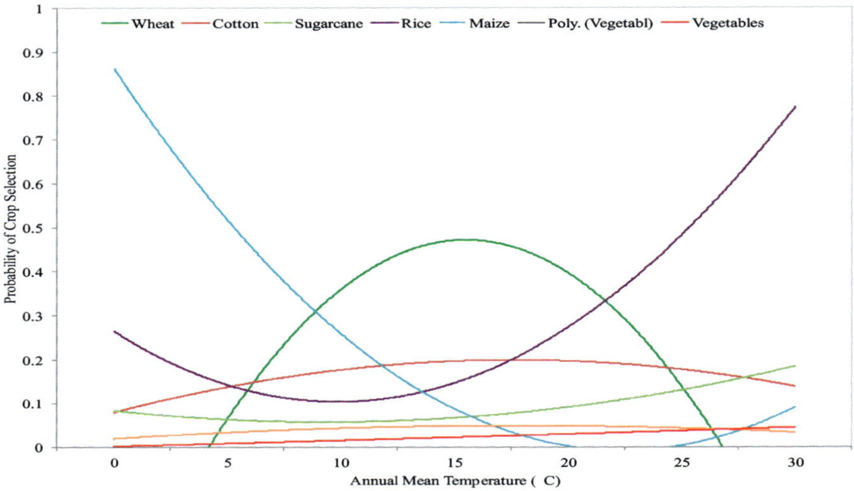

Note: the graphs present trends calculated from the parameter estimates. The seasonal model has been recalculated given annual averages.

Figure 7.1 Estimated probabilities given annual temperature averages

Source: own illustration based on parameter estimates (2012)

The above figure reveals interesting insights with regard to the optimal temperature range certain crops grow in. The probability of selecting wheat declines with higher temperatures, the same is observed for all other crops except rice and sugarcane, which shows an increasing likelihood of being selected when temperatures begin to increase, maize also reappears at higher mean temperatures. Rice for instance is highly adaptable to its environment and can be grown under various different climatic conditions, provided that man modifies the environment to the liking of the rice crop. Rice generally shows different responses to varying temperatures depending on its growth stage, nine such stages can be distinguished: germination, seedling emergence and establishment, rooting, leaf elongation, tillering, initiation of panicle primordia, panicle differentiation, anthesis and ripening. Critical temperatures can be grouped into low, high and optimum. For rice in the different stages (read from left to right) the critical low temperatures range from 10-20°C and the high threshold fluctuates between 30-45°C (45° only for Germination and Leaf Elongation), whereas the optimum levels for each stage are considered to be around 20°C-33°, depending on the region and variety (SHOUICHI, 1981). To this background, the crop seems to be suitable for higher temperatures in Pakistan, where it is grown under irrigated systems only. Moreover the varieties grown in Pakistan are particularly heat tolerant, such as the IRRI type medium long grain heat tolerant tropical rice in the south of N.W.F.P., Sindh and Balochistan (NARC, 2012). At mean temperatures wheat and cotton already show a decreasing trend, whereas others show a slight trend towards increase. The crops seem to have preferred temperature ranges. At the coolest end of the sample the probability of growing maize and wheat is high, whereas at the warmer end of the sample the preferred crops are sugarcane, rice and again maize. Some crops have a wider range for the optimum, whereas others have narrower ranges. Wheat to start with, has a much wider range for its optimum growing temperature, as Figure 7.1 shows, the highest probability that wheat is certainly chosen is estimated for an annual mean temperature range between approximately 12°C and 18°C. Temperatures in excess of 25°C clearly seem to be harmful for wheat. A flatter curve is observed for cotton, indicating a wider range the crop can be grown in. Rice clearly is selected at the warmer end of the sample. The crops clearly seem to have preferred temperature ranges, which farmers in addition to other factors seem to consider when deciding on which crop to grow.

Furthermore, Figure 7.2 illustrates the sensitivity of crop choice in Pakistan to annual mean precipitation. The annual mean precipitation amounts to 28.8 mm. The conditional probabilities plotted against annual mean precipitation clearly indicate a hill shaped relationship between precipitation and crop cultivation. With increasing precipitation farmers foremost select wheat, rice, cotton and sugarcane. Wheat is selected over all other crops when precipitation increases, however the curves for rice, cotton and sugarcane are flatter, indicating that even in case of further increases in precipitation, these crops are likely to be chosen, whereas additional increases clearly seem to effect the selection of wheat, such that at precipitation levels higher than 45 mm the probability of choosing wheat shows a clear decline.

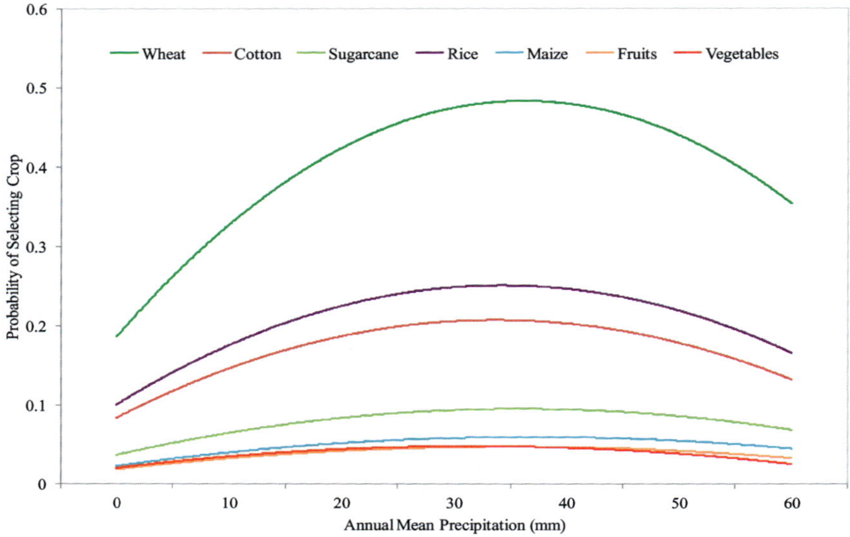

Note: the graphs present trends calculated from the parameter estimates. The seasonal model has been recalculated given annual averages.

Figure 7.2 Estimated probabilities given annual precipitation averages
Source: own calculations based on parameter estimates (2012)

Because the coefficients of the choice model (linear and squared terms) and the non-linear conditional probability plots which present trends can be tedious to interpret and present, the conditional marginal impacts (Ricardian climate sensitivities) have been calculated to understand the effect that the changes in the respective climate variable would have on the choice of a crop. For the multinomial choice model the marginal impacts on the probability of crop choice can be computed by differentiating Equation (21) (chapter 5.5) in the following way.

EQ (36) $$\frac{\partial P_i}{\partial Z_c} = P_i[\gamma_i - \sum_{k=1}^{I} P_k \gamma_k]$$

With P_i denoting the probability of crop i to be chosen, γ_i representing the model parameters and k specifying the exogenous farm characteristics. Table 7.9 reports the calculated marginal effects of a slight change in temperature (1°C) and precipitation (1 mm) on the probability of choosing a certain crop. The change is estimated for a uniform increase of 1°C in Temperature and a uniform increase in precipitation of 1 mm over all season that have been modeled.

Table 7.9 Marginal effects of Climate Change on crop choice in Pakistan

Crop	Temperature (°C)	Precipitation (mm)
Wheat	-1.290%	-0.012%
Cotton	0.026%	-0.011%
Sugarcane	0.071%	0.012%
Rice	0.774%	-0.004%
Maize	0.031%	0.007%
Fruits	0.010%	0.001%
Vegetables	0.369%	0.007%

Note: Marginal Impacts have been estimated at the mean of the corresponding climate variable.

Source: own calculations based on parameter estimates from the choice model.

The marginal effects are evaluated at the mean climate. As temperature marginally increases, farmers clearly choose wheat less often, whereas the likelihood of selecting all other crops increases. The rainfall effects are weaker. When the average annual mean rainfall increases by 1 mm, farmers less often select wheat and cotton, whereas they clearly prefer to grow maize and vegetables.

7.4 Second stage income regressions- Interpretation of results

In a second stage of the empirical analysis conditional net revenue functions are estimated by regressing the net crop revenue per hectare for each selected crop type on climate, soil, water and socio economic variables. As already explained, to correct for sample selection bias Dubin-McFadden selection bias correction is used (BOURGUIGNON et al., 2004). The second stage conditional net revenue model also relies on the same seasonal climate specification that was used for the first stage choice model regressions. The parameter estimates are presented in table 7.10. The findings confirm the climate sensitivity of the incomes for the six crops wheat, cotton, sugarcane, rice, vegetables and fruits. Many of the seasonal climate variables are highly significant, especially the squared terms, indicating a non-linear relationship between net revenues and climate. As in the second stage model each sample is treated individually and sample selection bias correction is applied, a much smaller sample size is left over for analysis, which explains a lower frequency of significant coefficients as compared to the choice analysis. The selection terms or selection bias coefficients hint towards interactions among the different crop types. Precisely, if the selection term is positive, this implies that the same factors that make the choice of one crop type attractive would make the choice of the other respective crop at study even more attractive and vice versa. For instance the selection term for vegetables in the wheat income regression is highly positive and significant, implying that farmers who the choice model predicted would choose vegetables (however, they selected wheat) have higher than expected incomes from growing wheat (compare SEO AND MENDELSOHN, 2008). Vice versa, if the selection term is negative such as in the case of cotton, sugarcane and rice in the wheat income regression, this implies that if the choice model predicted the farmer would have chosen these three crops but the farmer actually selected wheat, that the farmer would have lower income from wheat than if he had chosen cotton, rice or sugarcane. In the cotton income regressions the cotton selection term is positive, indicating that farmers have made the choice that was also predicted by the choice model. For the sugarcane income regression, the selection terms for wheat, cotton and sugarcane are all positive significant. In the rice income regressions the selection bias coefficients are positive significant for cotton, sugarcane, rice and fruits, whereas in the fruits income regressions the maize selection term is negative significant. For vegetables, the selection term on wheat is positive significant.

Again, as the model includes seasonal and quadratic terms that are not straightforward to interpret in terms of a single effect, marginal impacts of a change in climate o crop net revenues are estimated and presented in table 7.11.

Table 7.10 Conditional Income by crop regression for Pakistan

Independent Variables	Wheat	Cotton	Sugarcane	Rice	Maize	Fruits	Vegetables
	Coefficient	Coefficient	Coefficient	Coefficient	Coefficient	Coefficient	Coefficient
Summer Temperature	0.0744***	0.0908	0.266***	0.126**	-0.156	0.00440	0.193**
Summer Temperature²	0.0105*	-0.0716	-0.190**	-0.0240	0.00405	0.0835	0.00814
Fall Temperature²	-0.0250**	-0.194*	-0.108**	0.0845*	-0.0310	-0.125	0.0822
Winter/Spring Temperature²	0.0299***	0.154*	0.0861**	-0.0730**	0.00555	0.104*	-0.0737
Spring Precipitation	0.00317**	-0.00713	-0.00447	0.00349**	-0.000484	-0.0150	0.0259**
Spring Precipitation²	-3.55e-06	0.000256	-3.52e-06	4.81e-05**	7.21e-06	1.08e-05	-7.49e-05
Monsoon Precipitation	0.00274***	0.0278	0.00307	0.000474	0.00325	0.0190**	0.00323
Monsoon Precipitation²	-5.26e-06	0.000442	-0.000376**	9.75e-06	-3.00e-05	-0.000392	4.87e-05
Irrigated Land	0.0330***	0.172**	0.286***	0.0925**	0.145	0.0319	-0.0226
Crop Value (Price)	-0.0140*	0.00856**	-0.00939	0.00476	0.0588	-	-
Household size	0.104***	0.00502	-0.0112	0.0437	0.00227	-0.0802	-0.257
Education	0.00329	0.00572	0.0130	0.000929	0.0228	0.0218	-0.0172
Soil Cambisol	0.000393	-0.536**	0.000941	0.000497	-0.00450	-0.0297*	0.00535
Weak Soils	-0.407***	0.571**	0.339	0.303*	-3.073	1.543	-0.257
Farmland Area	-0.0393***	-0.180***	-0.306***	-0.104**	-0.150	-0.0468	-0.0178
Selection Terms							
Wheat-Selection	-0.211	-0.958	0.907**	0.483	-0.579	-1.662	3.571**
Cotton-Selection	-0.477*	0.393*	1.434**	1.635***	1.828	1.255	-1.590
Sugarcane-Selection	-0.561**	0.426	0.394***	0.842*	-0.683	0.0447	-0.149

Continued on next page…

Rice-Selection	-0.580*	-0.895	0.384	0.401**	0.843	-1.226	-1.132
Maize-Selection	0.0506	-5.498	-2.227	5.722	0.381	-7.727*	5.741
Fruits-Selection	0.635	-1.058	1.490	3.250**	-0.303	0.119	-0.874
Vegetables-Selection	1.794***	-1.791	-0.689	-1.357	-1.170	-2.382	-1.154
Anciliary							
Sigma2	0.726	3.846	3.625	1.436	1.306	6.594	5.575
rho1_Wheat	-0.247	-0.626	0.611	0.403	-0.649	-0.647	1.513
rho2_Cotton	-0.560	0.257	0.966	1.365	2.051	0.489	-0.674
rho3_Sugarcane	-0.658	0.278	0.266	0.703	-0.766	0.0174	-0.0631
rho4_Rice	-0.680	-0.585	0.258	0.334	0.946	-0.477	-0.480
rho5_Maize	0.0593	-3.595	-1.500	4.775	0.428	-3.009	2.431
rho6_Fruits	0.745	-0.692	1.004	2.712	-0.340	0.0463	-0.370
rho7_Vegetables	2.105	-1.171	-0.464	-1.133	-1.313	-0.928	-0.489
Adjusted R^2	0.3591	0.2772	0.1493	0.3024	0.4621	0.6038	0.5896
Observations	876	408	178	507	71	78	91
Constant	9.460***	7.506***	10.38***	10.00***	8.158*	8.044***	14.90***

Standard Errors are presented in the ANNEX; *** p<0.01, ** p<0.05, * p<0.1

Note: The dependent variable is the log of net farm revenue $/ha

Source: own calculations (2012)

Table 7.11 Marginal effects of climate on conditional net farm revenue in Pakistan

Crop	Temperature (°C)							Precipitation (mm)						
	Coef.		Std. Err.	z	P>z	[95% Conf.Interval]		Coef.		Std. Err.	z	P>z	[95% Conf.Interval]	
Wheat	0.0794	***	0.0214	3.71	0	0.0375	0.1214	0.0059	***	0.0011	5.25	0	0.0037	0.0081
Cotton	0.0509		0.0617	0.82	0.41	-0.07	0.1719	0.0207		0.0427	0.48	0.63	-0.063	0.1045
Sugarcane	0.2445	***	0.0616	3.97	0	0.1238	0.3651	-0.001		0.0089	-0.2	0.88	-0.019	0.0161
Rice	0.1371	**	0.061	2.25	0.03	0.0176	0.2567	0.004	*	0.0022	1.83	0.07	-3E-04	0.0082
Maize	-0.181		0.2825	-0.6	0.52	-0.7352	0.3722	0.0028		0.0185	0.15	0.88	-0.033	0.039
Fruits	-0.018		0	-0.1	0.91	-0.333	0	0.004		0.0081	0.49	0.62	-0.012	0.0198
Vegetables	0.201	**	0.0941	2.14	0.03	0.0166	0.3854	0.0292	***	0.0086	3.39	0	0.0123	0.046

Significance level denoted by: *** p<0.01, ** p<0.05, * p<0.1; values are presented in percent.

Source: own calculations using estimates from the conditional income regressions in table 7.10 (2012)

Table 7.11 shows that if climate marginally changes (a uniform change of 1°C in Annual Mean Temperature and 1 mm in Annual Mean Precipitation is calculated) from the current mean state, then gains in income (statistically significant) for wheat and particularly sugarcane, rice and vegetables can be expected. Losses in income are revealed for maize and fruits with a marginal warming. As far as the marginal change in precipitation is concerned, statistically significant benefits are found for wheat, rice and vegetables. Almost all coefficients for the marginal precipitation change are positive, as expected for a dry country like Pakistan. To check the consistency of the first and second stage estimations, the marginal effects for the choice regression (Table 7.9) and the income regression (Table 7.11) can be compared.

Except for wheat, the income effects for sugarcane, rice and vegetables are clearly consistent with farmers' choices. For wheat, which was found to be less selected if a uniform temperature increase of 1°C would occur, comparatively to the other crops that have been found to have income increases with a marginal temperature increase, the beneficial income effect from a slight warming is weaker. These findings confirm the consistency of both models. The precipitation effects in both stages are very small and make it difficult to draw conclusions about the directions of the impacts from a marginal precipitation change, for this to be facilitated further simulations are necessary. For a temperature increase, the results indicate that farmers substitute sugarcane, rice and vegetables for wheat. For precipitation the results are not clear and inferences are difficult to draw from a marginal change only, as mentioned before this shortcoming will be addressed in the subsequent chapter on climate change simulations.

8 Climate Change Scenarios and Simulations using the Ricardian Approach

The following chapter simulates the impacts of climate change on agriculture in Pakistan by using the parameter estimates obtained from both Ricardian regressions. By utilizing the estimated response functions the simulations incorporate the expected changes in climate delivered by an ensemble of Global Circulation Models on a district scale and calculate possible impacts for the agriculture in the country. Further investigations towards changes in future water availability and the estimation of the value of adaptation are made.

8.1 Specification of future Climate Change for Pakistan

To predict the impact of future climate scenarios this study makes use of the dataset which was prepared by the Oxford School of Geography and the Environment. Under the umbrella of the UNDP (United Nations Development Programme) the Oxford School of Geography and the Environment at the University of Oxford has developed the project Climate Change Country Profiles to address the climate change information gap in many developing countries. Funding for the project was received from the United Nations National Communications Support Programme (NSCP) and the UK Department for International Development (DfID). The project aims at making up-to-date data on real observations and simulated observations available to the wider research community. By using the latest data from various renowned sources such as the Climate Research Unit (CRU) and by additionally utilizing the multi-model projections from the WCRP CMIP3 (World Climate Research Program Coupled Model Intercomparison Project 3) the country profiles are created for a number of Asian and other developing countries (MCSWEENEY et al., 2010; MEEHL et al., 2007).

Table 8.1 Data Sources used for the UNDP Climate Change Country Profiles project

Monthly Temperature				
Climatic Research Unit (*CRU*)	New *et al.* (2002)	Gridded station data	1961-2000	0.5x0.5°
University of Delaware (*UDel*)	Matsuura and Willmott (2007a)	Gridded station data	1961-2006	0.5x0.5°
NCEP	Kalnay *et al.* (1996)	Re-analysis data	1960-2006	0.5x0.5°
ERA40	Uppala, et al. (2005)	Re-analysis data	1960-2001	0.5x0.5°
Monthly Precipitation				
Climatic Research Unit (*CRU*)	New *et al.* (2002)	Gridded Station data	1961-2000	0.5x0.5°
University of Delaware (*UDel*)	Matsuura and Willmott (2007b)	Gridded station data	1961-2006	0.5x0.5°
Global Precipitation Climatology Centre (GPCC)	Adler *et al.* (2003)	Merged station and Satellite data	1960-1979	2.5x2.5°

Source: MCSWEENEY et al., 2010

For modeling climate change in Pakistan, various data maps and illustrations on observed and modeled climate under the three SRES (Special Report on Emissions Scenarios) were used. The data set contains information for the emission scenarios A2, A1B, and B1. These emission scenarios were developed by the IPCC to assess the relationship between the forces that cause the emissions and their development. Thus the scenarios cover a wide range of technological, economic and demographic factors that influence future emissions. Different storylines or assumptions are at heart of the different scenarios or scenario families (when based on the same assumptions/storyline). In the context of the political situation in the developing countries, the SRES scenarios are not modeled under the assumption of an implementation of the UNFCCC or the emissions targets proposed by the Kyoto Protocol. However, the storylines nevertheless do take into account existing policies that are implemented in related fields such as the prevention of air pollution, technological advancement, population growth, which can directly or indirectly influence the emission of greenhouse gases (NAKICENOVIC et al., 2000). In following the different SRES scenario families are presented and the most realistic scenario for Pakistan is selected. For a detailed overview on the different SRES scenarios refer to the IPCC special report on emissions scenarios (NAKICENOVIC et al., 2000).

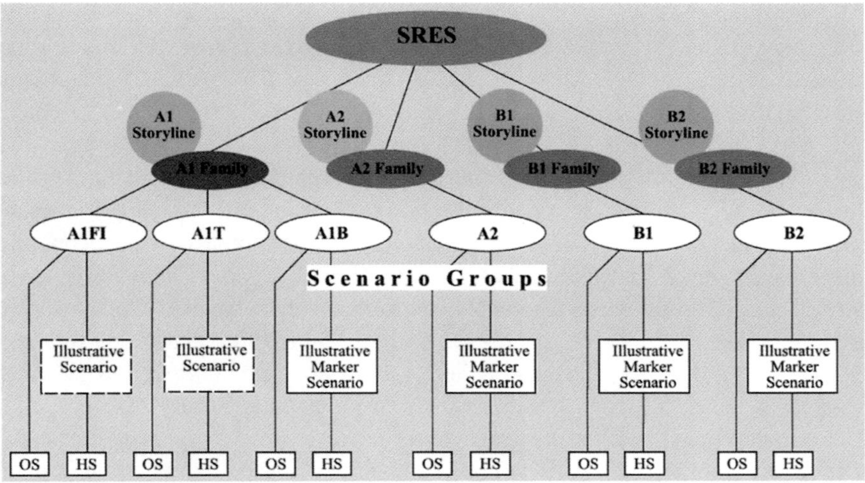

Figure 8.1 Characteristics of the four SRES storylines and scenario families
Source: NAKICENOVIC et al., 2000

Figure 8.1 shows the different SRES scenarios, which are constructed based on four different qualitative storylines. Scenarios based on the same storyline belong to a scenario family. Thus, illustration depicts the four storylines and the corresponding scenario families A1, A2, B1 and B2. In a next step out of the scenario families six different groups are drawn, with three groups from the A1 family and one group each from the A2, B1 and B2 scenario families. The three groups in the A1 scenario family are based on a differential characterization of the developments in the energy sector, with A1F1 depicting a fossil fuel intensive future, A1B assuming a more balanced use of different energy sources, and A1T referring to a future with a dominance of non-fossil energy sources. Within the different groups another distinction is made regarding assumptions on world population, energy and gross world product. When these so-called "driving forces" are harmonized amongst the scenarios the abbreviation "HS" is applied, and vice versa if the driving forces are treated as uncertain (beyond the HS case) the abbreviation "OS" is applied.

The A1 scenario family assumes a future world with fast paced economic growth and technological advancement and long term population decline (after peaking in 2050). Furthermore, regions are expected to converge culturally with an increase in social interactions and capacity building, causing income disparities to narrow. As mentioned above, the A1 family of scenarios is distinguished further by alternative developments in the energy sector. The A2 scenario family is a sort of counterpart of the A1 scenario storyline and family, outlining a rather heterogeneous world. Instead of converging, regions rely on themselves and preserve their very own identities. The trend of population growth is steadily increasing and economic development is regionally oriented. The per capita changes in economic growth are not uniform throughout the region and progress slowly. The B1 scenario family stands for an adapting world and presents a rather optimistic future, where without additional climate initiatives the world will have implemented solutions for achieving economic and environmental sustainability and equity through a rapid change towards a service and information economy. Population growth is handled the same way as in the scenario family A1. In the B2 scenario family the focus is on regional responses to climatic change, where solutions, including environmental protection and social equity, will be implemented rather locally than globally. The population is assumed to grow, however on a lower rate than in the A2 scenarios. Economic development is medium paced and technological development is seen to be more diverse. The following illustrations present the predicted climate change for Pakistan under different scenario families on an annual basis. Following the illustrations, the choice of the appropriate scenario family for Pakistan will be elaborated on and the amended dataset will be presented using a GIS for the study site.

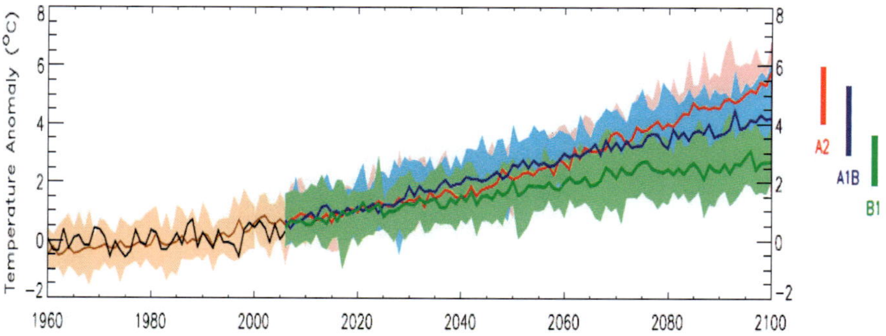

Note: "All values shown are anomalies, relative to the 1970-1999 mean climate. Black curves show the mean of observed data from 1960 to 2006, Brown curves show the median (solid line) and range (shading) of model simulations of recent climate across an ensemble of 15 models. Colored lines from 2006 onwards show the median (solid line) and range (shading) of the ensemble projections of climate under three emissions scenarios. Colored bars on the right-hand side of the projections summarize the range of mean 2090-2100 climates simulated by the 15 models for each emissions scenario (MCSWEENEY et al., 2010)".

Figure 8.2 Trends in annual mean temperature - past and projected

Source: MCSWEENEY et al., 2010

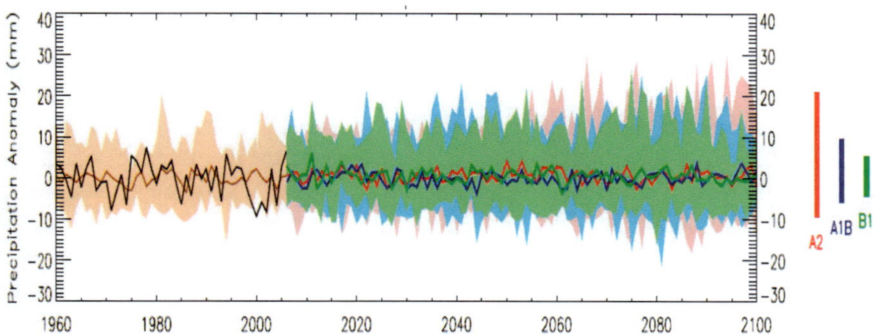

Figure 8.3 Trends in annual mean precipitation - past and projected
Source: MCSWEENEY et al., 2010

For the entire country the trends suggest that annual mean temperature will increase in the range between 1.4°C to 3.7°C by 2060 and between 1.9°C to 6°C by the 2090. Amongst all ensembles the predicted range of temperature change is estimated around 1.5 °C to 2°C. Precipitation changes are somewhat more uncertain. The range of the precipitation change prediction, in any direction, is large. Thus, changes in precipitation are predicted to range between -9 mm and +20mm per month by the 2090s. The ensemble means are close to zero, indicating a rather compensatory effect for precipitation, where one region's decrease is compensated by another regions precipitation increase. Pakistan is a developing country with various development related issues including high population growth rate, poverty, inequality, high national indebtedness, low institutional capacity, corruption and persistent political unrest, all weakening the countries capacity to respond to environmental changes and promoting economic growth. To this background, the SRES scenario family A2 seems to fit best to the situation in Pakistan, where population can be expected to rather grow than decline. Moreover, inside Pakistan the regions even today proclaim their own identities and can be expected to even more rigorously do so in future. The middle class in Pakistan is very thin, if existent at all, and the gap between rich and poor is widening, therefore providing additional evidence to opt for the A2 scenario for modeling the economic impacts resulting from a change in climate.

Country level climate change scenarios depicting changes for a ten year average for 2030, 2060 and 2090 under the SRES A2 scenario family are used. The dataset by the UNDP climate change country profile project reports observed trend and projected changes averaged over the whole country and provides spatial variations in change across the country on a 2.5 x 2.5 ° grid level. These grids contain the ensemble range and ensemble median. The median is used as the reference value for creating a district level dataset using a Geographic Information System (GIS). For this, the predicted temperature change in °C for 2030, 2060 and 2090 is added to a baseline temperature in each district. In the baseline scenario, the farmers are assumed to continue making the same choices that they make today, provided the climate does not change. This implies that other reasons that might inflict a change in the farmer's choice behaviors over the next eighty years, is not modeled. The focus is on examining the impacts of climate change. Furthermore, the percentage change in precipitation that is predicted by the AOGCM's summarized in the UNDP climate change country profiles, is multiplied by the baseline precipitation values in each district. This procedure ensures that a new climate prediction is available for every modeled district in Pakistan. The following maps present the district wise amended dataset for the study site on a district scale. In total six different maps are presented covering the three simulation periods 2030, 2060 and 2090, which are distinguished by temperature and precipitation.

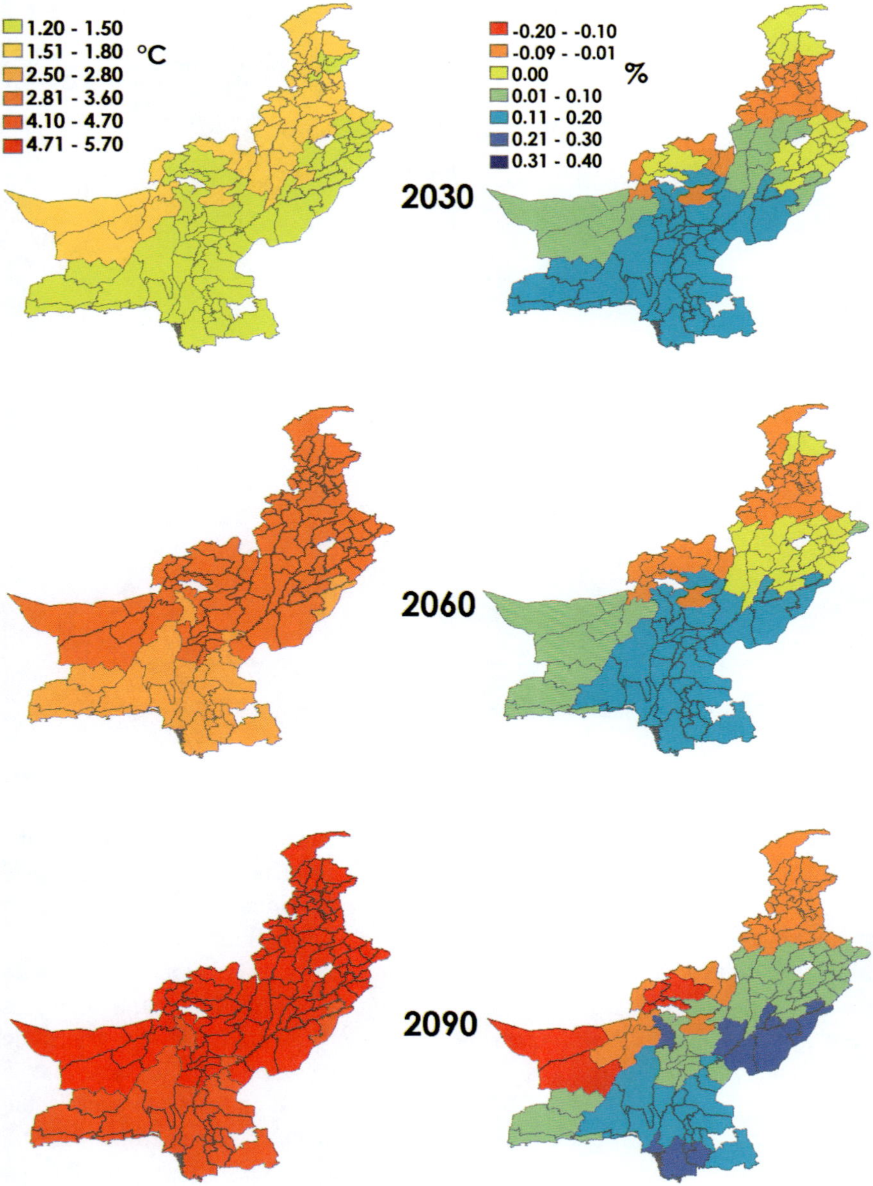

Figure 8.4 Spatial patterns of projected Climate Change in Pakistan (mean values)
Source: own illustration using Arc GIS (ESRI, 2011)

Table 8.2 Pakistani average annual and seasonal Climate Change scenarios 2030-2090

	Obs	Temperature (°C)	Precipitation (%)
2030			
Annual	2369	1.521992	4.89%
JFM	2369	1.652469	-15.50%
AMJ	2369	1.644407	2.71%
JAS	2369	1.457282	3.91%
OND	2369	1.599831	-4.89%
2060			
Annual	2369	3.086661	6.91%
JFM	2369	3.14943	-15.47%
AMJ	2369	3.094344	6.67%
JAS	2369	2.952216	17.20%
OND	2369	3.089194	-10.11%
2090			
Annual	2369	5.075728	3.75%
JFM	2369	5.236809	-32.66%
AMJ	2369	5.025791	7.58%
JAS	2369	4.446813	18.86%
OND	2369	5.098227	2.78%

Note: values denote deviations from the normal climate!

Source: own calculations based on amended dataset using McSWEENEY et al., 2010

Figure 8.4 shows that in the short term (2030) temperatures on a district level are expected to increase by a minimum of 1.2 °C in most of the south and east, and by a maximum of 1.8°C in most of the northern and western parts of the country, comprising districts like Chitral and Upper Dir. On a provincial scale the province N.W.F.P. is one with the highest increase in temperatures throughout its borders, followed by Punjab Balochistan and Sindh. In the mid-term (2060) the general trend more or less continues, with minimum temperature increase found in the Sindh province valuing 2.5°C and maximum increases predicted for the Northern parts of the country with 3.4°C to 4°C. The Balochistan province in this scenario is expected to warm with the same rate as the Punjab province, touching 3°C and above. Drastic changes are seen in the 2090s, where the entire country is expected to warm by 5°C on average, with 4.7 °C in the Southern parts and up to 5.7°C for the northern most parts. As table 8.2 shows, once again the importance of specifying seasonal climate scenarios to be used in Ricardian models is revealed, as seasonal patterns differ, at times considerably, from the overall annual pattern. In the 2030s for instance, the hottest months from July to September are on average expected to experience an increase in temperature by 1.4°C, whereas for the coldest months from January to March temperatures changes of plus 1.6°C are predicted. These differential seasonal warming trends continue in the mid-term and long-term scenarios.

As far as precipitation is concerned clear patterns of change are observed for the country. In the short- term for the most northern parts of the country and the western districts of the Punjab province the ensembles of AOGCMs do not predict any changes in precipitation. The eastern and southern parts of the country are expected to receive higher precipitation between 1% and 15%. In the mid-term precipitation shortages are shown for the entire north of the country and the northern parts of the Balochistan province. By 2060 the climate models predict no changes for the most central part of the country, whereas the amount of precipitation that is received compared to the long term averages from 1960-2000 increases significantly in the south. A lot more districts are predicted to receive increases in precipitation between 15% and 20%. In the long-term scenario (2090) compared to the mid-term scenario, the biggest change is that precipitation is expected to change over the entire country. The central parts in the 2090's scenario also show a positive trend in precipitation between 1% and 10%. However, the decreases in the northern parts and parts of the Balochistan province are also much stronger, such as minus 20% in the Quetta district. Table 8.2 shows the change in monthly precipitation for ten year periods in the future under the SRES A2 scenario family, all values are percentage deviations from the mean climate observed between 1970 and 1999. Seasonally, precipitation is expected to decrease in the months from October to December by almost 5% and with minus 15% even more so between January and March. Spring and summer precipitation shows an increasing trend. For the 2060s the changes in precipitation for the fall and winter season remain similar to the 2030 trends, however the increases in spring and summer precipitation are much stronger, with +6.67% and +17.2% , respectively. For the long term-scenario, on a country level, table 8.2 shows a drastic decline in precipitation in the winter months; however the fall period does not show decreases in precipitation anymore, increases are found for the three seasons spring, summer and fall. The maps and the table show the differences between overall country predictions and district level assessments; the number of districts that are expected to receive more precipitation by the 2090s is clearly larger as in 2030 and 2060. Moreover, the reader might be tempted to perceive the seasonal averages to average into the annual averages of the precipitation variable in table 8.2, as they only vaguely reflect the seasonal patterns, this is due to the fact that the absolute amounts of precipitation that are received in the spring and summer months represent the bulk of the country's annual precipitation, thus although relatively negative changes are quite large, the positive changes overcompensate them by some extent as far as absolute amounts of precipitation measured in millimeters are concerned. In following the results of the simulations shall be presented.

8.2 Climate Change Simulations

Using the specified climate change scenarios the original dataset was amended on the district level. The comparative static nature of the Ricardian model calls for a cautious approach for interpreting the findings from the simulation runs. As discussed in the literature review on Ricardian modeling other factors such as carbon dioxide, surface water, technology and prices can change over the simulated period. However, as i.e. carbon dioxide levels are not district specific, they can be only considered exogenously. To this backdrop, assumptions regarding e.g. surface water are only hypothetically made to account for the effect of variables other than climate changing in the course of the next eighty years. The assumption of the Ricardian model vis-à-vis future climate simulations therefore is that farmers will adjust to future climates just as they have adjusted to their current climates on the district levels. The technologies that so far have been developed for certain crops and climatic settings in the country will also be used in the future.

8.2.1 Simulations using the Standard Ricardian Model

The maps in figure 8.5 show the spatial patterns of the projected climate scenarios' impacts on land values across the country for three different farm types, mixed, rainfed and irrigated. The rainfed and irrigated farm samples do not sum up to the mixed sample. The results of the simulations clearly reveal that the country is highly vulnerable to climate change if the change is of a uniform nature, specifically rainfed farms are severely affected, with declines in land values and thus capitalized net revenues between -30% and -10% in the southern most parts of the country in the Sindh province. Only selected districts in the more or less humid northern most parts of the country and some parts in the western moderately humid parts reveal gains from a uniform change in climate. The SRES A2 scenarios however, show a slightly different picture. With increasing temperature and increasing precipitation, the losses are reduced over the course of the three scenarios from 2030 to 2090, showing that, if farmers keep on doing what they have done so far and provided that the other things modeled remain unchanged over the next 80 years, then in most of the districts agriculture can benefit from climate change. Common in all scenarios is the significantly high negative impact from climate change in the southern parts of the country, specifically in the Indus Delta.

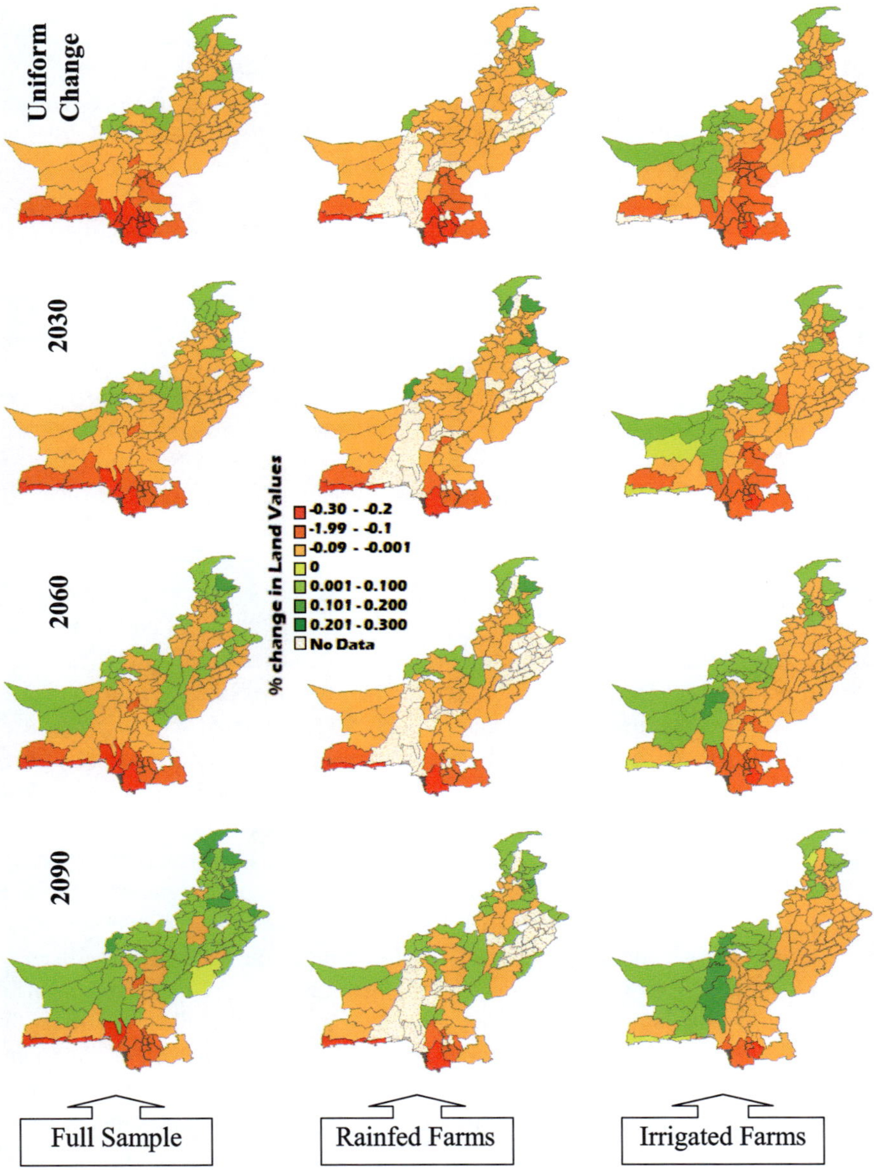

Figure 8.5 Spatial patterns of Climate Change impacts on land values – SRES A2
Source: own illustration in Arc GIS 10.1 (ESRI, 2011)

These maps present the percentage change in land values resulting from a uniform change in climate (1°C and 1 mm – first row), a predicted change by SRES A2 ensemble 2030 (second row), ensemble 2060 (third row) , and ensemble 2090 (fourth row). Further differentiation is applied with respect to farm type: first column (full sample), second (rainfed farms) and third column (irrigated).

Of key importance is the seasonal pattern of impacts which are shown in table 8.3. As far as temperature only (first aggregated row) is concerned, clearly in the first three scenarios the summer season damages are higher than the winter season benefits, in the 2090 scenario however, this constellation changes and the winter season benefits from climate change (temperature only) overcompensate the losses in the summer season. In the mixed or full farm model this is true for all provinces except for the Sindh province, where throughout losses are expected. In case of the rainfed farms the impacts are more severe initially, as expected, they also span over the northern parts. They are negative over all scenarios for the southern provinces Balochistan and Sindh, which are the hottest and driest of the country. For the irrigated farms, which present the most realistic picture with respect to current agricultural structure in the country (2/3 of agricultural activity is supported by irrigation), over all scenarios the adverse effects from climate change overcompensate the positive effects. The summer season temperature impacts are negative for all provinces and over all modeled scenarios. For precipitation (second aggregated column) the impacts are negative for summer and positive for winter, here as well the dry nature of the country comes into play, as the parameter estimates have shown the relationship between summer or monsoon precipitation and land values to be U-shaped. This shows that a precipitation deficiency is true for all provinces. Winter precipitation benefits in the case of precipitation do not compensate for the missing precipitation in summer. The third aggregated row presents the total impacts of climate change on land values (or capitalized net revenues) from a joint climatic change (precipitation and temperature) for the uniform scenario and the different SRES A2 scenarios for 2030, 2060 and 2090. Over all farm types and provinces the impacts from climate change in summer are found to be negative. Whereas in the mixed farm sample the winter season benefits in the long-term SRES A2 scenario (2090) manage to compensate summer season losses, for rainfed and irrigated farms the total impacts are continuously negative, especially for rainfed farms the district average losses are -169566$ in the summer season, whereas they amount to -153000$ for irrigated farms. In the mixed farm sample the highest losses from a temperature change in summer by 2090 are found for the N.W.F.P. and Balochistan, followed by Punjab and

Sindh. Rainfed farms are expected to suffer the most severe losses from warming in summer in the provinces of N.W.F.P. and Punjab, followed by Balochistan and Sindh. Irrigated farms are expected to experience high losses from a change in summer temperature throughout the provinces, however on a relatively smaller scale as compared to rainfed farms. This reveals that irrigated farms might be better equipped for buffering the adverse effects from warming in summer through the application of additional water, provided it is available. That irrigation might act as an adaptation measure and that irrigation could compensate for a certain share of the negative climate change impacts has been also find in other Ricardian studies e.g. for Africa (KURUKULASURIYA AND MENDELSOHN, 2008). The winter season benefits are particularly high for the N.W.F.P. and the Balochistan province. An indication of the importance of modeling of climate using a seasonal pattern is clearly revealed, as the disaggregated effects manage to reveal that winter benefits are capable of compensating for the adverse impacts in summer, at least in the mixed farms model. Moreover, rainfed farms are found to be more vulnerable to climate change and irrigation seems to serve as a compensatory factor. This stems from the fact that the growing season length in winter is positively altered and farmers have more options available. Currently the major crop that is cultivated in the winter season is wheat. The results in this sense reveal that to date farmers have more or less adapted well to their local circumstances and if their available resource base does not change they will continue to flourish in certain regions, whereas regardless of unchanged resource availability they will suffer in other districts.

Table 8.3 Welfare impacts from a change in climate

Scenario			Mixed Farms				
			Punjab	Sindh	N.W.F.P.	Balochistan	Pakistan
Temperature	Uniform Change	Summer	-43014.58	21073.09	-159435.4	-149907.6	***-82807.09***
		Winter	4384.72	-71314.58	136700.3	103333.4	***43300.64***
	SRES A2	Summer 2030	-91048.71	3850.427	-200401.6	-188847.6	***-120822***
		Winter 2030	71742.12	-42623.18	199652.3	160544.9	***99381.07***
		Summer 2060	-125848.4	-14551.02	-231073.4	-231235.7	***-153372.9***
		Winter 2060	132547.4	-16323.09	252420.7	216582	***150318.1***
		Summer 2090	-186105.6	-36000.2	-287919.2	-277988.6	***-201686.8***
		Winter 2090	219776	20340.25	326166.4	286869.6	***220179.6***
Precipitation	Uniform Change	Summer	-11969.16	-5266.052	-15276.57	-13485.06	***-11736.02***
		Winter	11798.64	3121.591	21849.45	11049.91	***12065.55***
	SRES A2	Summer 2030	-12168.89	-5274.293	-15847.95	-13589	***-11954.16***
		Winter 2030	12944.8	3215.015	22587.37	11697.14	***12785.09***
		Summer 2060	-12375.37	-5467.826	-15645.77	-13866	***-12086.83***
		Winter 2060	12470.6	3219.864	22656.6	13037.89	***12987.25***
		Summer 2090	-12583.6	-5394.819	-17089.51	-13853.3	***-12452.15***
		Winter 2090	367.7745	1354.853	-7983.831	1042.246	***-1085.295***
Total	Uniform Change	Summer	-54983.74	15807.04	-174711.9	-163392.7	***-94543.11***
		Winter	16183.36	-68192.99	158549.7	114383.3	***55366.19***
	SRES A2	Summer 2030	-103217.6	-1423.866	-216249.6	-202436.6	***-132776.1***
		Winter 2030	84686.92	-39408.16	222239.7	172242	***112166.2***
		Summer 2060	-138223.7	-20018.84	-246719.2	-245101.6	***-165459.8***
		Winter 2060	145018	-13103.22	275077.3	229619.9	***163305.4***
		Summer 2090	-198689.2	-41395.01	-305008.7	-291841.9	***-214138.9***
		Winter 2090	220143.7	21695.1	318182.5	287911.8	***219094.4***

Note: Uniform Scenario assumes a change in precipitation by 1 mm and by 1°C in temperature. Values are reported in Dollars/district, thus represent district averages on province scale.

…Table 8.3 continued

Scenario			Rainfed Farms				
			Punjab	Sindh	N.W.F.P.	Balochistan	Pakistan
Temperature	Uniform Change	Summer	-74292.68	3666.934	-154079.7	-82315.91	***-82965.09***
		Winter	51594.65	-44813.74	131180.5	55618.87	***55349.3***
	SRES A2	Summer 2030	-109864.6	-10054.64	-181781.3	-105517.9	***-108504.7***
		Winter 2030	98720.57	-23084.07	169824.4	85577.41	***90251.4***
		Summer 2060	-135632.6	-24946.52	-202141.1	-129228.1	***-129962.3***
		Winter 2060	141523.3	-3240.807	201569	115098.9	***121649.8***
		Summer 2090	-180045.9	-42379.19	-238420.8	-156603.5	***-161959.9***
		Winter 2090	201712.6	24602.01	248256.5	151504.8	***164992.5***
Precipitation	Uniform Change	Summer	-6114.36	-860.0175	-13557.65	-3488.757	***-6429.043***
		Winter	-3043.993	-8164.459	20722.83	-5190.902	***2031.103***
	SRES A2	Summer 2030	-6286.815	-817.8688	-14409.61	-3655.639	***-6747.631***
		Winter 2030	-1065.747	-7952.985	22493.15	-4217.373	***3317.79***
		Summer 2060	-6448.937	-898.2496	-14466.45	-3796.003	***-6857.433***
		Winter 2060	-1571.313	-7956.768	22592.4	-2383.71	***3715.534***
		Summer 2090	-6994.895	-827.0717	-16604.59	-3897.323	***-7606.165***
		Winter 2090	4614.764	5523.405	-8724.581	5600.702	***1304.217***
Total	Uniform Change	Summer	-80407.04	2806.917	-167637.3	-85804.67	***-89394.14***
		Winter	48550.66	-52978.2	151903.3	50427.97	***57380.4***
	SRES A2	Summer 2030	-116151.4	-10872.51	-196190.9	-109173.5	***-115252.3***
		Winter 2030	97654.82	-31037.05	192317.6	81360.03	***93569.19***
		Summer 2060	-142081.5	-25844.77	-216607.6	-133024.1	***-136819.7***
		Winter 2060	139952	-11197.57	224161.4	112715.2	***125365.3***
		Summer 2090	-187040.8	-43206.26	-255025.4	-160500.8	***-169566.1***
		Winter 2090	206327.3	30125.41	239531.9	157105.5	***166296.7***

...Table 8.3 continued

Scenario			Irrigated Farms				
			Punjab	**Sindh**	**N.W.F.P.**	**Balochistan**	**Pakistan**
Temperature	**Uniform Change**	Summer	-117076.9	-62996.47	-168904.5	-123649.9	***-119201.4***
		Winter	43209.14	3202.121	92963.59	112846.2	***63338.99***
	SRES A2	Summer 2030	-123286.3	-65058.53	-183526.4	-134364.6	***-127542***
		Winter 2030	58397.72	13339.32	115054.4	137723.8	***81427.81***
		Summer 2060	-135606.9	-67287.86	-199795.5	-146167.4	***-138566.9***
		Winter 2060	75691.36	19804.67	144512	170089.8	***102947.7***
		Summer 2090	-139186.9	-68667.19	-204004.2	-156659.1	***-143553.3***
		Winter 2090	95691.22	33542.67	167621.7	202943.4	***125552.4***
Precipitation	**Uniform Change**	Summer	-8036.039	-3785.218	-13159.8	-11455.06	***-9153.155***
		Winter	15296.64	6355.37	28250.76	19763.33	***17450.69***
	SRES A2	Summer 2030	-8110.215	-3788.468	-13436.08	-11514.87	***-9253.959***
		Winter 2030	15775.09	6396.331	28930.07	20118.76	***17854.59***
		Summer 2060	-8183.638	-3865.019	-13429.64	-11633.49	***-9321.623***
		Winter 2060	15575.41	6398.573	28902.92	20960.73	***17991.68***
		Summer 2090	-8253.365	-3837.065	-14169.48	-11631.81	***-9501.496***
		Winter 2090	1036.42	926.5691	-3103.87	1635.764	***251.086***
Total	**Uniform Change**	Summer	-125112.9	-66781.69	-182064.3	-135105	***-128354.5***
		Winter	58505.78	9557.491	121214.4	132609.6	***80789.68***
	SRES A2	Summer 2030	-131396.5	-68847	-196962.4	-145879.5	***-136795.9***
		Winter 2030	74172.81	19735.65	143984.5	157842.5	***99282.4***
		Summer 2060	-143790.5	-71152.87	-213225.2	-157800.9	***-147888.5***
		Winter 2060	91266.77	26203.24	173414.9	191050.5	***120939.4***
		Summer 2090	-147440.3	-72504.25	-218173.7	-168290.9	***-153054.7***
		Winter 2090	96727.64	34469.24	164517.8	204579.2	***125803.5***

Source: own illustration based on parameter estimates in STATA (2013)

The standard Ricardian model does not model specific adaptation, but it implicitly includes adaptation, which shows that the adaptations that were made by the farmers over the course of the last 30 to 50 years have been well directed and can be expected to benefit the farmers for adaptation to future risks. Obviously, the positive trend in the maps reveals adaptations, however how farmers adapt is not captured and no changes can be made to key variables to test the impact their change may inflict, for this reason the structural model has been formulated to answer the black box questions not addressed in the standard model. Here one can also see whether a reduction in surface water availability will leave incomes from agriculture unaffected or will negatively influence them. Moreover, adaptation cannot be quantified in the standard model, as it is implicit in the changes, which depend on all other variables in the model as well. For this reason in following the idea is to simulate the results from the structural model to quantify the value of adaptation and also test the assumption of reduced surface water availability, this in order to counter the ceteris paribus assumption of the standard model.

8.2.2 Simulations using the Structural Ricardian Model

The idea is to first measure the change in welfare that results from a change in climate from one point in time to the other, where the base period represents long term normal climate for the country. The following equation presents the measurement of the climate change induced welfare change.

EQ (37)
$$\Delta W = [V(C_1, Z_i) - V(C_0, Z_i)] \cdot L_i$$
$$= \sum_{k=1}^{K} \left[P_k(C_1) \cdot \pi_k(C_1) - \sum_{k=1}^{K} P_k(C_0) \cdot \pi_k(C_0) \right] \cdot L_i$$

Where C_1 denotes the new climate and C_0 represents the initial climate, L_i is the amount of land at each farm i, π_k is the conditional land value of a particular farm type and P_k stands for the probability of each choice the farmer faces. The major difference between the standard model's welfare estimates and the structural model estimates is that the structural model calculations assess the welfare change by combining the choice and conditional income results, whereas the standard model's welfare estimates depend on a single equation. Hence, in case of the structural model the welfare change captures two different effects. On one hand as the climate changes, the probabilities of crop choice are likely to alter and on the other hand climate change will thereby also impact on the conditional incomes from the chosen crops. To assess the value and thus importance of adaptation, the expected income (equation 37) of climate change without adaptation can be compared to the expected income in the case with adaptation. Equation 38 presents the case when farmers do not adapt.

EQ (38) $$W(C) = \sum_{k=1}^{K}[P_k(C_1) \cdot \pi_k(C_0) - \sum_{k=1}^{K} P_k(C_0) \cdot \pi_k(C_0)] \cdot L_i$$

Without adaptation measures farmers will continue to make the same choices that they were making under the initial climate. However, as the incomes of each choice also change, farmers will be made worse off than if they adapted. The value of adaption is given by the difference between equations 37 and 38.

Using the parameter estimates from the structural Ricardian regressions the consequences of climate change are simulated. Basically, the initial model is re-estimated using the new climate dataset for each time period. From these new estimates the probabilities of selecting a certain crop were extracted to compare them with the probabilities of the initial model. Table 8.4 shows the changes in the probabilities of selecting a certain crop for each climate scenario.

Table 8.4 Predicted probability change of selecting each crop under different scenarios

Probability	Obs	Current Climate	2030	2060	2090
Wheat	2369	41.16%	0.1579%	-0.3169%	-0.6416%
Cotton	2369	17.22%	-0.0312%	-0.0129%	-0.0687%
Sugarcane	2369	7.51%	0.3340%	0.2832%	-0.0141%
Rice	2369	21.40%	-0.4679%	0.1406%	0.5830%
Maize	2369	4.56%	0.0128%	0.0497%	0.1032%
Fruits	2369	3.97%	-0.0113%	-0.0134%	-0.0167%
Vegetables	2369	4.18%	0.0057%	-0.1302%	0.0549%

Note: probabilities have been obtained from the multinomial logit regression estimates.

Source: own calculations (2013)

Although the predicted changes seem to be weak, they to some extent highlight the importance of examining agro-climatic zone specific scenarios as precipitation increases can possibly offset, at least to a certain extent, damages and pressure from additional warming. The model does not claim to be correct on the decimal and absolute value level, however it does claim to accurately reflect the direction of the expected changes. In spirit of this claim, the probability results are interpreted. As climate changes in the short term (2030) (combined temperature and precipitation change) farmers are more likely to select wheat and sugarcane, whereas they are slightly also more likely to select maize and vegetables. On the other hand they are reluctant to choose rice, cotton and fruits. Rice especially, as has been explained earlier in the interpretations for the structural Ricardian model, has a U-shaped relationship with summer precipitation (rice growing season), thus minimum levels of precipitation are essential for the crop to flourish. By 2030 the change in temperature is evident over most of the country, however in the rice growing regions of the Punjab, precipitation levels do not change. Thus from a relative point of view, under the 2030 scenario farmers are more likely to switch from rice, cotton and fruits to wheat, maize, and foremost sugarcane. Under the mid-term scenario farmers clearly move away from wheat, vegetables, fruits and cotton towards maize, sugarcane and foremost rice. In the mid-term scenario, the precipitation gains in the Punjab province and the even stronger precipitation increases in the Sindh province, both major rice producing provinces of the country, seem to benefit rice cultivation, finally exceeding the minimum required precipitation amount for the cultivation of rice. Sugarcane, although also more selected in 2060s, when compared to 2030s shows a weaker positive effect (on probability), whereas the negative effect for cotton is relatively also smaller. In the long-term scenario farmers clearly switch away from wheat, cotton, fruits and newly also sugarcane to rice, vegetables and maize, clearly showing the effect of the strong increase in precipitation and at the same time temperature. Thus when it's hot and precipitation is available, other things remaining constant (see parameter estimates for structural model for the other variables), farmers select rice, vegetables and maize. The long term scenario results also hint towards increasing adaptations as warming progresses. How things change when other variables change, such as surface water availability, will be discussed in combination with the following deliberations on welfare and the value of adaptation. The above findings are highly consistent with the earlier presented conditional probability plots of crop choice and annual mean temperature and precipitation (see Figures 7.1 and 7.2).

As the structural model is a two stage model, the changes in net income are also shown in table 8.5. For obtaining these results, all models were re-estimated using the AOGCM scenarios for 2030, 2060 and 2090.

Table 8.5 Predicted change in conditional net crop revenue under different CC scenarios

Crop	Temperature (°C)							Precipitation (mm)						
2030														
	Coef.		Std. Err.	z	P>z	[95% Conf.Interval]		Coef.		Std. Err.	z	P>z	[95% Conf.Interval]	
Wheat	0.0297		0.0217	1.37	0.17	-0.012747	0.0721619	0.0059	***	0.0011	5.24	0	0.0037002	0.0081246
Cotton	0.1198		0.2095	0.57	0.567	-0.29079	0.5303555	0.0508		0.0374	1.36	0.174	-0.022455	0.1240899
Sugarcane	0.0952		0.2117	0.45	0.653	-0.319657	0.5100227	0.0563		0.0376	1.5	0.134	-0.017334	0.1300015
Rice	0.1779	*	0.077	2.31	0.021	0.0269235	0.3288821	0.0041	**	0.0022	1.9	0.057	-0.000124	0.0084135
Maize	-0.112		0.1721	-0.65	0.514	-0.449638	0.2248681	0.0029		0.0184	0.16	0.874	-0.0332	0.0390262
Fruits	-0.202		0.1722	-1.17	0.242	-0.539171	0.1359511	0.0055		0.0077	0.71	0.479	-0.009682	0.0206384
Vegetables	0.1579		0.1325	1.19	0.234	-0.101856	0.4176148	0.0287	***	0.009	3.19	0.001	0.0110526	0.0462628
2060														
Crop	Temperature (°C)							Precipitation (mm)						
	Coef.		Std. Err.	z	P>z	[95% Conf.Interval]		Coef.		Std. Err.	z	P>z	[95% Conf.Interval]	
Wheat	-0.017		0.028	-0.61	0.544	-0.071878	0.0378816	0.006	***	0.0012	5.09	0	0.0036652	0.0082481
Cotton	0.197		0.3944	0.5	0.618	-0.576122	0.9700224	0.0441		0.0354	1.24	0.213	-0.025319	0.1134379
Sugarcane	0.1483		0.3986	0.37	0.71	-0.632971	0.9295934	0.0494		0.0356	1.39	0.165	-0.020301	0.1191895
Rice	0.2165	*	0.1301	1.66	0.096	-0.038621	0.4715443	0.004	*	0.0022	1.78	0.075	-0.000406	0.0083702
Maize	-0.049		0.0949	-0.51	0.608	-0.234709	0.1372916	0.0031		0.0185	0.17	0.867	-0.033191	0.0394154
Fruits	-0.387	*	0.2351	-1.65	0.099	-0.84822	0.0732548	0.0082		0.0079	1.04	0.296	-0.007213	0.0236854
Vegetables	0.1094		0.1925	0.57	0.57	-0.268006	0.4867473	0.0285	***	0.0095	2.99	0.003	0.0098263	0.0470904
2090														
Crop	Temperature (°C)							Precipitation (mm)						
	Coef.		Std. Err.	z	P>z	[95% Conf.Interval]		Coef.		Std. Err.	z	P>z	[95% Conf.Interval]	
Wheat	-0.073	**	0.0354	-2.06	0.039	-0.142548	-0.003656	0.0059	***	0.0011	5.15	0	0.0036677	0.0081678
Cotton	0.2783		0.5931	0.47	0.639	-0.884074	1.44	0.0466		0.0385	1.21	0.226	-0.028784	0.121977
Sugarcane	0.2023		0.5993	0.34	0.736	-0.97237	1.377013	0.0524		0.0387	1.36	0.175	-0.02338	0.1281856
Rice	0.2534		0.1975	1.28	0.199	-0.133662	0.6404024	0.0046	**	0.0022	2.07	0.039	0.0002418	0.0089842
Maize	0.0407		0.1347	0.3	0.763	-0.223347	0.3046672	0.0033		0.0182	0.18	0.858	-0.032458	0.0389787
Fruits	-0.565	**	0.2994	-1.89	0.059	-1.15	0.0217059	0.009		0.0082	1.1	0.27	-0.00703	0.0251272
Vegetables	0.0625		0.2648	0.24	0.813	-0.456502	0.5814645	0.0274	***	0.0103	2.65	0.008	0.0071501	0.0475827

Significance level denoted by: *** p<0.01, ** p<0.05, * p<0.1; values are presented in percent.

Source: own calculations (2013)

Most of the income simulations are consistent with the choices that are made by farmers with a changing climate. Under all scenarios wheat incomes are declining. With changing climate from 2030 to 2090 a declining trend is also observed fruits and maize, whereas growing trends are found for cotton, sugarcane, rice and vegetables. Significant positive effects are found for rice income, also indicated by the increasing probability of choosing rice over other crops when climate changes from the current normals. After initially declining, by 2090 maize incomes are predicted to increase by 4%. Precipitation effects are all positive for income, showing the importance for of additional water for all crops, particularly for rice, wheat and vegetables. By 2090 rice incomes are expected to increase by 25%, whereas income from fruits is expected to decline by 56% triggered by warming, whereas precipitation changes a likely to lead to income increases for vegetables to the tune of 2.7% and 0.5% for rice and wheat. After combining the results of both modeling stages the expected change in net income is shown for the selected climate scenarios in table 8.6

Table 8.6 Welfare change in Pakistan resulting from different climate scenarios

Farmer response	2030	2060	2090
	Welfare US$/farm	*Welfare US$/farm*	*Welfare US$/farm*
Crops unchanged	-0.966690756	0.142661891	1.261022014
Crop switiching	71.08248059	74.92259725	79.82895438
	%-change	*%-change*	*%-change*
Crops unchanged	-0.14%	0.02%	0.18%
Crop switiching	10.08%	10.62%	11.31%
Value of Adaptation	**70.11578983**	**74.77993536**	**78.56793237**

Note: The expected income at the current climate is 705$/farm.
Crops unchanged assumes no change in cropping patterns, whereas crop switching assumes farmers to adjust cropping patterns to future climate.

Source: own calculations (2013)

For the short-term scenario income from crop farming in Pakistan is expected to drop, whereas in the mid-term and long-term slight increases are predicted. If the farmers stick to their current portfolio of crops, revenues are likely to decrease marginally by 2030. In the mid-term and long term scenarios a slightly positive effect is observed. The second and fourth rows calculate the situation, when farmers do adjust their cropping patterns to match future climate. If they adapt, welfare increases by 10% in the short-term scenario, 10.6% in the mid-term scenario and 11.3% in the long-term scenario. As has been mentioned earlier, the simulations model the change in climate only, assuming other factors to remain stable over the course of the next 80 years. This also includes surface water availability. Under these assumptions table 8.6 shows the value and importance of crop switching as an adaptation measure. Crop switching substantially increases farmers' revenues and thus provides a cushion for other factors that might also change with climate change. Moreover, the fact that farmers, if they do not adapt, are found to have slightly positive revenues in the long-term climate, might be misleading, as the underlying assumption is that everything else remains the same. To see what implications can be expected from a situation, where these other factors change, the following simulation repeatedly exemplifies the importance of adaptation. In the following scenario the assumption is that, farmers keep on growing the same crops that they grow under the current climate, however besides climate also new-equilibriums are assumed for surface water availability. The World Bank predicts Pakistan's water resources to be depleted by 30% - 40% in the long-run (WORLD BANK, 2005). Taking this forecast, the surface water availability proxy variable "mean annual runoff" is reduced by the proposed percentage. Table 8.7 presents the results

Table 8.7 Welfare change in Pakistan - assuming reduced surface water availability

Farmer response	2030	2060	2090
	Welfare US$/farm	*Welfare US$/farm*	*Welfare US$/farm*
Crops unchanged	-2.366609966	-1.525773588	-0.821749152
Crop switiching	69.0792512	72.31127141	76.16965436
	%-change	*%-change*	*%-change*
Crops unchanged	-0.34%	-0.22%	-0.12%
Crop switiching	9.80%	10.26%	10.80%
Value of Adaptation	**66.71264124**	**73.837045**	**76.99140351**

Note: The expected income at the current climate is 705$/farm.
Crops unchanged assumes no change in cropping patterns, whereas crop switching assumes farmers to adjust cropping patterns to future climate.

Source: own calculations (2012)

As expected, reduced surface water availability leads to declines in welfare for all three scenarios. Again, the absolute values do not claim to represent the real and true amount of changes, they however do claim to accurately predict the directional changes. This is specifically important to understand, as the surface water availability is modeled using a proxy variable with its dimension of measurement specified in millimeters, whereas in real world conditions output from hydrological models would provide the data for the true magnitude of flows and their changes. Nevertheless, the estimates approximate the surface water availability variable well enough as the underlying tests of the econometric model have shown. Thus on the basis of these findings it can be inferred that adaptation is of paramount importance to cope with the adverse effects of climate change in future. This becomes clear, when assessing the results in the second simulation, with reduced surface water availability. Considering the five million farms of the country, the estimates reveal the value of adaptation to be as high as 330 and 380 million dollars (2030/2090) for the country. If adaptation does not take place, the losses for the country range between 4 and 12 million dollars. The reduced surface water simulations also confirm the apprehension which was hinted at in the deliberations on the climate change only simulations, namely that although the welfare change in the mid and long- term scenarios was not find to be negative, the changes were very close to zero (no change), thus leaving a very thin room for error or changes in factors other than climate, such as surface water availability. Thus indeed, reducing surface water availability reveals that adaptation is pivotal to cope with climate change, the results in table 8.7 show that although less surface water is available, the adaptations made by the farmers cushion the effects of climate change. Thus although the positive welfare effects are weaker, they are significantly positive. Although the simulations in general assume a change in climate only, the SRES A2 scenarios account for this shortcoming. It is in general hard to imagine the status of Pakistan in 2090; however the IPCC storylines are formulated to capture distinctly different future directions. They try to account for underlying uncertainties by including a range of key future characteristics, including demographic change, economic development, and technological change. "For this reason, their plausibility or feasibility should not be considered solely on the basis of an extrapolation of current economic, technological, and social trends" (NAKICENOVIC et al., 2000). In following the conclusions shall be drawn and possible ways forward shall be elicited.

The following chapter is devoted to conclude the study by providing key findings of the study alongside limitations and future research needs. Where applicable and appropriate, policy implications shall be elicited.

9 Conclusions and Policy Implications

The economic assessment of climate change impacts poses a serious challenge as great temporal and spatial variation can be expected. This was also revealed by the present study. Research with particular emphasis on the economic impacts of climate change on agriculture has been scarce, especially for developing countries, including Pakistan. To fill this gap this study not only contributes to the literature by modeling climate change impacts on the farming sector on a full country scale, but also by endeavoring to depict the adaptations that farmers are most likely to make in the crop sector under differential climatic conditions. The quantification of adaptations plays a central role. A major contribution is made towards the analysis of spatial effects across Agro-Ecological Zones, districts and provinces using a multi-seasonal approach. For the first time both, standard and structural models are applied for a developing country, with the advantage to have a comparison between both approaches and to test for the robustness of the results from a differential modeling perspective. The Standard Ricardian Model for the analysis of climate change impacts on agricultural land values is constructed by using the Pakistan Social and Living Standards Measurement Survey (PSLM) for the farming season 2008/2009. Ultimately, 21.910 agricultural households, which were selected on the basis of land ownership from an original sample of 75.188 households, were aggregated into 110 districts with further differentiation into rainfed and irrigated farms. For the Structural Ricardian Model, which in contrast to the Standard Ricardian Model analyzes net crop revenues, the study uses the community level PSLM survey (2007/2008) covering detailed information on the farming related activities of 2.369 farm households in 79 districts.

The rationale behind using two different dependent variables and datasets lies in testing the sensitivity and consistency of the results. Another advantage of modeling both approaches is that the results of the structural model can be further broken down or disaggregated to understand how sub-vectors behave if altered. For instance, the structural model can help to uncover farmer behavior when certain factors other than climate also vary over time, such as water availability for instance. To the shortcoming of previous studies, surface water availability is included using two different hydrological components, namely runoff and a unique river density index on a district scale. To account for soil heterogeneity a dataset is constructed for each district in the country. Moreover, with specific reference to the country's vulnerability to flooding a metric is introduced to account for the relevance and impact of extreme events on agricultural performance. To capture the demand side effects in the hedonic approach for the first time additionally to the population density variable the Human Development Index (HDI) is introduced as a control. The interesting additional advantage of the HDI over simple income as a demand side indicator is that the HDI also indicates the policy choices of the district (in the modeled case) to some extent and also hints towards a higher capacity of the policy makers towards investing in sustainable undertakings such as education. Additionally, the study contributes to previous research by accounting for infrastructure and technological abundance, in this way reducing the possible bias towards overestimation of the climate sensitivity. By applying a semi-logarithmic function using iteratively reweighted least-squares estimation and specifying land values as the explanatory variable, adaptations outside agriculture are also accounted for in the Standard Ricardian Model. The study also tests the interaction between surface water availability and temperature and models a proxy for solar radiation. As a key contribution, the study provides insights to policy makers of how farmers can be expected to adapt to climate change, this by understanding past adaptations. To account for separate effects for irrigated and rainfed farms, the study uses two separate samples to unveil the specific impacts for each farm type. For the first time a structural Ricardian model is estimated for a South Asian country. Prices, education, household size and irrigation are used to identify the choice equations. The choice model is not based on a choice survey as such, but follows a revealed preference approach using multinomial logit estimation. Another contribution of the study is that it does not only analyze staple crops; by including multiple crops a greater fraction of the agriculture sector is represented. The study covers all major cash crops and food crops of the country. Furthermore, the study only analyzes farms that own their land, this with the intention to reveal the importance of property rights.

The models confirm the high sensitivity of farming in Pakistan to climate, the results also confirm the non-linear relationship between climate and farm income. Thus, increases in climate variables seem to be beneficial to farming up to a certain extent, but beyond a certain limit the impacts will be clearly negative, as suggested by the results. The land value analysis using the Standard Ricardian Model also affirms the importance of estimating separate models for rainfed and irrigated farms. Marginal impact analysis, i.e. assuming a temperature increase by 1°C, for the summer farming season computes a -14% loss in land values for rainfed farms, whereas -12.8% reductions are found for irrigated farms. The gains from a marginal warming in the winter farming season amount to 6.1% for rainfed farms (increase in land value or capitalized net revenue) and 5.5% for irrigated farms. Most importantly the separate analysis of rainfed and irrigated farms reveals that rainfed farms are more vulnerable to climate change, a finding that has also been obtained for Africa (KURUKULASURIYA AND MENDELSOHN, 2008). The marginal precipitation effect in the country's rainy summer season is found to be negative, with -1.77% decreases in rainfed land value and -1.03% in irrigated land value. These results are in line with the functional relationship between precipitation and land values, which is U-shaped, indicating that with every additional millimeter of rainfall the losses will become smaller up until the minimum extreme point, after which land values will start actually increasing.

At the Agro-Ecological-Zone level, the results indicate that in general all zones except for the Northern Dry Mountains Zone and the Wet Mountain/Northern Dry Mountain Zone will experience losses in agricultural revenue. Clearly those zones with the lowest average land values (below 500$ per hectare), are also found to suffer the highest losses from additional warming by 1°C. The dry and hot regions of Sindh and Balochistan comprising the zones Dry Western Plateau, Sulaiman Piedmont and Indus Delta, are found to suffer annual losses in land values of -15% (62 $/ha) , -17.1% (130 $/ha) and -27.76% (132 $/ha), respectively. At the province level, the results show the highest negative impacts from a joint increase in temperature and precipitation for the Sindh province (-15% or -110$/ha annual reduction in land value), followed by the Balochistan province with -8% or -62$/ha, followed by the Punjab province (-4% or -90$/ha) and the NWFP with -1% or -37$/ha. Higher accumulated precipitation in the spring period benefits the country; hence gains in land value from a marginal increase in precipitation range from an additional 0.305% in land value/ha in the Sindh province and 0.139% in the Balochistan province to merely 0.041% in the Punjab province. The highest summer rainfall deficiencies are found for the Balochistan province, followed by the Sindh and Punjab provinces. Different soil types can be important in climate research, as better soils tend to compensate for adverse climatic impacts. Sowing, growing and harvesting periods show different responses to climatic change. Marginally higher temperatures in the earlier parts of the respective growing season (or spring and fall) are found to be beneficial for farming. Longer sowing seasons seem to benefit farmers. Higher technological abundance (tractor density) seems to be relatively more beneficial for land values in rainfed areas. Increasing surface water availability, ceteris paribus, has a significant positive effect on land values, which i.e. are raised by 5 % in rainfed areas with a one unit increase in river density. The sensitivity of the land market to climatic extreme events is confirmed by the results for the flood risk variable. Contrary to the expectations, in Pakistan additional surface water availability does not seem to compensate for higher temperatures in the case of rainfed and mixed farms, this is confirmed by the interaction term between annual mean runoff and annual mean temperature. For the irrigated sample a separate interaction term between irrigated acreage and temperature was introduced. The results indicate that additional irrigation, provided water supplies are available, can compensate for higher temperatures. When warming is simulated further, interesting results are revealed. Winter season benefits by 2090 over-compensate summer season losses, however not in the dryer parts of the country. This is only true if everything else remains the same, when tested with the advanced structural model, a reduction in surface water availability by 30% would clearly be detrimental for farming. In general farmers in Pakistan have well adapted to the current climatic conditions, given climate change and if nothing else changes, they would not have to suffer great losses, however other factors such as water for instance do change. An important result is also that those areas with the lowest land values (capitalized net revenues) are also the most vulnerable ones. This could have serious implications for designing policies for the rural poor.

Farmers are found to adjust their crop choices as warming progresses by considering the expected income resulting from the switch. Thus for instance the two (selection-bias-corrected) estimates for crop choice probability and income effects for the rice crop are analogous; rice income by 2090 is expected to increase by 25% and at the same time the probability model reveals an increasing likelihood for rice to be chosen by the farmers. That farmers decide considering profit has been confirmed by the model. In a World Bank study on Pakistan's water resources it is clearly stated that the major goal of farming in Pakistan also is to have profitability (WORLD BANK, 2005). As climate warms, farmers in Pakistan are more likely to choose rice, vegetables and maize, whereas they move away from wheat, sugarcane, cotton and fruits. These findings are consistent with other study's findings i.e. vegetables and rice in other studies have also been found to be an option for warmer regions (see chapter 7). Rice specifically seems to be of relevance as climate changes. Rice cultivation is characterized by high productivity and sustainability, therefore in the Asian monsoon region it can be expected to play a key role in future (KYUMA, 2009). Different crops are revealed to have specific preferred climate ranges, where they grow best. If the optimal ranges are altered crop productivity either falls or increases, dependent on the change in climate. To this background the importance of accounting for adaptation is crucial when analyzing the impacts of a change in climate on the agriculture sector. Studies that do not model adaptation or consider it endogenous with respect to crop, livestock, farm type choice will overestimate the negative impacts on farming as they assume farmers to continue growing what they have been used to grow. Other Ricardian studies did not find objective characteristics such as education, to play a role in determining farm net revenues (MENDELSOHN AND DINAR, 2009). However for Pakistan these characteristics seem to be of relevance. The results reveal that educated farmers prefer to grow fruits and maize, both of which require incentives by the farmer himself, especially for availing beneficial marketing options. Moreover, larger households are found to favor sugarcane and cotton cultivation, both of which are labor intensive undertakings.

The study reveals that farmers will react to a change in climate by altering crop choices that have the potential to make them economically better off. The welfare calculations based on the selected sample reveal the value of adaptation (through crop switching only) to be as high as 330-380 million dollars over all five million farms. On the other hand, if adaptations in the crop sector do not take place and water availability decreases by 30%, the estimates reveal losses between 4 and 12 million dollars (2030/2090). Again the estimates hint towards the well-directed adaptations of farmers in Pakistan. Moreover, they reveal that adaptation in any case will benefit the farmers, as by adapting they can cushion the adverse effects from unprecedented extreme weather events. These welfare estimates only consider crops and do not include livestock effects, which could be of high relevance as the livestock sector is an important part of agriculture in Pakistan. An additional shortcoming of the study is that due to a comparative static nature of the approach dynamic effects are not accounted for. Furthermore, when farmers switch or adapt from an initial state to different future state, the costs of this transition are not included. As the dataset was of a cross sectional nature the hypothesis whether the estimated Ricardian climate response functions would behave well over time and whether they are stable or not, could not be tested. If panel data is available a study covering inter-annual variation could be conducted in a panel framework. Another weakness of the study is that it could not directly account for the impacts of electricity on land values. From personal communication it was found that electricity shortage constraints not industrial but also agricultural practice in Pakistan. Moreover, global warming is expected to benefit certain regions of the world, while negatively affecting others, in this context trade can play a key role. Detailed models in this direction are clearly needed, especially for assessing the impact of price increases and decreases on a global scale. The frequency of extreme events should also be modeled, when panel data is available. Moreover, detailed hydrological models are needed to accurately assess the development of water availability in future.

In spite of the aforementioned shortcomings, the welfare calculations based on the structural model capture the alteration in the probabilities of choice as well as the impact on conditional income from the chosen crops. From a policy perspective, it is important for policy makers that these adjustments are considered when designing policies. The risk of overestimating the damages predicted by models that do not incorporate adaptations is thereby reduced. Policy should be designed to facilitate these adjustments. Moreover, as the spatial impacts reveal, location specific impacts will require location specific policies. Pakistan not only has a diverse climate, it is also home to a number of different ethnic groups. The provinces are culturally diverse and resource abundance also differs substantially. Intra-provincial water treaties do not exist, therefore water use issues have been of concern lately. On top of that, the provinces claim autonomy from the federal state on various cultural and political disputes. The government therefore has to be careful to design different policies for the diverse regions within the country. Different provinces, different people different sub-cultures, and different institutional capacities of the provincial governments, demand a prudent approach to location specific policy design. Not necessarily farmers exposed to similar climates have to be affected in the same manner, clearly other factors i.e. economic and institutional conditions they face might be of high relevance; this has been partially revealed by including the level of human development in the model, which also sheds light on possible policy choices that have been made in the respective regions. The study only analyzes farms that are owner cultivated, thus property rights for the modeled farmers are secured. The results reveal that farmers who own their land also adapt. To this background policy also has to ensure that landless laborers and tenants are provided land under secure property rights. Land reforms have to be actively enforced to give secure access to the rural poor who are deprived of land ownership. That secured property rights benefit investments directed towards adaptation measures, has been shown in other studies (DEININGER AND JIN, 2006). Policy has to use the existing local traditional knowledge in communities to facilitate location specific adaptation. The design of adaptation strategies has to be based upon local knowledge. Systems that are in place since centuries like the "warabandi" (water sharing) and the "chat" (water storage) have to be included as low cost options for rural communities chapter 5.4), especially considering that not many poor farmers will be able to pay higher prices for water. Having this in mind, this study focused on autonomous adaptation with the objective to use the estimations as a guide for designing policies for a long-term planned ex-ante adaptation. Irrigation is likely to be an important adaptation option, where water is available. In the past the country has successfully used the Indus River for fostering agricultural growth. Policy has contributed to this success over the last fifty to sixty years, however possibly tougher challenges are expected in future, especially for sustaining the benefits from the major life line of the country (DOROSH et al., 2007). The policy role in this context can be to take measures that lead to a more efficient provision of water. For glacial-melt-water fed river systems, the differentiation between long-and short term effects will have to be undertaken. Adaptation can include new cropping-patterns, introduction of heat and drought resistant crop varieties, water harvesting, especially in years where weather developments are indicating water shortages from precipitation. The government can also contribute by providing information to the farmers and rural community about climate change and extreme events, in this way awareness could be raised amongst the wider farming community. For the dissemination of information and successful implementation of adaptation policies, respective agencies, such as agricultural extension services will have to be strengthened, their outreach will have to be significantly increased. A closer cooperation between government agencies and research institutions will have to be established, especially to the background that climate change, agriculture, water availability are all interlinked. Policy makers are called upon to take advantage from benefits and limit negatives, i.e. in some regions clearly winter benefits compensate for summer losses, and this relation has to be considered when designing adaptation policies. The country has shown in the past and more recently that it possesses the potential to adapt to changes in the environment. For instance in parts of the Punjab and Sindh provinces cotton cultivated fields have been replaced by food crops, especially vegetables to counter the effects of extended dry seasons (PINCC, 2003). Several projects are ongoing in the country, that are focusing on the reduction of vulnerability of the rural poor to a multiplicity of adverse environmental impacts (Food Security Programme, Rural Access Programme, Food Security and Early Warning System- for a detailed overview, see WORLD BANK, 2011). Knowledge that is generated from these projects has to be mainstreamed into the wider development agenda for the country; in this sense the findings of this study can provide a useful starting point for understanding the (climate change) vulnerability dimensions of the agriculture sector within and amongst the provinces and the country's diverse agro-climatic settings.

10 German Summary

Die ökonomische Bewertung von Klimaveränderungen stellt eine beachtliche Herausforderung dar, da beträchtliche zeitliche und räumliche Variation zu erwarten sind. Dies kommt auch in der vorliegenden Studie zum Ausdruck. Forschungsarbeiten mit einem spezifischen Schwerpunkt auf die ökonomischen Auswirkungen einer Klimaänderung auf den Landwirtschaftssektor sind rar, vor allem für Entwicklungsländer, einschließlich Pakistan. Um diese Forschungslücke zu schließen, leistet die vorliegende Arbeit nicht nur durch die Modellierung der Auswirkungen von Klimaänderungen auf den Agrarsektor (über alle Kern-Distrikte) einen Beitrag zur Literatur, sondern auch durch die Bestimmung der möglichen Anpassungen der Landwirte, die diese als Reaktion auf eine Klimaänderung in dem Ackerbausektor durchführen. Der Quantifizierung von Anpassungsstrategien kommt dabei eine zentrale Rolle zu. Einen Hauptbeitrag leistet die Studie hinsichtlich der Agrar-Ökologischen Zonen-, Distrikte- und Provinzen übergreifenden Analyse von räumlichen Effekten mittels eines multi-saisonalen (Jahreszeiten) Ansatzes. Erstmalig kommen beide Modelle, standard- und strukturell, für ein Entwicklungsland zum Einsatz, mit dem Vorteil, einen Vergleich zwischen den beiden Ansätzen zu erhalten und die Ergebnisse unter unterschiedlichen Modellierungs-perspektiven Robustheits- und Sensitivitätstests unterziehen zu können. Das Standard Ricardian Modell zur Analyse der Klimaauswirkungen auf Preise von Landwirtschaftsflächen wird mittels der Pakistan Social and Living Standards Measurement Befragung (PSLM) für das Erntejahr 2008/2009 konstruiert. Letztendlich wurden 21910 landwirtschaftliche Haushalte, die auf Basis von Landeigentum aus einer ursprünglichen Stichprobe von 75188 Haushalten ausgewählt wurden, in 110 Distrikte zusammengefasst, mit einer zusätzlichen Unterscheidung zwischen Beregnungs- und Bewässerungsfeldbau. Für das Strukturelle Ricardian Modell, das im Gegensatz zum Standard Ricardian Modell Deckungsbeiträge analysiert, verwendet die vorliegende Arbeit die PSLM Befragung auf Gemeindeebene für das Jahr 2007/2008, welche detaillierte Informationen zu landwirtschaftlichen Aktivitäten von 2369 Haushalten in 79 Distrikten erfasst.

Die Verwendung zweier unterschiedlicher abhängiger Variablen und Datensätze liegt darin begründet, dass sich so die Möglichkeit bietet, die Ergebnisse Sensitivitäts- und Konsistenztests zu unterziehen. Ein weiterer Vorteil, der aus der Modellierung beider Ansätze/ Methoden hervorgeht, liegt darin, dass die Ergebnisse des strukturellen Modells weiter herunter gebrochen bzw.

disaggregiert werden können und somit die Möglichkeit eröffnen nachzuvollziehen, wie einzelne Sub-Vektoren sich verhalten, wenn sie verändert werden. So kann das strukturelle Modell beispielsweise zum Verständnis des Verhaltens von Landwirten auch in solchen Fällen, in denen sich zusätzlich zum Klima auch andere Faktoren über die Zeit verändern, wie z.B. die Wasserverfügbarkeit, beitragen. Zur Ergänzung früherer Studien wird Oberflächenwasserverfügbarkeit in das Modell mittels zweier unterschiedlicher hydrologischer Komponenten, Oberflächenabfluss und einem eigens berechneten Gewässer-Dichte-Index auf Distrikt Ebene, mit-eingebaut. Um die Heterogenität des Bodens zu erfassen, verwendet die Studie einen spezifisch für das Modell konstruierten Datensatz auf Distriktebene. Darüber hinaus, geht in Anbetracht der Vulnerabilität des Landes gegenüber Flutkatastrophen auch eine Variable für Flutrisiko in das Modell mit ein, um so die Relevanz und die Auswirkungen von Extremwetterereignissen auf die landwirtschaftliche Produktion bzw. den Landwert zu bestimmen. Um die nachfrageseitigen Effekte im vorliegenden hedonischen Ansatz zu berücksichtigen, fließt neben der Bevölkerungsdichte auch der Human Development Index (HDI) als Kontrollvariable in die Berechnungen mit ein. Ein deutlicher Vorteil, den der HDI gegenüber einer einfachen Einkommensvariable hat, liegt darin, dass der HDI in gewissem Maße auch Aufschluss über die verschiedenen Politikentscheidungen der einzelnen modellierten Distrikte gibt und gleichzeitig auch auf die unterschiedlichen Handlungskapazitäten der politischen Entscheidungsträger hinsichtlich der Investition in nachhaltige Bereiche, wie z.B. Bildung, hinweist. Des Weiteren leistet die vorliegende Arbeit einen ergänzenden Beitrag zu vorangegangenen Studien, indem sie bei den Berechnungen auch Faktoren wie z.B. Infrastruktur und Technologie mitberücksichtigt und somit einer möglichen Verzerrung hinsichtlich einer Überbewertung der Klimasensitivität vorbeugt.

Durch Anwendung einer semi-logarithmischen Funktion und der Verwendung der Methode der iterativ-neugewichteten kleinsten Quadrate und den Wert des Ackerlandes als abhängige Variable bestimmend, werden auch Anpassungen außerhalb des Agrarsektors im Standard Ricardian Modell mitberücksichtigt. Die Studie testet auch die Interaktion zwischen Oberflächenwasserverfügbarkeit und Temperatur und modelliert auch eine Annäherungsvariable für Sonneneinstrahlung. Als Kernbeitrag versucht die Studie politischen Entscheidungsträgern einen Überblick über die möglichen Anpassungsstrategien, die die Landwirte als Reaktion auf eine Klimaveränderung auf Basis von in der Vergangenheit getätigten Anpassungen in Betracht ziehen könnten, zu verschaffen.

Um die unterschiedlichen Auswirkungen für Beregnungs- und Bewässerungsfeldbau zu analysieren, nutzt die Studie zwei verschiedene Stichproben. In dieser Arbeit findet erstmalig für die Region Süd-Asien ein strukturelles Modell zur Analyse haushaltsspezifischer Anpassungsmaßnahmen eine Anwendung. Landwirtschaftliche Preise, Bildung, Haushaltsgröße und Bewässerung sind dabei Variablen, die zur Schätzung des sogenannten „revealed preference" choice Ansatzes verwendet werden. Dabei basiert das Choice-Modell nicht auf einer choice-Befragung, sondern auf Rekodierung der meist-präferierten Alternative in unterschiedliche Dummy-Variablen. Einen weiteren Beitrag leistet die Studie, indem Sie nicht nur Hauptgetreide analysiert, sondern durch die Einbeziehung von mehreren Feldfrüchten einen größeren Anteil des landwirtschaftlichen Sektors abbildet. So werden sämtliche bedeutende Marktfrüchte (Cash- und Food Crops) des Landes berücksichtigt. Mit der Absicht, die Bedeutung von Eigentumsrechten hervorzuheben, werden bei den Berechnungen in der vorliegenden Arbeit nur Landwirte mit Landeigentum berücksichtigt.

Die Modellberechnungen bestätigen die hohe Klima-Vulnerabilität der pakistanischen Landwirtschaft und verdeutlichen auch den nicht-linearen Zusammenhang zwischen Klima und landwirtschaftlichem Einkommen. Folglich zeigen die Berechnungen, dass Klimaveränderungen sich bis zu einer bestimmten Schwelle vorteilhaft für die Landwirtschaft erweisen können, jedoch über diese Schwelle hinaus, zu deutlichen Einschränkungen führen. Die Ergebnisse aus dem Structural Ricardian Modell bestätigen die Notwendigkeit, separate Analysen für Beregnungs- und Bewässerungsfeldbau durch zu führen. Die Grenzbetrachtung der Klimaveränderung, beispielsweise unter der Annahme eines Temperaturanstiegs in Höhe von 1°C im Mittel für die Sommer-Anabausaison kommt zum Ergebnis, dass landwirtschaftliche Einkommen in Regionen mit ausschließlich Beregnungsfeldbau um -14% einbrechen, wobei sich die Verluste in Regionen mit ausschließlich Bewässerungsfeldbau auf -12.8% belaufen. In der Winter-Anabausaison zeigt das Modell, dass positive Einkommenseffekte bzw. Landpreiseffekte in Höhe von + 6.1% im Durchschnitt für Beregnungsfeldbau und +5.5% für Bewässerungsfeldbau durch einen marginalen Temperaturanstieg möglich sind. Insbesondere zeigt die separate Analyse von beregneten und bewässerten Betrieben, dass der Beregnungsfeldbau klimasensibler reagiert; ein Ergebnis, das auch schon für den afrikanischen Kontext bestätigt wurde (KURUKULASURIYA AND MENDELSOHN, 2008). Der marginale Niederschlagseffekt in den Sommermonaten ist negativ mit -1.77% Verlust im Wert der Landwirtschaftsflächen für Beregnungsbetriebe und -1.03% für Bewässerungsbetriebe. Diese Ergebnisse sind komplementär zu dem geschätzten funktionalen Zusammenhang zwischen Niederschlag und dem Wert der Ackerfläche, welcher U-förmig ist und somit indiziert, dass mit jedem zusätzlichen Millimeter Niederschlag die Verluste der Landwirte marginal betrachtet kleiner werden (wenn auch absolut negativ) bis zu dem Punkt, ab dem sie auch absolut zu Gewinn werden.

Auf der Ebene der Agrar-Ökologischen-Zonen zeigen die Ergebnisse, dass im Allgemeinen in allen Zonen die Landwirte Einkommenseinbußen durch eine marginale Klimaveränderung werden hinnehmen müssen. Auch wird deutlich, dass Agrar-Ökologische-Zonen mit dem durchschnittlich niedrigsten Wert der Ackerfläche (<500$/ha) auch gleichzeitig die größten Verluste durch einen marginalen Temperaturanstieg hinnehmen müssen. Die trockenen und heißen Regionen der Sindh- und Balochistan Provinzen, die die Agrar-Ökologischen-Zonen Dry Western Plateau, Sulaiman Piedmont und Indus Delta umfassen, werden den Berechnungen zu Folge mit entsprechenden Verlusten in Höhe von -15% (62 $/ha), -17.1% (130 $/ha) and -27.76% (132 $/ha), rechnen müssen. Eine marginale Klimaveränderung (1°C Temperaturanstieg und 1mm Anstieg des Niederschlags im Mittel) zeigt die größten negativen Effekte für die Sindh Provinz (-15% bzw. -110 $/ha jährliche Reduktion des Wertes der Ackerfläche), gefolgt von den Provinzen Balochistan mit -8% bzw. -62 $/ha, Punjab mit -4% bzw. -90 $/ha und NWFP mit -1% bzw. -37 $/ha. Zusätzlicher akkumulierter Niederschlag in der Frühlingssaison wirkt sich positiv auf das Land aus, so sind Anstiege des durchschnittlichen Wertes der Ackerfläche je Hektar in Höhe von +0.305%, +0.139% und +0.041% in den Provinzen Sindh, Balochistan und Punjab möglich. Marginal höhere Temperaturen in den Anfangsphasen der jeweiligen Anbausaison sind vorteilhaft für die Landwirtschaft in Pakistan. Längere Aussaatperioden liefern den Landwirten mehr Optionen. Technologische Abundanz (gemessen als Traktoren-Dichte) wirkt sich vor allem auf Regionen mit ausschließlich Beregnungsfeldbau positiv aus. Höhere Oberflächenwasserverfügbarkeit, ceteris paribus, hat gemäß der Modellergebnisse einen positiven Effekt auf die Preise von Landwirtschaftsflächen und folglich auch auf das Einkommen. Weiterhin wird durch die Schätzungen unter Einbeziehung der Flut-Risiko-Variablen auch die Reagibilität der landwirtschaftlichen Flächenmärkte auf Extremwetterereignisse bestätigt. Wie der Interaktionsterm zwischen Oberflächenwasserverfügbarkeit und Temperatur zeigt, führt eine höhere Oberflächenwasserverfügbarkeit nicht zum Ausgleich der negativen Temperatureffekte, insbesondere für den Beregnungsfeldbau und gemischte Anbauformen. Für die Stichprobe der Betriebe mit Bewässerungsfeldbau wurde ein separater Interaktionsterm zwischen bewässerter Anbaufläche und der Jahresdurchschnittstemperatur eingeführt. In diesem Fall zeigen die Ergebnisse, dass zusätzliches Wasser, das zum Bewässern genutzt werden kann, wenn es denn verfügbar ist, den negativen Temperatureffekt kompensiert.

Wird die Erwärmung weiter simuliert (2030, 2060, 2090), ergeben sich zusätzliche hoch interessante Ergebnisse. So überkompensieren z.B. die positiven Effekte im Winter die negativen Effekte aus der Sommer-Anbausaison. Dies ist jedoch nur unter der restriktiven Annahme, dass alles Andere (u.a. Wasser, sozio-ökonomische Faktoren) außer dem Klima konstant bleibt, der Fall. Wird mittels des strukturellen Modells z.B. die Oberflächenwasserverfügbarkeit um 30% (laut Weltbank Szenario) reduziert, hat dies durchweg negative Effekte für die Landwirtschaft zur Folge. Ferner zeigen die Ergebnisse auch, dass sich die Landwirte in Pakistan über die Jahre verhältnismäßig gut an die vorherrschenden klimatischen Gegebenheiten angepasst haben. Sollte sich nur das Klima verändern und die restlichen modellierten Variablen blieben konstant, müssten die Landwirte keine großen Verluste hinnehmen. Dass die anderen modellierten Faktoren, wie z.B. Wasserverfügbarkeit sich ändern, daran besteht laut aktuellem wissenschaftlichen Stand kein Zweifel. Auch bestätigen die Ergebnisse, dass die ärmsten Regionen, die am stärksten betroffenen sind. Dieses Ergebnis ist von hoher Relevanz für politische Entscheidungsträger und den Entwurf von Politikmaßnahmen für die wirtschaftlich Schwächeren in den ländlichen Regionen.

Die Ergebnisse des Choice-Modells bestätigen die Annahme, dass Landwirte unter Berücksichtigung des zu erwartenden Einkommens, das aus einem Fruchtwechsel resultieren könnte, ihre Anbauentscheidungen treffen bzw. anpassen. So sind beide Ergebnisse (Auswahlwahrscheinlichkeit einer Fruchtart und resultierender Einkommenseffekt-korrigiert hinsichtlich Selektionsfehler zur Gewährleistung unabhängiger Schätzungen) z.B. für die Reisfrucht analog. Das Einkommen aus Reisanbau steigt laut den Simulationen in 2090 um +25% an; gleichzeitig zeigt das Choice-Modell, dass auch die Wahrscheinlichkeit, dass Landwirte Reis auswählen, ebenfalls ansteigt. Zusätzlich bestätigt das Modell, dass Landwirte auch im Entwicklungsland Pakistan nach Profitabilitätskriterien entscheiden. Dies wird auch in einer von der Weltbank veröffentlichten Studie deutlich (WORLD BANK, 2005). Mit einer zusätzlichen Erwärmung präferieren Landwirte in Pakistan Reis, Gemüse und Mais als Hauptfrüchte, wobei sie sich weniger für Weizen, Zuckerrohr, Baumwolle und Obst entscheiden. Ähnliche Ergebnisse finden sich auch in Studien für andere Länder und Kontinente (siehe

Kapitel 7). Vor allem Reis scheint von hoher Bedeutung zu sein, wenn das Klima sich erwärmt. Der Reisanbau ist durch hohe Produktivität und Nachhaltigkeit gekennzeichnet, daher ist zu erwarten, dass er in naher Zukunft für die asiatische Monsoon-Region eine noch bedeutendere Rolle spielen wird (KYUMA, 2009).

Verschiedene Feldfrüchte zeigen unterschiedliche klimatische Präferenzen, bei denen sie am besten gedeihen. Wenn sich diese optimalen Bedingungen jedoch ändern, fällt bzw. steigt auch die Frucht-Produktivität in Abhängigkeit von dem Ausmaß und der Richtung der Veränderung. Vor diesem Hintergrund ist es essentiell, Anpassungen bei der Analyse der Auswirkungen von Klimaveränderungen auf den Landwirtschaftssektor zu berücksichtigen. Solche Studien, bei denen autonome Anpassungen (z.B. Wahl der Ackerfrucht, Spezies etc.) nicht in die Berechnungen der ökonomischen Auswirkungen des Klimawandels auf den Agrarsektor einfließen, tendieren dazu, die negativen Effekte einer Klimaveränderung über zu bewerten, da sie annehmen, dass Landwirte auch in Zukunft weiterhin das gewohnte anbauen werden. In anderen Studien konnte ein Zusammenhang zwischen objektiven Charakteristika, wie z.B. Bildung, Haushaltsgröße und landwirtschaftlichem Einkommen nicht nachgewiesen werden (MENDELSOHN AND DINAR, 2009).

Für Pakistan scheinen diese Charakteristika jedoch von hoher Relevanz zu sein. So zeigen die Ergebnisse, dass gebildete Landwirte eher Mais und Obst anbauen. Gerade Mais- und Obstanbau in Pakistan verlangen hohe Eigeninitiative durch den Landwirt, um Zugang zu vorteilhaften Vermarktungskanälen zu erhalten. Größere Haushalte wählen vor allem Zuckerrohr oder Baumwolle, beide sehr arbeitsintensive Feldfrüchte, als Hauptfrucht. Die Studie zeigt, dass Landwirte in Pakistan ihre Anbauentscheidungen unter Berücksichtigung der ökonomischen Faktoren treffen und eher solche Optionen vorziehen, die das Potenzial besitzen eine Verbesserung der wirtschaftlichen Lebenssituation zu bewirken. Die Wohlfahrtsberechnungen zeigen, dass der monetäre Wert der Anpassung (nur durch Fruchtwechsel) sich in etwa auf 330-380 Mill.$ (2030-2090) über alle fünf Millionen pakistanischen landwirtschaftlichen Betriebe beläuft. Findet keine Anpassung statt und die Wasserverfügbarkeit sinkt um 30%, dann entstehen Wohlfahrtsverluste zwischen 4 und 12 Mill.$. Auch anhand dieser Berechnungen auf Basis des strukturellen Modells wird deutlich, dass die Landwirte in Pakistan sich an die teils extremen Bedingungen im Land gut angepasst haben. Gleichwohl wird deutlich, dass die Landwirtschaft von Anpassungen stark profitieren und sich somit auch eine finanzielle Sicherheit schaffen kann, um auf unvorhersehbare Extremwetterereignisse reagieren zu können. Es muss auch berücksichtigt werden, dass in die Berechnungen nur der Feldfruchtbau einfließt und keine Betrachtung des Nutztiersektors, welcher einen wichtigen Teil der pakistanischen Landwirtschaft ausmacht.

Ein Schwachpunkt der Studie ist, dass aufgrund der komparativ-statischen Natur des Modellierungsansatzes keine dynamischen Effekte simuliert werden können. Darüber hinaus werden auch keine Transaktionskosten, die entstehen, wenn der Landwirt seine Anbauentscheidung von Situation A (z.B. Weizen) zur Situation B (z.B. Obst) anpasst, berücksichtigt. Da der Datensatz einen Querschnitt zeigt und kein Panel verfügbar ist, konnte die berechnete Klima-Resonanz-Funktion nicht auf ihre Robustheit über die Jahre getestet werden. Wenn Panel-Daten zur Verfügung stehen, kann das Modell auch interannuelle Variationen berechnen. Des Weiteren konnte auch nicht der Einfluss von einer eingeschränkten Elektrizitätsverfügbarkeit auf den Landpreis als Kontrollvariable, mangels Datenverfügbarkeit, in das Modell einfließen. Durch persönliche Gespräche mit Wissenschaftlern vor Ort, wurde in Erfahrung gebracht, dass die schlechte Elektrizitätsinfrastruktur nicht nur die industrielle- sondern auch die landwirtschaftliche Produktion in Pakistan stark einschränkt. Da der Klimawandel manche Regionen besser stellen kann und somit auch positive Preiseffekte neben negativen anderenorts entstehen können, ist es essentiell, internationale Handelsströme in Hinblick auf zukünftige Entwicklungen zu analysieren, damit auch der Schwachpunkt des Modells, dass Preise in der Simulation konstant bleiben, gelöst werden kann. Auch gilt es, in weiteren Arbeiten die Häufigkeit von Extremwetterereignissen und deren Auswirkungen über die Zeit zu modellieren, auch hierfür sind Panel-Daten unerlässlich. Zur Analyse der Entwicklung der Wasserverfügbarkeit in Zukunft bedarf es detaillierter hydrologischer Modelle. Trotz der angesprochenen Schwachpunkte ist das Modell in der Lage, bei den Berechnungen der Wohlfahrteffekte auf Basis des strukturellen (choice) Ansatzes sowohl die Änderungen in den Auswahlwahrscheinlichkeiten einer bestimmten Ackerfrucht als auch die Auswirkungen auf das konditionale Einkommen in Abhängigkeit der gewählten Anbaufrucht zu berücksichtigen. Hinsichtlich der Formulierung von Handlungsstrategien zur Anpassung and den Klimawandel ist es essenziell, diese unterschiedlichen autonomen Anpassungen seitens der Landwirte einzubeziehen. Insbesondere vor dem Hintergrund, dass das Modell durch Einbeziehung des Anpassungsverhaltens bei der ökonomischen Bewertung der Auswirkungen von Klimaveränderungen das Risiko einer Überbewertung des Einflusses der Klimavariablen reduziert. Politikmaßnahmen sollten demnach auf die Förderung solcher Anpassungsstrategien zugeschnitten werden. Darüber hinaus ist es essenziell, wie auch die Modellergebnisse zeigen, ortsspezifische bzw. regionsspezifische Maßnahmen zu entwickeln. Nicht nur hat Pakistan ein diverses Klima, es beheimatet auch zahlreiche teils sehr unterschiedliche Ethnien. Die Provinzen sind kulturell divers und auch ihre Ausstattung mit natürlichen Ressourcen variiert sehr stark. Zwischen den Provinzen existieren keine Wasserabkommen, daher sind Wasserkonflikte im Land keine Seltenheit. Hinzu kommt, dass sich die Provinzen nach Unabhängigkeit vom Staat sehnen und aufgrund zahlreicher meist historisch bedingter kultureller und politischer Differenzen einen verstärkten Drang nach Autonomie hegen. Diese Umstände verlangen, dass Anpassungsmaßnahmen in vielerlei Hinsicht auf die Regionen zugeschnitten sind. Die Studie verdeutlicht auch, dass nicht alle Landwirte, die ähnlichen klimatischen Verhältnissen ausgesetzt sind, auch die gleiche Klima-Vulnerabilität zeigen; vor allem sozio-ökonomische und institutionelle Grundbedingungen scheinen beim Ausmaß der Vulnerabilität eine entscheidende Rolle zu spielen. Dies wird auch durch die Modellergebnisse unter Berücksichtigung der unterschiedlichen regionalen Entwicklungsstufen mittels des HDI bestätigt.

Die Studie bestätigt, dass sich Landwirte mit Landbesitz an den Klimawandel anpassen. Dabei spielen gesicherte Eigentumsrechte eine zentrale Rolle. In Anbetracht dieses Modellergebnisses gilt es, Maßnahmen zur Zuteilung von Land an Wanderarbeiter und Sharecropper noch zielgerichteter als bisher voran zu treiben. Dabei muss sichergestellt werden, dass die andauernden Landreformen noch stärker greifen und dazu beitragen, dass den landlosen Bevölkerungsteilen im Landwirtschaftssektor sicherer Zugang zu Land gewährt wird. Dass sichere Eigentumsrechte sich auch positiv auf Investitionen in Anpassungsmaßnahmen auswirken, wurde auch in anderen Studien deutlich (DEININGER AND JIN, 2006). Um Orts- und regionsspezifische Maßnahmen zu entwickeln, müssen Entwicklungsländer noch intensiver auf vorhandenes Wissen zurückgreifen. Traditionelles Wissen bzw. langjährig erprobte Maßnahmen wie z.B. das „Warabandi“ (Wasser teilen) und das Chatt (Wasser speichern) System, müssen als low-cost Optionen in den Anpassungsmaßnahmen-Katalog aufgenommen werden. Dies ist wichtig, da nicht alle Landwirte unter gegebenen finanziellen Verhältnissen in der Lage sein werden, höhere Preise z.B. für Wasser zu zahlen. Die vorliegende Arbeit leistet somit einen Beitrag zum Verständnis autonomer Anpassung, welcher dazu nutzen soll, langfristige ex-ante Anpassungsstrategien für die einzelnen Regionen zu entwickeln. Bewässerung scheint eine wichtige Anpassungsmaßnahme zu sein, insbesondere in Regionen, die auf ausreichende Wasservorräte zurückgreifen können. In der Vergangenheit hat das Land den Indus erfolgreich für die

Förderung der Landwirtschaft genutzt. Auch die Politik hat in den letzten fünfzig bis sechzig Jahren sicherlich einen Beitrag dazu geleistet, jedoch stehen in Zukunft weitaus härtere Zeiten bevor, insbesondere hinsichtlich der nachhaltigen Nutzung der Lebenslinie des Landes (Indus) (DOROSH ET AL., 2007). Die Rolle der Politik ist es in diesem Kontext, Maßnahmen für eine effizientere Wassernutzung zu entwickeln. Vor allem für derartige Gletscher-gespeiste hydrologische Systeme wird es umso wichtiger sein, zwischen kurz- und langfristigen Effekten zu differenzieren. Im Allgemeinen kann Anpassung auch die Einführung neuer Fruchtfolgen, hitzetoleranter Sorten, Water-Harvesting etc. beinhalten. Die Politik hat auch zur Aufgabe, die Bevölkerung über Extremwetterereignisse und mögliche Gefahren in Kenntnis zu setzen und einen kontinuierlichen Informationsfluss zu gewährleisten. So gilt es langfristig zu einem ausgeprägteren Klimabewusstsein innerhalb der Bevölkerung beizutragen. Zur Kommunikation und Umsetzung der Maßnahmen müssen landwirtschaftliche Beratungsinstitutionen gestärkt- und deren Reichweite verbessert werden. Vor allem sollte es Ziel sein Politik und Wissenschaft zu vernetzen, damit die multidimensionalen Erkenntnisse zur Entwicklung von gebündelten Maßnahmenpaketen (Wasser, Landwirtschaft, Klima etc.) genutzt werden können. Ferner sind politische Entscheidungsträger aufgerufen, Nutzen aus den positiven Effekten einer Klimaveränderung zu ziehen und negative Auswirkungen zu begrenzen bzw. diesen vorzubeugen. Diesbezüglich zeigen die Ergebnisse beispielsweise, dass in manchen Regionen die nachteiligen Auswirkungen einer Klimaveränderung im Sommer durch positive Auswirkungen im Winter kompensiert bzw. teils auch überkompensiert werden.

Pakistan hat in der Vergangenheit, aber auch jüngst, gezeigt, dass es über das nötige Potential verfügt sich an Umweltveränderungen anzupassen. So wurde z.B. in den Provinzen Sindh und Punjab der Baumwollanbau durch den Anbau von Marktfrüchten zur Nahrungsmittelproduktion ersetzt, vor allem durch Gemüse, um dadurch den Auswirkungen von längeren Dürreperioden entgegenzuwirken (PINCC, 2003). Zusätzlich laufen derzeit zahlreiche Projekte im Land, die eine Reduktion der Vulnerabilität der ärmeren Landbevölkerung gegenüber Umweltveränderungen zum Ziel haben (Food Security Programme, Rural Access Programme, Food Security and Early Warning System- siehe WORLD BANK, 2011 für einen detaillierten Überblick). Wissen, das durch derartige Projekte generiert wird, sollte in die weitere Entwicklungsagenda für das Land einfließen. In diesem Sinne können die Ergebnisse der vorliegenden Studie zu einem besseren Verständnis der landes-, Agrar-Ökologische-Zonen- und regionsspezifischen Vulnerabilitätsausmaße im Agrarsektor (speziell Ackerbau) beitragen.

List of References

AC (2000): Agricultural Census 2000, Pakistan Report, Government of Pakistan, Statistics Division, Agricultural Census Organization, Lahore, Pakistan, 2000

AC (2011): Agricultural Statistics of Pakistan 2010-2011. Government of Pakistan, Pakistan Bureau of Statistics, Islamabad

Adams, R.M.; Hurd, S.L. and Leary, N. (1998): Effects of global climate change on agriculture: an interpretative review. Climate Research, Vol. 11, pp. 19-30

Adams, R.; McCarl, B.; Segerson, K.; Rosenzweig, C.; Bryant, K.; Dixon, B.; Conner, R.; Evenson, R.; Ojima, D. (1999): The Economic Effects of Climate Change on US Agriculture- in Mendelsohn, R. and Neumann, J. (eds) (1999): The Impact of Climate Change on the United States Economy, Cambridge University Press, Cambridge, UK, 1999

Adler, R.F.; Huffman, G.J.; Chang, A. et al. (2003): The Version-2 Global Precipitation Climatology Project (GPCP) monthly precipitation analysis (1979-present). Journal of Hydrometeorology, Vol. 4, No. 6, pp. 1147-1167

Agrawal, A. (2008): The role of local institutions in adaptation to climate change. Paper prepared for the Social Dimensions of Climate Change, Social Development Department, The World Bank, Washington D.C.

Ahmad, I.; Zhaobo,S.; Weitao, D.; Ambreen, R. (2003): Trend Analysis of January Temperature in Pakistan over the Period of 1961-2006: Geographical Perspective, Pakistan Journal of Meteorology Vol. 7, Issue 13, pp. 11-22

Ahmad, M. (1985): Soils of Pakistan. Proceedings of XII International Forum on Soil Taxonomy and Agrotechnology Transfer, October 9- 23, Pakistan 1985, pp. 47-57

Ahmad, N. (2005): Sustainable crop productivity through balanced and integrated plant nutrient management: Pakistan experience. 2005 IFA regional conference for Asia and the Pacific Singapore, 6-8 December 2005

Ahmed, M.N. (2010): TFP Estimation for Pakistan-the Importance of the Collective Infrastructure to Feed the Poor – Poster selected for presentation at the Tropentag — World Food System —A Contribution from Europe, September 14-16, 2010, Zurich

Ahmed, M. N. and Schmitz, P. M. (2011a): Economic Assessment of the Impact of Climate Change on the Agriculture of Pakistan. Business & Economic Horizons, BEH, Vol.4, Issue 1, Prague, 2011, pp. 1-12

Ahmed, M. N. and Schmitz, P. M. (2011b): Using the Ricardian Technique to estimate the impacts of Climate Change on Crop Farming in Pakistan-European Association of Agricultural Economists 2011 International Congress, August 30-September 2, 2011, Zurich, Switzerland, published online on AgEconSearch, http://purl.umn.edu/114217

Ainsworth, E.A. and Rogers, A. (2007): The response of photosynthesis and stomatal conductance to rising [CO2]: mechanisms and environmental interactions- Plant, Cell and Environment, Vol. 30, pp. 258-270

Ainsworth, E.A.; Long, S.P. (2005): What have we learned from 15 years of free-air CO2 enrichment (FACE)? A meta-analytic review of the responses of photosynthesis, canopy properties and plant production to rising CO2. New Phytologist Vol.165, pp. 351–371

Alam, M.T. (2009): Impact of Climate Change over Pakistan. WMO Technical Conference preceding CAS XV. Environmental Prediction into the Next Decade: Weather, Climate, Water and the Air We Breathe Incheon, Republic of Korea, 16-17, November 2009

Ali, G., Hassan, S. and Khan, A.M. (2009): Climate Change: Implications and Adaptation of Water Resources in Pakistan, Research Report No.GCISC-RR-13, Global Change Impact Studies Centre, Islamabad

Allen (Jr.), L.H. (1991): Managing Water Resources in the West Under Conditions of Climate Uncertainty: A Proceedings. Committee on Climate Uncertainty and Water Resources Management, National Research Council. Washington, DC: The National Academies Press, pp.101-147, 1991

Allen, R. G.; Pereira, L. S.; Raes, D. and Smith, M. (1998): Crop evapotranspiration: Guidelines for computing crop water requirements. FAO Irrigation and Drainage Paper No. 56, Rome

Amemiya, T. (1981): Qualitative response models: A survey. Journal of Economic Literature, Vol. 19, pp. 1483–1536

Amiraslany, A. (2010): The Impact of Climate Change on Canadian Agriculture: A Ricardian Approach. PhD Thesis, University of Saskatchewan, Canada

Anderegg, W.R.L.; Prall, J.W.; Harold, J. and Schneider, S.H. (2010): Expert credibility in climate change. Social Sciences - Environmental Sciences: PNAS- Proceedings of the National Academy of Sciences of the USA, Vol. 107, No. 27, pp. 12107-12109

Anselin, L.; Syabri, I. and Kho, Y. (2006): GeoDa: An Introduction to Spatial Data Analysis. Geographical Analysis, Vol. 38, No. 1, pp. 5-22

APO (2008): Total Factor Productivity –a note by the Asian Productivity Organization, Japan, www.apo-tokyo.org , 2008

Aquastat (2011): Country profile Pakistan. Rome: Food and Agriculture Organization

Archambault, D.J. (2007): Efficacy of Herbicides under Elevated Temperature and CO2. Agroecosystems in a Changing Climate.CRC, Press, pp.336-336

Archer, D. R.; Forsythe, N.; Fowler, H. J. and Shah, S. M. (2010): Sustainability of water resourcesmanagement in the Indus Basin under changing climatic and socio economic conditions. Hydrology and Earth System Sciences, European Geosciences Union Copernicus Publishing, Vol. 14, pp. 1669–1680, 2010

Arriagada, L.E.S. (2005): Optimal Crop Choice: Farmer Adaptation to Climate Change. Conference Paper for the American Economic Association Annual Meeting, 6-8 January, Boston, USA

Ashfaq, M.; Abid, M.; Bakhsh, K. and Fatima, N. (2012): Analysis of Resource use efficiencies and return to scale of Medium sized Bt cotton farmers in Punjab, Pakistan. Sarhad Journal of Agriculture, Vol.28, No.3, pp. 493-498

Aslam, M. and Prathapar, S. A. (2006): Strategies to mitigate secondary salinization in the Indus Basin of Pakistan: A selective review. Research Report 97. Colombo, Sri Lanka: International Water Management Institute (IWMI).

Aujla, K.M.; Abbas, M.; Mahmood, K. and Saadullah, S. (2007): Marketing System of Fruits, Margins and Export Potential in Pakistan. Pakistan Journal of Life and Social Sciences, Vol. 5, pp. 34-39

Badar, H.; Mohy-ud-Din, Q. and Ali, T. (2007): An analysis of domestic support to agriculture sector in Pakistan under WTO regime. Pakistan Journal of Agricultural Sciences, Vol 44, No. 4, 2007, pp. 679-683

Bakhsh, K. (2007): An Analysis of technical efficiency and profitability of growing potato, carrot, radish & bitter gourd: a case study of Pakistani Punjab. PhD-Thesis. Department of Environmental & Resource Economics, University of Agriculture, Faisalabad, Pakistan, 2007

Banuri, T. (1998): Pakistan: Environmental Impact of Cotton Production and Trade. International Institute for Sustainable Development, Manitoba, Canada, 1999

Bardhan, P. (1989): The Economic Theory of Agrarian Institutions. Oxford University Press, 1989

Barnett, V. and Lewis, T. (1978): Outliers in Statistical Data. John Wiley & Sons, New York

Barry, R.G. and Chorley, R.J. (1987): Atmosphere. Weather and Climate. 5th ed. New York, Routledge

Barry, S.; Burton, I.; Klein, R. and Wandel, J. (2000): An Anatomy of Adaptation to Climate Change and Variability. Climatic Change, Vol. 45, pp. 223-51

Benhin, J. (2008): South African Crop Farming and Climate Change-an economic assessment. Global Environmental Change, Vol.18, pp. 666-678

BOI (2007): Pakistan Agriculture Sector. Investment information Guide, Board of Investment Government of Pakistan, Islamabad, 2007

Bourguignon, F.; Fournier, M. and Gurgand, M. (2004): Selection Bias Corrections Based on the Multinomial Logit Model: Monte-Carlo Comparisons, DELTA Working Papers, No.20

Bourguignon, F.; Fournier, M. and Gurgand, M. (2007): Selection Bias Corrections Based On The Multinomial Logit Model: Monte Carlo Comparisons. Journal of Economic Surveys, Wiley Blackwell, Vol. 21, No. 1, pp. 174-205

Bretschger, L. (1999): Knowledge Diffusion and the Development of Regions. Annals of Regional Science, Vol. 33, pp. 251-268

Budyko, M. I. (1974): Climate and Life. Elsevier, New York, pp. 321

Burke, J.J. and Wanjura, D.F. (2009): Plant responses to temperature extremes. pp. 123-128. In:J.M. Stewart, D.M. Oosterhuis, J.J. Heitholt and J.R. Mauney (eds.) Physiology of Cotton, Springer, New York

CAIT (2011). Climate Analysis Indicators Tool version 8.0.,Washington, DC: World Resources Institute, 2011. Available at http://cait.wri.org

Calzadilla,A.; Rehdanz, K.; Betts, B.; Falloon,,P.; Wiltshire, A. and Tol, R.S.J. (2010): Climate Change Impacts On Global Agriculture. Working Paper FNU-185, Forschungsstelle Nachhaltige Umweltentwicklung, Universität Hamburg, April 2010

CBSE (2006): Natural Hazards and Disaster Management. Supplementary Textbook in Geography. UNIT 11 : Natural Hazards and Disasters. Central Board of Secondary Education, Preet Vihar, Delhi, pp.1-10

CEEPA (2002): Centre for Environmental Economics and Policy in Africa http://www.ceepa.co.za/organization.html

Centritto, M. and Loreto, F. (2005): Preface. Photosynthesis in a changing world: Photosynthesis and Abiotic Stresses. Agriculture Ecosystems & Environment, Vol. 106, pp. 115-117

Chambwera, M. and Stage, J. (2010): Climate change adaptation in developing countries: issues and perspectives for economic analysis, International Institute for Environment and Development, UK, May 2010

Chan, N.Y.; Ebi, K.L.; Smith, F.; Wilson, T.F. and Smith, A.E. (1999): An Integrated Assessment Framework for Climate Change and Infectious Diseases. Environmental Health Perspectives, Vol. 107, pp. 329-337

Chatterjee, S. and Price B. (1991): Regression Analysis by Example. 2nd ed.,Wiley, New York

Chaudhry, M.G. and Chaudhry, G.M. (1997): Pakistan's Agricultural Development since Independence: Intertemporal Trends and Explanations – Pakistan Development Review, Vol. 36, No.4, Part II, pp. 593-612

Chaudary Q.Z., Sheik M.M.,Bari A.,Hayat A. (2001): History's Worst Drought Conditions Prevailed over Pakistan, Pakistan Metrological Department, Islamabad, PMD

Chaudhary, Q. Z. (1994): Pakistan summer monsoon rainfall's association with global and regional circulation features and its seasonal prediction. Proceeding of International Conference on Monsoon Variability and Prediction, Triest, Italy, May 9-13, pp. 136-145

Chaudhry, Q. Z.; Mahmood, A.; Rasul, G. and Afzaal, M. (2009): Climate Change Indicators of Pakistan, Pakistan Meteorological Department, Technical Report No. PMD. 22/2009, Islamabad

Chow, G. (1983): Econometrics. McGraw-Hill, New York

CIA (2011): The World Factbook 2011. Washington, DC: Central Intelligence Agency, 2011

Cline, W. R. (1996): The Impact of Global Warming on Agriculture: Comment. American Economic Review Vol. 86, pp. 1309-1312

Cline, W. R. (2007): Global Warming and Agriculture: Impact Estimates by Country. Center for Global Development and Peterson Institute for International Economics, Washington

Collier, P.; Conway, G. and Venables, T. (2008): Climate Change and Africa. Oxford Review of Economic Policy, Vol. 24, No. 2, pp. 337-53

Cook, R. D. and Weisberg, S. (1982): Residuals and Influence in Regression. Chapman and Hall, London

Cox, M. (2002): Land tenure and rural Development. FAO Land Tenure Studies Vol. 3 , Food and Agriculture Organization of the United Nations, Rome, Italy

Critchley , W. and Siegert, K. (1991): A Manual for the Design and Construction of Water Harvesting Schemes for Plant Production. Food and Agriculture Organization of the United Nations - Rome, 1991

Cruz, R.V., Harasawa H., Lal M., Wu S., Anokhin Y., Punsalmaa B., Honda Y., Jafari M., Li C. and Huu Ninh N. (2007): Asia. Climate Change 2007: Impacts, Adaptation and Vulnerability. Contribution of Working Group II to the Fourth Assessment Report of the Intergovernmental Panel on Climate Change, Parry M L, Canziani O F, Palutikof J P, van der Linden, P. J. and Hanson, C. E. (eds). Cambridge University Press. Cambridge, UK. pp. 469 – 506

Curran, C. (2010): Sample Selectivity Bias. The Connexions Project, module m34547, www.cnx.org

Darwin, R. (1999): The Impact of Global Warming on Agriculture: A Ricardian Analysis: Comment" American Economic Review Vol. 89, pp. 1049-1052

Darwin, R.; Tsigas, M.; Lewandrowski, J.; Raneses, A. (1995): World agriculture and climate change: economic adaptations. Agricultural Economic Report No. 703. Natural Resources and Environmental Division, Economic Research Service, U.S. Department of Agriculture, Washington, DC

Deininger, K. and Jin, S. (2006): Tenure security and land related investment: evidence from Ethiopia. European Economic Review, Vol. 50, No. 5, pp. 1245-1277

Deke, O.; Hooss, K.G.; Kasten, C. ; Klepper, G. and Springer, K. (2001): Economic Impact of Climate Change: Simulations with a Regionalized Climate-Economy Model. Kiel Working Papers 1065, Kiel Institute for World Economics

Deschenes, O. and Greenstone, M. (2007): The Economic Impacts of Climate Change: Evidence from Agricultural Output and Random Fluctuations in Weather, American Economic Review, Vol. 97, No. 1, pp. 354-385

Dinar, A.; Hassan, R. and Mendelsohn, R. (2008): Climate Change and Agriculture in Africa: Impact Assessment and Adaptation Strategies. EARTHSCAN, Taylor and Francis Publishing

Dore, M.H.I. (2005): Climate change and changes in global precipitation patterns: What do we know? Environment International, Vol.31, pp. 1167–1181

Dorosh, P. and Salam, A. (2007): Distortions to Agricultural Incentives in Pakistan. Agricultural Distortions Working Paper 33, The World Bank, Washington D.C., 12/2007

Dorosh, P. and Valdes, A. (1990): Effect of Exchange Rate and Trade Policies on Agriculture in Pakistan. International Food Policy Research Institute, IFPRI Research Report No. 84, Washington D.C., USA

Dorosh, P. and Abdul, S. (2006): Wheat Markets and Price Stabilisation in Pakistan: An Analysis of Policy Options, MPRA Working Paper No. 2244, November 2007, Munich, Germany

Dowlatabadi, H. and Morgan, M.G. (1995): A Model Framework for Integrated Assessment of theClimate Problem. Mmimeo, Department of Engineering and Public Policy, Carnegie Mellon University 1995

Dubin, J. and McFadden, D. (1984): An Econometric Analysis of Residential Electric Appliance Holdings and Consumption, Econometrica Vol. 50, pp. 345-362

Eitzinger, J.; Utset, A.; Trnka, M.; Zalud, Z.; Nikolaev, M. and Uskov, I. (2007): Weather and climate and optimization of farm technologies at different input levels. In: Sivakumar, M.V.K. and R.P. Motha (Eds.). Managing Weather and Climate Risks in Agriculture, pp. 141-170, Springer

Ekeland, I.; Heckman, J.J. and Nesheim, L. (2004): Identification and Estimation of Hedonic Models. Journal of Political Economy. University of Chicago Press, Vol. 112, pp.60-109

EM-DAT (2012): The OFDA/CRED International Disaster Database, www.emdat.be - Université catholique de Louvain - Brussels – Belgium, Created on: Jan-30-2012. Data version:v.12.07

ESP (2011): Economic Survey of Pakistan 2010-2011. Islamabad,: Ministry of Finance, Government of Pakistan, pp.15-33

ESP (2007). Economic Survey of Pakistan, www.accountancy.com.pk , 2006-2007

ESRI (2011): ArcGis Desktop 10.1. New York Street, Redlands, CA, USA

ESRI (1999): ESRI Data and Maps. Rome: Food and Agriculture Organization

Evenson, R.E.; Flavio, A. and Avila, D. (2004): Total Factor Productivity Growth in Agriculture: The Role of Technological Capital- Economic Growth Center, Yale University, New Haven, 2004

Fagerland, M. W.; Hosmer, D.W. and Bofin, A.M. (2008): Multinomial goodness-of-fit tests for logistic regression models. Statistics in Medicine, Vol. 27, pp. 4238–4253

Fankhauser, S.; Smith, J.B.; Tol, R.S.J. (1999): Weathering climate change: some simple rules to guide adaptation decisions. Ecological Economics, Vol. 30, pp. 67–78

FAO (1996): Agro-Ecological Zoning: Guidelines - FAO Soils Bulletin - 731996, W2962/E, Natural Resources and Environment Department of the Food and Agriculture Organization, Rome, Italy, 1996

FAO (2001): Lecture notes on the major soils of the world. World Soil Resources Reports, No. 94, Food and Agriculture Organization of the United Nations, Rome, Italy

FAO (2003): The digital soil map of the world (DSMW) CD-ROM. Food and Agriculture Organization, Italy, Rome.

FAO (2004): CROPWAT Version 7.0 software for Windows in English, French and Spanish, Water Resources, Development and Management Service, Food and Agriculture Organization of the united Nations, Rome, Italy

FAO (2004): Fertilizer use by crop in Pakistan. Land and Plant Nutrition Management Service, Land and Water Development Division. Food and Agriculture Organization of The United Nations, Rome, 2004

FAO (2005): New LocClim (incl. CD-ROM), Environment and Natural Resources Working paper No. 20

FAO and IIASA (2000): Global Agro-Ecological Zones (GAEZ)-Global Agro-Ecological Zones (GAEZ) CD-ROM, Laxenburg, Austria, 2000; http://www.iiasa.ac.at/Research/LUC/GAEZ/index.htm; accessed on September 30th 2010

FAO-ERCU (2010): Flood affected districts - Pakistan 1950 – 2011. Prepared by: Information Management Unit, FAO-ERCU, Pakistan

FAOSTAT (2011): FAO LAND STATISTICS. Rome, 2011

Farooq, M.; Majid and A., Pazir, M. (1992): Effect of tillage techniques on some physical characteristics and yield of sugarcane. Pak. J. Agric. Res., Vol. 13, No. 3, pp. 251-254

Farooqi A.; Khan, A.; Mir, H. (2005): Climate change perspective in Pakistan. Pakistan Journal of Meteorology Vol. 2, No.3, pp. 11–21

FAS USDA (2009): Pakistan Agriculture Economy and Policy Report. Foreign Agricultural Service United States Department of Agriculture, 02/2012

FBS (2009): Pakistan Living Standards Measurement Survey 2007-2008, Federal Bureau of Statistics, Government of Pakistan, Islamabad

FBS (2010): Pakistan Living Standards Measurement Survey 2008-2009, Federal Bureau of Statistics, Government of Pakistan, Islamabad

Feldman, I. and Kahan, J. (2007): Preparing for the day after tomorrow: Frameworks for climate change adaptation. Sustainable Development Law & Policy, Vol.8

Fisher, A.; Hanemann, M.; Roberts, M. and Schlenker, W. (2007): Potential Impacts of Climate Change on Crop Yields and Land Values in U.S. Agriculture: Negative, Significant, and Robust. University of California Energy Institute Seminar, Berkley, 2009

Fleischer; A.; Lichtman, I. and Mendelsohn, R. (2008): Climate Change, Irrigation, and Israeli Agriculture:Will Warming Be Harmful? Ecological Economics, Vol. 65, No.3, pp. 508-515

Fox, J. and Long, S. J. (1990): Modern Methods of Data Analysis. Sage Publications, Inc., Newbury Park, CA

Fuhrer, J. (2003): Agroecosystem responses to combination of elevated CO2, ozone, and global climate change. Agriculture, Ecosystems and Environment, Vol. 97, pp. 1-20

Gan, J. (2004): Risk and damage of southern pine beetle outbreaks under global climate change. Forest Ecological Management, Vol. 191, pp. 61-71

Garcia, R.M. (2009): Pakistan's Food and Agriculture Systems. Nathan Associates Inc., The United States Agency for International Development, 2009

Gavian, S.; Tamer E., Lamia B. and Ender, G. (2002): The Importance of Agricultural Growth to SME Development and Rural Employment in Egypt. MVE Unit, APRP, Special Study No. 5, Abt Associates, 2002

Gazdar, G. and Quan, J. (2004): Poverty and Access to Land in South Asia. DFID-NRI, UK, May 2004

Gazdar, H. (2007): Determinants and Drivers of Poverty Reduction and ADB's Contribution in Rural Pakistan: Rural Economy and Livelihoods in Pakistan. ADB TA-4319-PAK. Islamabad: ADB.

Gelman, A. and Hill, J. (2007): Data Analysis Using Regression and Multilevel-Hierarchical Models. Cambridge University Press, Cambridge

Ghannoum, O.; von Caemmerer, S.; Ziska, L.H. and Conroy, J.P. (2000): The growth response of C4 plants to rising atmospheric CO2 partial pressure: a reassessment. Plant, Cell and Environment Vol. 23, pp. 931–942

Gifford, R.M. (2004): The CO2 fertilizing effect - does it occur in the real world? New Phytologist Vol. 163, pp. 221-225

Gillig, D. and McCarl, B.A. (2002): Introduction to Computable General Equilibrium Model (CGE). College Station, Texas, USA

GIZ (2011): Support for the Federal Bureau of Statistics Pakistan. Gesellschaft für Internationale Zusammenarbeit, Project 2005-2011, German Federal Ministry for Economic Cooperation and Development (BMZ)

GoB (2012): Overview on the Districts of Balochistan. Government of Balochistan, Balochistan.gov.pk, obtained via personal contact in October 2012.

Gommes R.; Grieser J. and Bernardi, M. (2004): FAO agroclimatic databases and mapping tools-ESA Newsletter No.26, FAO, Rome

GoN (2009): Development Statistics of the NWFP 2009. Bureau of Statistics NWFP, Government of the NWFP, Peshawar

Gonzalez, F.J.; Basson, T. and Schultz, B. (2005): Final Report of the international Panel of Experts for review of Studies on Water Escapage below Kotri Barrage, Delft, the Netherlands.

GoS (2008): Development Statistics of Sindh 2008. Bureau of Statistics Sindh, Government of Sindh, Karachi, Pakistan

Goulder, L. H., and Pizer, W. A. (2006): The Economics of Climate Change. Resources for the Future, Discussion Paper

Greco, A. J. (2005): Cross Elasticity of Supply: Seldom Heard of and Seldom Taught. Journal of Economic Educators, Vol. 5, No. 1, pp. 1-5

Greene, W. (2008): Econometric Analysis, 6th ed., Prentice Hall

Gujarati, D.N. (1995): Basic Econometrics. McGraw-Hill, New York, USA

Haider and Kuhnen (1974): Land Tenure and Rural Development In Pakistan. Reprint from Land Reform, Land Settlement and Cooperatives 1974 - No.1, Issue 2. Food and Agriculture Organization of the United Nations, Rome, Italy, pp.62-67

Hair, J. F. Jr.; Anderson, R. E.; Tatham, R. L. and Black, W. C (1995): Multivariate Data Analysis, 3rd ed., Macmillan, New York, USA

Hamilton, L.C. (1992): Resistant normality check and outlier determination. STATA Technical Bulletin Reprints, Vol.1, pp. 86-90

Hanif, U.; Syed, S.H.; Ahmad, R. and Malik, K.A. (2010): Economic impact of climate change on Agricultural sector of Punjab. 25th Annual Meeting of the Pakistan Institute of Development Economics

Hansen, J., Sato,M.; Kharecha,P.; Beerling,D.; Berner,R.; Masson-Delmotte,V.; Pagani, M.; Raymo,M; Royer,D.L. and Zachos,J.C. (2008): Target atmospheric CO2: Where should humanity aim? Open Atmospheric Science Journal, Vol. 2, pp. 217-231

Harmeling, S. (2011): Global Climate Risk Index 2012-Who suffers most from extreme weather Events? Weather-related loss events in 2010 and 1991 to 2010. Germanwatch Briefing Paper, Bonn, Germany

Hausman, J., and McFadden, D. (1984): Specification Tests for the Multinomial Logit Model, Econometrica Vol. 52 No. 5, pp. 1219-1240

Heckman, J.J. (1979): Sample selection bias as a specification error. Econometrica, Vol. 47, No.1, pp. 153–162

Heldt, H.W.; Piechulla B. (2011): Plant Biochemistry. Elsevier, London, UK, pp. 113–131

Helms, S.; Mendelsohn, R. and Neumann, J. (1996): The Impact of Climate Change on Agriculture. Climatic Change, Vol. 33, pp. 1-6

Hosmer, D. W. Jr. and Lemeshow, S. (1980): Goodness-of-fit tests for the multiple logistic regression model. Communications in Statistics: Theory and Methods, Part A, Vol.9, pp. 1043–1069

Houghton, J. T.,;Y. Ding, D. J.; Griggs, M. N.; Van der Linden,P.J. and Xiaosu,D. (2001): Climate Change 2001: The Scientific Basis: Contributions of Working Group I to the Third Assessment Report of the Intergovernmental Panel on Climate Change. Cambridge University Press

Huesca, L. and Camberos, M. (2010): Selection-bias correction based on the multinomial logit: An application to the Mexican labor market. Mexican Stata Users' Group Meetings 2010, Stata Users Group

Hulme, M.; Osborn, T. J. and Johns, T. C. (1998): Precipitation sensitivity to global warming: comparison of observations with HadCM2 simulations. Geophysical Research Letters, Wiley,Vol. 25, pp. 3379–3382

Hussain, A. and Thapa, G.B (2012): Smallholders' access to agricultural credit in Pakistan. Journal Food Security Volume 4, Issue 1, Springer Netherlands, pp. 73-85

Hussain, S.I. (2002): Economy of an Elitist State. Oxford University Press

ICIMOD (2009): Local Responses to Too Much and Too Little Water in the Greater Himalayan Region. International Centre for Integrated Mountain Development, Kathmandu, Nepal

IFRC (2009): World Disasters Report 2009. International Federation of Red Cross and Red Crescent Societies. ATAR Roto Presse, Satigny/Vernier, Switzerland

IPCC (2007): Climate Change 2007: Impacts, Adaptation and Vulnerability. Contribution of Working Group II to the Fourth Assessment Report of the Intergovernmental Panel on Climate Change, M.L. Parry, O.F. Canziani, J.P. Palutikof, P.J. van der Linden and C.E. Hanson (eds).Cambridge University Press, Cambridge, pp. 976

Iqbal, M.; Ahmad, M. and Abbas, K. (2003): The impact of institutional credit on agricultural production in Pakistan. The Pakistan Development Review, Volume 42, Issue 4, pp. 469-485

IWMI (2007): Water for Food,Water for Life: A Comprehensive Assessment of Water Management in Agriculture. London: Earthscan, and Colombo: International Water Management Institute, 2007

Jamal, H. and Khan, A.J. (2007): Trends in Regional Human Development Indices. Social Policy and Development Centre, Research Report No. 73, Pakistan

Javonillo, C.J. (2010): Indus Valley Civilization: Enigmatic, Exemplary, and Undeciphered. ESSAI, Vol. 8, Article 21

Jorgenson, D.W.; Goette, R. J.; Hurd, B.H., Smith, J.B. (2004): US market consequences of global climate change. Pew Center on Global Climate Change

Junejo, M., Rohra, Ch., Kanasro, H. (2007): Analyzing the impact of Sindh Rice Industries on economy of Pakistan. Australian Journal of Basic and Applied Sciences, Vol.1(4), pp.853-59.

Kalnay, E., Kistler, R., Collins, W. et al. (2001): The NCEP-NCAR 50-year reanalysis: Monthly means CD-ROM and documentation. Bulletin of the American Meteorological Society, Vol. 82, pp. 247-267

Kaufmann, R. K. (1998): The Impact of Climate Change on US Agriculture: A Response to Mendelsohn et al. (1994). Ecological Economics, Vol. 26, No.2, pp. 113-119

Kelly, D.L. and Kolstad, C.D. (1999): Integrated assessment models for climate change control. In: Folmer H, Tietenberg T (eds): International yearbook of environmental and resource economics 1999/2000: a survey of current issues. Edward Elgar, pp 171–197

Kemal, A.R. (1988): Commercial policy, international trade and structure of protection in Pakistan. Pakistan Institute of Development Economics, Islamabad. Mimeo.

Kempton, W. (2001): Cognitive anthropology and the environment. In: New Directions in Anthropology and Environment – Intersections. (ed. Carole L. Crumley with A. E. van Deventer and J.J. Fletcher) Altamira Press, California, USA, pp. 49-71

Kennedy, P. (1992): A Guide to Econometrics. Blackwell Publishing, Oxford, UK

Kenoyer, J.M. and Meadows, R.H. (2000): The Indus Valley mystery. Discovering Archaeology, April Issue, pp. 38-43

Kersebaum,K.C.; Hecker,J.M.; Mirschel,W.;Wegehenkel, M. (2010): Modeling Water and Nutrient Dynamics in Soil-crop Systems. Springer, 2010

Khan, A.G. (2004): The Characterization of the Agro Ecological Context in Which Fangr (Farm Animal Genetic Resource) are Found. Technical Report. Animal Nutrition, Animal Sciences Institute National Agricultural Research Centre Islamabad, Pakistan

Khan, M.H. (2000): Pakistan's Dual Economy. Dawn. In: USAID- Property Rights & Resource Governance, Pakistan—Land Tenure and Property Rights Profile, U.S. Agency for International Development Washington, D.C., 2010

Khan, M.H. (1981): Underdevelopment and agrarian structure in Pakistan. Boulder: Westview Press.

Khan, S.; Hasan, M.; Bari, A.; Khan, F.K. (2010): Climate Classification of Pakistan, Paper presented at the Balwois Conference 2010 on water observation and information systems for decision support – Ohrid, Republic of Macedonia –25, 29 May 2010

Khushk, A.M.; Memon, A. and Lashari, M. I. (2006): Marketing System of Selected Fruits in Pakistan. Bangladesh Journal of Agricultural Research, Vol. 31, No.1, pp.39-68

Kiani, A.K. (2008): Farm Size and Productivity in Pakistan. European Journal of Social Sciences – Vol. 7, No. 2, pp. 42-52

Kiani, K. (2005): Water-related crisis feared in 20 years. Dawn: the Internet Edition. 04 January 2005. The Dawn Group of Newspapers: Islamabad, Pakistan. http://www.dawn.com/2005/01/04/nat12.htm

Kim, D. (1999): A Standardization Technique to Reduce the Problem of Multicollinearity in Polynomial Regression Analysis. Bulletin of the International Statistical Institute, 52nd Session, Proceedings Tome LVIII, Finland

Kim, H.Y.; Lieffering, M.; Kobayashi, K.; Okada, M.; Mitchell, M.W. and Gumpertz, M. (2003): Effects of free-air CO2 enrichment and nitrogen supply on the yield of temperate paddy rice crops. Field Crops Research, Vol. 83, pp. 261-270

Kimball, B.A.; Mauney, J.R.; Nakayama, F.S. and Idso, S.B. (1993): Effects of increasing atmospheric CO2 on vegetation. Vegetatio Vol. 104, No.105, pp. 65-75

Koenker, R. (1981): A note on studentizing a test for heteroscedasticity. Journal of Econometrics, Vol. 17, pp. 107-112

Koenker, R. and Bassett, G. (1982): Robust Tests for Heteroscedasticity Based on Regression Quantiles. Econometrica, Vol. 50, pp. 43-61

Köppen, W. (1936): Das geographische System der Klimate, in: Handbuch der Klimatologie, edited by: Köppen, W. and Geiger, G., 1. C. Gebr, Borntraeger, S. 1–44, 1936.

Koti, S.; Reddy, K.R.; Kakani, V.G.; Zhao, D. and Reddy, V.R. (2005): Interactive effects of carbon dioxide, temperature and ultraviolet-B radiation on flower and pollen morphology, quantity and quality of pollen in soybean (Glycine max L.) genotypes. Journal of Experimental Botany, Vol. 56, pp. 725–736

Kramer P.J. (1981): Carbon dioxide concentration, photosynthesis, and dry matter production. Bioscience Vol. 31, pp. 29-33.

Kropko, J. (2010): A Comparison of Three Discrete Choice Estimators. University of North Carolina, Chapel Hill

Kuhnen, F. (1996): Land tenure in Asia: Access to land-access to income. Heidelberg, Germany: GTZ.

Kumar, K. and Parikh, J. (2001): Socio-economic Impacts of Climate Change on Indian Agriculture. International Review for Environmental Strategies, Vol. 2, No.2

Kumar, N.; Peters, T.; Nixon, V.; Sinha, K.; Jiang, X. and Ziegenhorn, S. (2009): An Optimal Spatial Configuration of Sample Sites for Air Pollution Monitoring. Journal of the Air and Waste Management Association, pp. 1308–1316.

Kurosaki, T. (2007): Land-use Changes and Agricultural Growth in India, Pakistan, and Bangladesh, 1901-2004. International Conference on Comparative Development, Indian Statistical Institute, December 18-20, 2007, ISI, Delhi

Kurukulasuriya, P. and Ajwad, M.I (2007): Application of the Ricardian Technique to Estimate the Impact of Climate Change on Smallholder Farming in Sri Lanka, Climatic Change, Volume 81, No. 1, March, 2007

Kurukulasuriya, P. and Ajwad, M.I (2007): Application of the Ricardian Technique to Estimate the Impact of Climate Change on Smallholder Farming in Sri Lanka. Climatic Change, Vol. 81, No. 1

Kurukulasuriya, P. and Mendelsohn, R. (2007): A Ricardian analysis of the impact of climate change on African cropland. World Bank Policy Research Working Paper No. 4305

Kurukulasuriya, P. and Mendelsohn, R. (2008a): Modeling Endogenous Irrigation: The Impact of Climate Change on Farmers In Africa, World Bank Policy Research Working Paper No. 4278, Washington D.C.

Kurukulasuriya, P. and Mendelsohn, R. (2008): Crop Switching as an Adaptation Strategy to Climate Change. African Journal Agriculture and Resource Economics, Vol. 2, pp. 105-126

Kurukulasuriya, P. and Rosenthal, S. (2003): Climate change and agriculture: a review of impacts and adaptation. Paper No. 91. The World Bank Environment Department, Washington D.C.

Kurukulasuriya, P.; Kala, N. and Mendelsohn, R. (2011): Adaptation and climate change impacts: a Structural Ricardian model of irrigation and farm income in Africa. Climate Change Economics, Vol. 2, No. 2, pp. 149–174

Kutner, N. (2004): Applied linear regression models. 4th ed., McGraw-Hill, Irwin

Kyuma, K. (2009): Nature and Agriculture in Monsoon Asia. Keynote at the MARCO Symposium 05th Oct.-7th Oct., Japan

Lal, M.; Singh, K. K.; Srinivasan, G.; Rathore, L. S.; Naidu, D. and Tripathi, C. N. (1999): Growth and yield response of soybean in Madhya Pradesh, India to climate variability and change, Agriculture and Forest Meteorology, Vol. 93, pp. 53–70

LP (2008): LEAD Climate Change Action Program- LEAD Pakistan, Islamabad, Internal Document

Lee, S.Y. (1998): Sozioökonomische Auswirkungen von Klimaveränderungen auf die Landwirtschaft von Südkorea – Einfluss der Klimavariablen Temperatur, Niederschlag und Sonneneinstrahlung. LIT Verlag Münster-Hamburg-London, Studien zur ländlichen Entwicklung, Vol. 56

Lempert, R.J.; Schlesinger, M.E. and Banks, S.C. (1996): When We Don't Know the Costs or Benefits: Adaptive Strategies for Abating Climate Change. Climatic Change, Vol. 33, pp. 235-274

Linsley, R.K.; Kohler, M.A. and Paulhus, J.L. (1958): Hydrology for Engineers. McGraw-Hill, New York, pp. 151-155

Lippert, C.; Krimly T.; Aurbacher, J. (2009): A Ricardian analysis of the impact of climate change onagriculture in Germany. Climatic Change, Vol. 97 (No.3-4), pp. 593-610

Long, J. S. and Freese, J. (2006): Regression Models for Categorical Dependent Variables Using Stata. 2nd ed., College Station, Stata Press, TX, USA

LUAOP (2009): Land Use Atlas of Pakistan. Islamabad: Ministry of Environment, Government of Pakistan, pp.19

Luo, Y.; Gerten, D.; Maire, G.L.; Parton, W.J.; Weng, E. and Zhou, X. (2008): Modeled interactive effects of precipitation, temperature and [CO2] on ecosystem carbon and water dynamics in different climatic zones. Global Change Biology, Vol. 14, 1986–1999

Maddison, D. (2000): A hedonic analysis of agricultural land prices in England and Wales. European Review of Agricultural Economics, Vol. 27, No. 4, pp. 519-532

Maddison, D.J.; Manley, M. and Kurukulasuriya, P. (2007): The Impact of Climate Change on African Agriculture: A Ricardian Approach. World Bank Policy Research Working Paper, No. 4306, Washington, D.C.

Majeed, S.; Bhatti, A.A. and Abbas, G. (2010): Productivity and Profitability of Barani Wheat under chemical and organic farming systems. Pakistan Journal of Agricultural Research, Vol. 23, No. 1-2

Mal, P., Manjunatha, A.V., Bauer S., Ahmed, M. Nomman, (2011): Technical Efficiency and Environmental Impact of Bt Cotton and Non- Bt Cotton in North India –AgBioForum Journal of Agrobitechnology Management and Economics,Vol. 14(3) 2011: 164-170, Columbia, USA

Malik, A.; Qin, X. and Smith, S.C. (2010): Autonomous Adaptation to Climate Change: A Literature Review. IIEP Working Paper 2010-24. Elliott School of International Affairs, George Washington University, Washington D.C.

Malik, S.B. (2010): The Culture of Technology in Barani (Rainfed) Areas of Potahar: An Analysis of Farming Tools and Technologies from Gender Perspective. International Journal of Business and Social Science Vol. 1, No. 3; pp. 146-157

Malik, S.J. and Randhawa, S.A.A. (2006): Pakistan Agriculture in Pakistan and the Doha Development Agenda-Challenges and opportunities, EU-TRTP-International Trade Centre (UNCTAD/WTO), September 2006

Margulis et al. (2008): The Economics of Adaptation to Climate Change. World Bank Methodology Report, The World Bank, Washington D.C.

Matsui, T.; Kobayasi, K.; Yoshimota, M. and Hasegawa, T. (2007): Stability of rice pollination in the field under hot and dry conditions in the Riverina region of New South Wales, Australia. Plant Production Science, Vol.10, pp. 57–63

Matsuura, K. and Willmott C.J. (2007a): Terrestrial Air Temperature: 1900-2006 Gridded Monthly Time Series (Version 1.01) http://climate.geog.udel.edu/~climate/html_pages/Global_ts_2007/README.global.t_ts_2007.html

Matsuura, K. and Willmott, C.J. (2007b): Terrestrial Precipitation: 1900-2006 Gridded Monthly TimeSeries (version 1.01) http://climate.geog.udel.edu/~climate/html_pages/Global_ts_2007/README.global.p_ts_2007.html

McFadden, D.L. (1981): Econometric models of probabilistic choice. In McFadden, D.L., Structural Analysis of Discrete Data and Econometric Applications. MIT press, Cambridge

McSweeney,C.; New, M. and Lizcano, G. (2010): The UNDP Climate Change Country Profiles: improving the accessibility of observed and projected climate information for studies of climate change in developing countries. Bulletin of the American Meteorological Society, Vol. 91, pp. 157–166

Mearns, L.O.; Mavromatis, T.; Tsvetsinskaya, E.; Hays, C. and Easterling, W. (1999): Comparativeresponses of EPIC and CERES crop models to high and low resolution climate change scenarios. Journal of Geophysical Research, Vol. 40 (D6), pp. 6623-6646

Meehl, G. et al. (2007): Global Climate Projections. In Solomon et al. (eds.), Climate Change 2007: The Physical Science Basis. Working Group I Contribution to the Fourth Assessment Report of the IPCC. Cambridge University Press: Cambridge

Memon, A.M. (2002): Sugar Industry in Pakistan. Pakistan & Gulf Economist, Issue September 2002, Economist Publications (Private) Limited, Karachi, Pakistan

Menard, S. (1995): Applied logistic regression analysis. Thousand Oaks, CA, Sage

Mendelsohn, R. and Dinar, A. (2003): Climate, water, and agriculture. Land Economics, vol. 79, No.3, pp. 328–341

Mendelsohn, R. and Dinar, A. (1999). Climate Change, Agriculture, and Developing Countries: Does Adaptation Matter? The World Bank Research Observer 14: 277-293

Mendelsohn R.; Nordhaus W.; Shaw D. (1994): The impact of global warming on agriculture: A Ricardian analysis-American Economic Review, Vol. 84: pp. 753–771

Mendelsohn, R. (2000): Efficient Adaptation to Climate Change. Climatic Change, Vol. 45, pp. 583-600

Mendelsohn, R. (2000a): Two essays on climate change and agriculture. FAO Economic and Social Development Paper, Version 145, Food and Agriculture Organization of the United Nations, Rome, Italy

Mendelsohn, R. (2001): Global Warming and the American Economy: A Regional analysis, Camberley, Surrey: Edward Elgar Publishing.

Mendelsohn, R. (2005): Forecasting the Impacts of Climate Change: Comparing the Effects in the Far East and the Rest of the World-Paper presented at the proceedings of the Annual Meeting of the Allied Social Science Associations, Session –Environment, Economy & Global Warming, Boston, MA,USA, January 6-8, 2006

Mendelsohn, R. (2006): The Role of Markets and Governments in Helping Society Adapt to a Changing Climate. Climatic Change Special Issue on Climate, Economy, and Society: From Adaptation to Adaptive Management, Vol. 78, pp. 203-215

Mendelsohn, R. and Dinar, A. (2009): Climate change and agriculture: An economic analysis of global impacts, adaptation and distributional effects. Northampton, MA: Edward Elgar Publishing, Inc.

Mendelsohn, R., Dinar, A.; Sanghi A. (2001): The Effect of Development on the Climate Sensitivity of Agriculture, Environment and Development Economics, 2001, Vol. 6, pp. 85-101.

Mendelsohn, R. and Nordhaus, W. (1996): The Impact of Global Warming on Agriculture: Reply. American Economic Review, American Economic Association, Vol. 86, No.5, pp. 1312-1315

Mendelsohn, R. and Nordhaus, W. (1999): The impact of global warming on agriculture Reply to Darwin. American Economic Review 89, pp. 1053–1055

Mendelsohn, R.; Bassist, A.; Kurukulasuriya, P. and Dinar, A. (2007): Climate and rural income. Climatic Change, Vol. 81, No. 1, pp. 101-118

Mendelsohn, R.; Kurukulasuriya, P.; Basist, A.; Kogan, F. and Williams, C. (2007a): Measuring Climate Change Impacts with Satellite versus Weather Station Data. Climatic Change, Vol. 81, pp. 71–83

Mendelsohn, R.; Neumann, J. (1999): The Impacts of Climate Change on the US Economy. Cambridge University Press, Cambridge

Mendelsohn, R.; Nordhaus, W.D.; Shaw, D. (1996): Climate impacts on aggregate farm value: accounting for adaptation. Agricultural and Forest Meteorology, Vol 80, pp. 55-66

Mendelsohn, R; Avila, A.F.; Seo, S.N. (2007b): Incorporation of Climate Change to Strategies of Rural Development: Synthesis of the Latin American Results PROCISUR/IICA. Montevideo, Uruguay

Mendelsohn R. and Williams L. (2004): Comparing Forecasts of the Global Impacts of Climate Change- *Mitigation and Adaptation Strategies for Global Change*, 2004, Vol. 9, pp. 315-333

Mian, M.A. and Syal M.N. (1985): Geomorphology of Pakistan. Proceedings of XII International Forum on Soil Taxonomy and Agrotechnology Transfer, October 9- 23, Pakistan 1985, pp. 26-42

MINFAL (2007): Agricultural Statistics of Pakistan 2006-2007. Ministry of Food Agriculture and Livestock, Islamabad, December 2007.

Molua, E.L. and Lambi, C.M. (2006): Assessing the impact of climate on crop water use and crop water productivity: The CROPWAT analysis of three districts in Cameroon. CEEPA Discussion Paper No. 33, Centre for Environmental Economics and Policy in Africa, University of Pretoria

Moorman, F. R. and Van Breemen, N. (1978): Rice: soil, water, land. International Rice Research Institute, Los Baños, Philippines

Morison, J.I.L. (1996): Global environmental change impacts on crop growth and production inEurope. Implications of global environmental change for crops in Europe. Aspects of Applied Biology Vol. 45, pp. 15-29

Morison, J.I.L. and Morecroft, M. D. (2006): Plant Growth and Climate Chang. Biological Sciences Series. Issue 1, Wiley, 2006

Muhammad, A. (2005): The Causes and Effects of Agricultural Mechanization and Labor Displacement in NWFP. PhD thesis, University of Peshawar, Peshawar

Nadeem, S.; Elahi, I.; Hadi, A. and Uddin, I. (2009): Traditional knowledge and local institutions support adaptation to water-induced hazards in Chitral, Pakistan. ICIMOD, International Centre for Integrated Mountain Development, Kathmandu, Nepal

Nakamura, A.O. and Diewert, E. (2002): The Measurement of Aggregate Total Factor Productivity Growth, paper written for the forthcoming Elsevier Science Handbook of Econometrics, Vol. 6, 2002

Nakicenovic , N., Alcamo, J., Davis, G. et al. (2000): Special Report on Emissions Scenarios: A SpecialReport of Working Group III of the Intergovernmental Panel on Climate Change, Cambridge University Press, Cambridge, U.K.

Naqvi, S. N. H. and Kemal, A.R. (1983): The structure of protection in Pakistan 1980-81. Pakistan Institute of Development Economics, Vol. 1, Islamabad, Pakistan

NARC (2012): National Coordinated Rice Programme. Pakistan Agricultural Research Council, PARC, Islamabad. http://www.parc.gov.pk/narc/riceprog/pages/intro.html

NCDC (2011): National Climatic Data Center, NESDIS, NOAA -U.S. Department of Commerce, Global Surface Summary of the Day – GSOD. Asheville, NC Dataset Publisher: National Climatic Data Center

New, M.; Lister, D.; Hulme, M. and Makin, I. (2002): A high-resolution data set of surface climate over global land areas. Climate Research Vol. 21

Niazi, T. (1997): Ecological bases of social violence in Pakistan. Ann Arbor, MI: University Microfilms International. In: Niazi, T. (2003): Land Tenure, Land Use, and Land Degradation: A Case for Sustainable Development in Pakistan. The Journal of Environment Development, Vol.12, 2003, pp. 275-295

Niazi, T. (2003): Land Tenure, Land Use, and Land Degradation: A Case for Sustainable Development in Pakistan. The Journal of Environment Development,Vol.12, 2003, pp. 275-295

NOAA (2011): National Climatic Data Center, State of the Climate: Global Analysis for Annual 2011, published online December 2011, retrieved on January 29, 2013 from http://www.ncdc.noaa.gov/sotc/global/2011/13.

NOAA Central Library (2010): Historical Meteorological and Climatological Data for Pakistan (1947-1975) - Climate Data Imaging Project, Silver Spring, MD, National Oceanographic Data Center, USA

Nordhaus, W.D. and Yang, Z. (1996): RICE: A Regional Dynamic General Equilibrium Model of Optimal Climate-Change Policy. American Economic Review, Vol. 86, No. 4, pp. 741-765

NSIDC (2012): Arctic Climatology and Meteorology-Latitude. The National Snow and Ice Data CenterAdvancing knowledge of Earth's frozen regions, University of Colorado Boulder, CO, www.nsidc.org, accessed July 2012

O'Brien, R. M. (2007): A Caution Regarding Rules of Thumb for Variance Inflation Factors. Quality & Quantity, Springer, Vol. 41, pp. 673–690

OECD (2009): Integrating Climate Change Adaptation into Development Co-operation. OECD Policy Guidance, Paris

OECD (2001): Measuring Productivity: Measurement of Aggregate and Industry-level Productivity Growth, Statistical Manual, 2001

Oosterhuis, D.M. and Snider, J.L. (2011): High temperature stress on floral development and Yield, pp. 1-24. In: D.M. Oosterhuis (ed.) Stress Physiology in Cotton. The Cotton Foundation, Cordova

Osakwe, E. (2009): Cotton fact sheet Pakistan. International Cotton Advisory Committee, Washington DC, USA, 2009

Ouraich, I. and Tyner, W.E. (2012): Agricultural Climate Change Impacts on Moroccan Agriculture and the Whole Economy Including an Analysis of the Impacts of the "Plan Maroc Vert (PMV) in Morocco. UNU-WIDER conference on Climate Change and Development Policy, Helsinki, 28–29 September 2012

Padgham, J. (2009): Agricultural Development under a Changing Climate: Opportunities and Challenges for Adaptation. Environment and Agriculture and Rural Development Department, Joint Discussion Paper, The World Bank, Washington D.C.

Pakistan Statistical Yearbook (2010): Pakistan Bureau of Statistics, Islamabad

Paradis, E. (2012): Moran's Autocorrelation Coefficient in Comparative Methods. The comprehensive R Archive Network (CRAN), http://www.icesi.edu.co/CRAN/web/packages/ape/vignettes/MoranI.pdf

PARC (1980): Agro-ecological Regions of Pakistan- Pakistan Agricultural Research Council, Ministry for Food and Agriculture, Islamabad, Pakistan, 1980

PCO (1998): District wise Reports on Basic Population and Housing Data by Union Councils. Pakistan Census Organization, Islamabad

PINCC (2003): Pakistan's Initial National Communication on Climate Change, Government of Pakistan, Islamabad, pp. 56

Plaster, E. J. (2008): Soil science and management (5th ed.). Albany, NY: Delmar Cengage Learning

PMD (1985): Climate and Seasons. Proceedings of XII International Forum on Soil Taxonomy and Agrotechnology Transfer, October 9- 23, Pakistan 1985, pp.43-46

Polsky, C. (2004): Putting Space and Time in Ricardian Climate Change Impact Studies: Agriculture in the U.S. Great Plains, 1969-1992- Annals of the Association of American Geographers, Vol. 94, No. 3, pp. 549-564

Polsky, C. and Easterling, W.E. (2001): Adaptation to climate variability and change in the US Great Plains: A multi-scale analysis of Ricardian climate sensitivities. Agriculture, Ecosystems, and Environment, Vol. 85, pp. 133–144

Possehl, G. L. (2002): The Indus Valley civilization: A contemporary perspective. Walnut Creek, CA:Altamira Press

Qazilbash I. A. (2005): Water and energy. Country water resources assistance strategy. Background Paper No.4 March, 2005

Quiggin, J. and Horowitz, J. (1999): The impact of global warming on agriculture: a Ricardian analysis: comment. American Economic Review, Vol. 89, pp. 1044–1045

Randhawa, H.A. (2001): Water development for irrigated agriculture in Pakistan: past trends, returns and future requirements. In FAO, Investment in Land and Water - Proceedings of the Regional Consultation (S. 261-282). Bangkok, Thailand: RAP Publication 2002/09, Food and Agriculture Organization of the United Nations, 2002.

Rao, C.H.H. (1975): Technological Change and Distribution of Gains in Indian Agriculture, Macmillan, Delhi, India

Rafique, S. (2009): Punjab Development Statistics 2009. Bureau of Statistics Government of the Punjab, Lahore, Pakistan

Reddy, K. R. and Hodges, H.F. (2000): Climate Change and Global. Crop Productivity. CAB International, UK, p.296 and p.340

Reddy, V.R.; Baker, D.N. and Hodges, H.F. (1991): Temperature effect on cotton canopy growth, photosynthesis and respiration. Agronomy Journal, Vol. 83, pp. 699–704

Reilly, J., Baethgen, W.; Chege, F.E.; Van de Geikn, S.C. Lin Erda, A. Iglesias, G. Kenny, D. Petterson, J. Rogasik, R. Rötter, C. Rosenzweig, W. Sombroek, and J. Westbrook (1996): Agriculture in a Changing Climate: Impacts and Adaptation. Chapter 5 in Intergovernmental Panel on Climate Change. Climate Change 1995: Impacts, Adaptations and Mitigation of Climate Change: Scientific-Technical Analysis. Cambridge University Press, Cambridge, England, pp. 427-467

Reilly, J., Hohmann, N. and Jane, S. (1994): Climate change and agricultural trade: Who benefits, who loses? Global Environmental Change, Vol. 4, No.1, pp. 24-36

Reinsborough, M. J. (2003): A Ricardian model of climate change in Canada. Canadian Journal of Economics, Revue canadienne d'économique, Vol.36, pp. 21–40

Revenga, C.; Brunner, J.; Henninger,N.; Kassem,K. and Payne, R. (2000): Pilot Analysis of Global Ecosystems (PAGE): Freshwater Systems. World Resources Institute. Washington DC.

Ricardo, D. (1817): On the Principles of Political Economy and Taxation. John, London

Ritchie, G.L.; Bednarz, C.W.; Jost, J.H. and Brown, S.M. (2007): Cotton growth and development. Bulletin 1253

Ritchie, J.T.; Singh, U.; Godwin, D.C. and Bowen, W.T. (1998): Cereal growth, development and yield. p. 79–98. In G.Y. Tsuji, G. Hoogenboom, and P.K. Th ornton (ed.) Understanding options for agricultural production. Kluwer Academic Publ., Dordrecht, the Netherlands.

Rivera-Ferre, M.G.; Di Masso, M.; Miele, M.; López-i-Gelats, F.; Gallar, D.; Vara, I. and Cuellar, M. (2012): Understanding the Role of Local and Traditional Agricultural Knowledge in a Changing World Climate: The case of the Indo-Gangetic Plains. CGIAR Report

Rogers, H. H. and Dahlman, R.C. (1993): Crop responses to CO2 enrichment. Vegetatio Vol.104, No. 105, pp. 117–131

Rolland, F; Baena-Gonzalez, E.; Sheen, J. (2006): Sugar sensing and signaling in plants: conserved and novel mechanisms. Annual Review of Plant Biology, Vol. 57, pp. 675–709

Rosen, S. (1974): Hedonic prices in Implicit Markets: Product Differentiation in Pure Competition. Journal of Political Economy, Vol. 82, No.1, pp. 34-55.

Rosenzweig, C. and Iglesias, A. (1994): Implications of climate change for international agriculture: crop modeling study. US Environmental Protection Agency, Washington D.C.

Rosenzweig, C. and Iglesias, A. (1998): The use of crop models for international climate change impact assessment in understanding options for Agricultural Production, G Y Tsuji, G Hoogenboom, P K Thornton (eds) Kluwer Academic Publishers, Dordrecht, The Netherlands, pp. 267-292

Rotter, R. and Van de Geijn, S.C. (1999): Climate change effects on plant growth, crop yield and livestock. Climatic Change, Issue 43, pp. 651-681

Rowntree, L.; Lewis, M. and Price, M. (2011): Diversity Amid Globalization: World Regions, Environment, Development. Prentice Hall; 5ed., New Jersey, 2011, p. 545

S4 GIS Institute (2006): Spatial Analysis with GeoDa- Spatial Autocorrelation. Spatial Structures in the Social Sciences, S4 GIS Institute, Brown University, RI, USA

Sajjad, S.H.; Khan, M.A. and Raza, A. (2009): Climate Variability and Drought Threats in Sindh and Balochistan Provinces of Pakistan. Sustainable Agriculture, Climate Change and Environment. 3rd International Conference on Agriculture Education and Environment. Education for Rural People FAO, Rome, Italy, 2009

Sanghi, A. (1998): Global Warming and Climate Sensitivity: Brazilian and Indian Agriculture. Ph.D. Dissertation, Dept. of Economics, University of Chicago, USA

Saseendran, A. S.; Singh, K. K.; Rathore, L. S.; Singh, S. V. and Sinha, S. K. (2000): Effects of climate change on rice production in the tropical humid climate of Kerala, India. Climatic Change Vol. 44, pp. 495–514

Schlenker, W. (2000): Estimating the Potential Cost from Global Climate Change on US Agriculture using the Hedonic Approach. Department of Agricultural and Resource Economics, University of California, Berkeley, Berkeley, CA

Schlenker, W.; Hanemann, W.M. and Fisher, A.C. (2005): Will U.S. Agriculture Really Benefit From Global Warming? Accounting for Irrigation in the Hedonic Approach. American Economic Review, March 2005, Vol. 95 (No. 1), pp. 395-406

Schlenker, W.; Hanemann, W.M. and Fisher, A.C. (2007): Water Availability, Degree Days, and the Potential Impact of Climate Change on Irrigated Agriculture in California. Climatic Change, Vol. 81 No.1, pp. 19-38

Schneider, M.K., Luscher, A., Richter, M., Aeschlimann, U. and Hartwig, U.A. (2004): Ten years of free-air CO2 enrichment altered the mobilization of N from soil in Lolium perenne L. swards. Global Change Biology, Vol. 10, pp. 1377–1388

Seo, N. and R. Mendelsohn (2008): A Ricardian Analysis of the Impact of Climate Change on SouthAmerican Farms. Chilean Journal of Agricultural Research, Vol. 68, pp. 69-79

Seo, N. and R. Mendelsohn (2008a): Measuring Impacts and Adaptation to Climate Change: A Structural Ricardian Model of African Livestock Management. Agricultural Economics, Vol. 38, pp. 150-165

Seo, S.N., McCarl, B. and Mendelsohn, R. (2010): From Beef Cattle to Sheep under Global Warming? An Analysis of Adaptation by Livestock Species Choice in South America. Ecological Economics, International Society for Ecological Economics ,Vol. 69, pp. 2486-2494

Seo, S.N.; Mendelsohn, R. and Munasinghe, M. (2005): Climate change and agriculture in Sri Lanka: a Ricardian valuation. Environment and development Economics, Vol. 10, pp. 581-596

Sharpley, A. N. and Williams, J. R. (1990): EPIC- Erosion-productivity impact calculator. 1. Model documentation. Technical Bulletin No. 1768, U.S. Department of Agriculture

Sheikh, M.M.; Iqbal, M.M.; Ali, G. and Khan A.M. (2011): Global Warming In The Context Of Pakistan: Major Concerns And Remedial Strategies. Symposium on “Changing Environmental Pattern and its impact with Special Focus on Pakistan”- in: Pakistan Engineering Congress publication Vol. 33, 2011, Paper No. 287, pp. 51-78

Sheikh, M.M.; Manzoor, N.; Adnan, M.; Ashraf, J. and Khan, A.M. (2009): Climate profile and past climate change in Pakistan. GCISC Research Report No. 01 (GCISC-RR-01), Global Change Impact Studies Centre (GCISC), Islamabad

Sillitoe, P. (2007): Local Science vs. Global Science: Approaches to Indigenous Knowledge in International Development. Bergahn Books, New York

Smit, B. and Pilifosova, O. (2001): Adaptation to climate change in the context of sustainable development and equity. In: McCarthy, J.J., Canzianni, O.F., Leary, N.A., Dokken, D.J.,White, K.S. (Eds.), Climate Change 2001: Impacts, Adaptation, and Vulnerability—Contribution of Working Group II to the Third Assessment Report of the Intergovernmental Panel on Climate Change. Cambridge University Press, Cambridge, pp. 1000

SPSS (2009): SPSS 17. SPSS Hong Kong Headquarters, Quarry Bay, HK

Squire, G. R. and Unsworth, M. H. (1988): Effects of CO2 and climatic change on agriculture. Contract Report to the Department of the Environment, Sheffield University, Sheffield, UK

Srinivasan, A. (2004): Local knowledge for facilitating adaptation to climate change in Asia and the Pacific, Policy Implications

SSP (1985): Soil Survey of Pakistan. Proceedings of XII International Forum on Soil Taxonomy and Agrotechnology Transfer, October 9- 23, Pakistan 1985

Staley, E. (1966): Arid Mountain Agriculture in Northern West Pakistan- A Geographical Study. PhD thesis, University of the Punjab, Lahore

StataCorp (2009): Stata Statistical Software: Release 11.2/SE. Stata Corporation, College Station, TX, USA

Stern, N. (2007): The Economics of Climate Change-The Stern Review. Cabinet Office - HM Treasury Paperback, Cambridge University Press, UK

Stratfor (2008): Pakistan River System Map. www.stratfor.com, accessed: 11th August 2011.

Strzepek, K. and McCluskey, A. (2006): District level hydroclimatic time series and scenario analyses to assess the impacts of climate change on regional water resources and agriculture in Africa. CEEPA Discussion Paper No. 13, Centre for Environmental Economics and Policy in Africa, University of Pretoria

Stuczyinski, T.; Demidowicz, G.; Deputat, T.; Górski, T.; Krazowicz, S. and Kus, J. (2000): Adaptation scenarios of Agriculture in Poland to future climate changes. Environmental Monitoring and Assessment, Vol. 61, pp. 133-144

Sundaramari, M. (2001): Adoption and perceived effectiveness of indigenous agricultural practices in different farming systems. Ph.D. Thesis, Gandhgram Rural Institute, Tamil Nadu, India

Tariq, M. and Iqbal, H. (2010): Maize in Pakistan – An Overview. Review Article. Kasetsart Journal (Nat. Sci.), Vol. 44, pp. 757-763

TFCC (2010): Task Force on Climate Change, Final Report, Planning Commission, Government of Pakistan, Islamabad

The World Bank, IRIN (2009): Convenient Solutions to an Inconvenient Truth: Ecosystem-basedApproaches to Climate Change. Environment Department, The World Bank, Washington D.C.

Timmins, C. (2003): Endogenous land use and the Ricardian valuation of climate change. Mimeo. Department of Economics, Yale University

TPR (2008): Trade Policy Review Report by the Secretariat Pakistan. Trade Policy Review Body of the World Trade Organization, Geneva, Switzerland, 2008

Trenberth, K. (2010): Russian Heat, Asian Floods May Be Linked-Interview given by the head of climate analysis department at the National Center for Atmospheric Research (USA) to Wired Science on the 10th August 2010, http://www.wired.com/wiredscience/2010/08/russian-heat-asian-floods/; accessed on October 13th 2010

Trewartha, G.T. (1954): An Introduction to Climate. 3rd Edition, Mc-Grawhill-Book Company. INC, p. 395

Tubiello, F.N.; Soussana, J.F.; Howden, M. and Easterling, W. (2007): Crop and pasture response to climate change; fundamental processes. Proceedings of the National Academy of Sciences, Vol. 104, 19686-19690

Tubiello, F.N. and Ewert, F. (2002): Modeling the effects of elevated CO2 on crop growth and yield: a review. European Journal of Agriculture, Vol. 18, (No. 1–2), pp. 57–74

UNFCCC (2007): Climate Change: Impacts, Vulnerabilities And Adaptation In Developing Countries- United Nations Framework Convention on Climate Change, Bonn, Germany

UN Comtrade (2011): Tade Profile Pakistan 2009. United Nations, http://comtrade.un.org/db/default.aspx, accessed December 2011

United Nations Population Division (2011): World Population Prospects: The 2010 Revision. New York.

Unsworth, M.H.; Hogsett, W.E. (1996): Combined effects of changing CO2, temperature, UVB radiation and O3 on crop growth. In: Bazzaz, F.; Somborek, W. (1996): Global climate change and agricultural production. FAO and John Wiley & Sons, New York, 1996, pp. 171-198

Uppala, S.M.; Kallberg, P.W.; Simmons, A.J. et al. (2005): The ERA-40 reanalysis. Quarterly Journal of the Royal Meteorological Society, Vol. 131, No.612, pp. 2961-3012

USAID (2009): Pakistan Food and Agriculture. USAID Project Report, Weidemann Associates, Inc., submitted to USAID Pakistan, 03/ 2009

USAID (2010): USAID- Property Rights & Resource Governance, Pakistan— Land Tenure and Property Rights Profile, U.S. Agency for International Development Washington, D.C., 2010

Verardi, V. and Croux, C. (2009): Robust regression in Stata. The Stata Journal, Vol.9, No.3

Vijverberg, W. P. (2011): Testing for IIA with the Hausman-McFadden Test. IZA Discussion Papers, No. 5826, Institute for the Study of Labor (IZA)

Waldman, D. (1983): A note on the algebraic equivalence of White's test and variant of Godfrey /Breusch-Pagan test for heteroskedasticity. Economic Letters, Vol. 13, pp. 197-200

Wang, J.; Mendelsohn, R.; Dinar, A.; Huang, J.; Rozelle, S. and Zhang, L. (2008): The Impact of Climate Change on China's Agriculture. Agricultural Economics, Vol. 40, pp. 323-337

Warren, D.M.; Slikerveer, L.J. and Brokensha, D. (1995): The Cultural Dimension of Development: Indigenous Knowledge Systems. Intermediate Technology Publication, London

Warrick, R.A. (1988): Carbon dioxide, climatic change and agriculture. The Geographical Journal, Vol. 154, pp.221-233

Washington, R. et al. (2006): African climate change: Taking the shorter route. Bulletin of the American Meteorological Soiety, Vol. 87, pp. 1355–1366

Watanabe, M.; Adams, R.M. ; Wu, J.; Bolte, J.P.; Cox, M.M.; Johnson, S.L.; Liss, W.J.; Boggess, W.G. and Ebersole, J.L. (2005): Toward Efficient Riparian Restoration: Integrating Economic, Physical, and Biological Models. Journal of Environmental Management, Vol. 75, pp. 93-104

Weber, M. and Hauer, G. (2003): A Regional Analysis of Climate Change Impacts on Canadian Agriculture. Canadian Public Policy, Vol. 29, No. 2, pp. 163-179,

Weyant, J.; Davidson, O.; Dowlatabadi, H.; Edmonds, J.; Grubb, M.; Parson, E.; Richels, R.; Rotmans, J.; Shukla, P.; Tol, R.; Cline, W. and Fankhauser, S. (1996): Integrated Assessment of Climate Change: An Overview and Comparison of Approaches and Results. In: J. P. Bruce, H. Lee, and E. F. Haites, eds., Climate Change 1995: Economic and Social Dimensions of Climate Change, Cambridge: Cambridge University Press, 1996

Wolfe, D.W. and Erickson, J.D. 1993. Carbon dioxide effects on plants: uncertainties and implications for modelling crop response to climate change. In: Agricultural Dimensions of Global Climate Change. H.M. Kaiser and T.E. Drennen (eds.). pp. 153-178

Woodley, E. (1991): Indigenous ecological knowledge systems and development. Agriculture and Human Values, Vol. 8, pp. 173–178

Wooldridge, J.M. (2001): Econometric Analysis of Cross Section and Panel Data. MIT Press Books, The MIT Press, 1st ed., Vol. 1

World Bank (2008): Climate Change Impacts in Drought and Flood Affected Areas: Case Studies in India. South Asia Region, Washington D.C.

World Bank (2008a): The Economics of Adaptation to Climate Change. Methodology Report, Washington D.C.

World Bank (2009): Impact of climate change on agricultural yields in Morocco. World Bank - Morocco study on the impact of climate change on the agricultural sector. Washington D.C.

World Bank (2005): Country Water Resources Assistance Strategy-Water Economy: Running Dry. Washington: Agriculture and Rural Development Unit-South Asia Region

World Bank (2011): Vulnerability, Risk Reduction, and Adaptation to Climate Change. Climate Risk and Adaptation Country Profile Pakistan, The World Bank, April 2011

Xiao,G.; Zhang, Q.; Yao,Y.; Zhao,H. and Wang, R. (2008): Impact of recent climatic change on the yield of winter wheat at low and high altitudes in semi-arid northwestern China. Agriculture, Ecosystems and Environment, Vol. 127, pp. 37-42

Yaseen, M.; Zakria,M.; Islam-ud-din-Shahzad, M.; Khan, I. and Javed, M.A. (2005): Modeling and Forecasting the Sugarcane Yield of Pakistan. International Journal of Agriculture & Biology, Vol. 7, No. 2, Friends Science Publishers, Faisalabad, Pakistan, 2005

Yin, X. and Kropff, M.J. (1996): The Effect of Temperature on Leaf Appearance in Rice. Annals of Botany, Vol. 77, pp. 215-221

You, S.C. (2001): Agricultural adaptation to climate change in China. Journal of Environmental Sciences, Vol. 13, No. 2, pp. 192-197

Zahid, M. and Rasul G. (2011): Thermal Classification of Pakistan. Atmospheric and Climate Sciences, Scientific Research Publishing, Vol.1 2011, pp. 206-213

Ziska, L.H. and Bunce, J.A. (2006): Plant responses to rising atmospheric carbon dioxide. In: Morison, J.I.L., Morecroft, M.D.: Plant growth and climate change. Oxford, UK: Blackwell Publishing, pp. 17–47

Annex

A.) Outlier analysis – studentized residuals:

	id	district	r
1.	36	Jhelum	-3.80319
2.	53	Larkana	-2.527366
3.	38	Karachi	-2.325032
4.	106	Upper Dir	-2.252052
5.	11	Bolan/Kacchi	-2.209584
6.	34	Jhal Magsi	-2.084457
7.	110	Zohb	-2.071434
8.	16	Chitral	-1.993827
9.	45	Kharan	-1.474471
10.	71	Narowal	-1.463139

	r	ln_lv_~t	CENTMAY	CENTMAY^2	CKHgrow	CKHgrow^2	NCENTra..	NCENTr..	NCENTra..
1.	-3.80319	11.32558	.59962	.3595441	-.0975776	.0095214	.6958476	.4842038	.7009372
2.	-2.527366	12.16405	4.59962	21.1565	2.802426	7.85359	4.162515	17.32653	4.200937
3.	-2.325032	11.27621	1.19962	1.439089	-.6642432	.4412191	5.329179	28.40015	6.13427
4.	-2.252052	13.36068	-7.80038	60.84592	-6.53091	42.65279	-5.304152	28.13403	-7.632397
5.	-2.209584	9.714183	3.249622	10.56004	4.635758	21.49025	1.752513	3.071301	2.24427
6.	-2.084457	9.690597	.3896209	.1518045	3.135758	9.832977	-.70082	.4911486	-.142397

	NCENTr..	CENTspr~m	CENTsp~^{2}	Cjunjul~P	Cjunj~P^2	RABCGROW	RABCGR~^{2}	CENTHDI	LNpop_~m	LOGele~n
1.	.491313	43.27627	1872.836	93.89825	8816.881	15.75863	248.3345	.1535	5.56452	5.811141
2.	17.64787	-65.72373	4319.609	-34.60175	1197.281	-24.5747	603.9158	-.0675	5.560682	3.871201
3.	37.62926	-60.72373	3687.371	-7.85175	61.64997	-20.24137	409.713	.1725	7.935587	4.49981
4.	58.25349	369.2763	136365	54.89825	3013.818	124.092	15398.82	-.0575	5.049856	7.453562
5.	5.036748	-63.45373	4026.375	-35.30675	1246.567	-22.31137	497.7971	-.1595	3.637586	4.624973
6.	.0202769	-62.72373	3934.266	-34.10175	1162.929	-21.90803	479.9619	-.1815	3.401197	4.043051

	CENTlat~e	SQRTru~f	SQRTrdi	SQRTFl~a	runoff~p	SQRTtr~v	roaddens	cl	fl	rk	rg
1.	2.58291	11.74734	6.900724	.5641942	3240.7	3.598795	3.53549	.485247	0	0	.01607
2.	-2.797091	1	13.81448	.7988586	26.84167	8.474857	4.45771	.694752	.305248	0	0
3.	-5.44709	1	4.679744	0	26.00833	2.66125	.485876	.273349	0	0	0
4.	4.762911	18.3303	0	0	5910.8	0	4.21486	0	0	0	0
5.	-1.717091	0	4.916299	.5958354	0	1.109601	.873239	.772979	0	0	0
6.	-1.84709	0	6.807349	.4906485	0	1.234035	1.19337	.957559	0	0	0

6.	r	ln_lv_~t	CENTMAY	CENTMAY²	CKHgrow	CKHgrow²	NCENTra..	NCENTr..	NCENTra..		
	-2.084457	9.690597	.3896209	.1518045	3.135758	9.832977	-.70082	.4911486	-.142397		
	NCENTr..	CENTspr~m	CENTsp~²	Cjunjul~P	Cjunj~P²	RABCGROW	RABCGR~²	CENTHDI	LNpop_~m	LOGele~n	
	.0202769	-62.72373	3934.266	-34.10175	1162.929	-21.90803	479.9619	-.1815	3.401197	4.043051	
	CENTlat~e	SQRTru~f	SQRTrdi	SQRTFl~a	runoff~p	SQRTtr~v	roaddens	cl	fl	rk	rg
	-1.84709	0	6.807349	.4906485	0	1.234035	1.19337	.957559	0	0	0
7.	r	ln_lv_~t	CENTMAY	CENTMAY²	CKHgrow	CKHgrow²	NCENTra..	NCENTr..	NCENTra..		
	-2.071434	10.76222	-4.30038	18.49327	-1.93091	3.728412	-4.37082	19.10407	-5.26573		
	NCENTr..	CENTspr~m	CENTsp~²	Cjunjul~P	Cjunj~P²	RABCGROW	RABCGR~²	CENTHDI	LNpop_~m	LOGele~n	
	27.72792	9.27627	86.04919	-19.60175	384.2286	-.5747	.3302801	-.0585	2.639057	7.326466	
	CENTlat~e	SQRTru~f	SQRTrdi	SQRTFl~a	runoff~p	SQRTtr~v	roaddens	cl	fl	rk	rg
	.9929101	1.732051	3.764306	0	57.525	1.743379	.16798	.015541	0	0	0
107.	r	ln_lv_~t	CENTMAY	CENTMAY²	CKHgrow	CKHgrow²	NCENTra..	NCENTr..	NCENTra..		
	2.026067	14.11065	-.8003796	.6406075	.5690904	.3238638	-.2708196	.0733433	-.6657296		
	NCENTr..	CENTspr~m	CENTsp~²	Cjunjul~P	Cjunj~P²	RABCGROW	RABCGR~²	CENTHDI	LNpop_~m	LOGele~n	
	.443196	83.27627	6934.937	-16.85175	283.9814	21.4253	459.0435	.0225	7.381502	5.817111	
	CENTlat~e	SQRTru~f	SQRTrdi	SQRTFl~a	runoff~p	SQRTtr~v	roaddens	cl	fl	rk	rg
	3.65291	3.741657	7.186098	.1434133	319.55	2.297816	3.10155	.929238	0	0	0
108.	r	ln_lv_~t	CENTMAY	CENTMAY²	CKHgrow	CKHgrow²	NCENTra..	NCENTr..	NCENTra..		
	2.366752	13.90985	-9.600381	92.16731	-6.53091	42.65279	-7.070821	49.99651	-8.36573		
	NCENTr..	CENTspr~m	CENTsp~²	Cjunjul~P	Cjunj~P²	RABCGROW	RABCGR~²	CENTHDI	LNpop_~m	LOGele~n	
	69.98544	429.3763	184364	65.57325	4299.851	120.7586	14582.65	-.0335	6.118097	6.628041	
	CENTlat~e	SQRTru~f	SQRTrdi	SQRTFl~a	runoff~p	SQRTtr~v	roaddens	cl	fl	rk	rg
	3.65291	4.358899	3.065942	.296314	288.6417	2.515185	3.06188	0	0	0	0
109.	r	ln_lv_~t	CENTMAY	CENTMAY²	CKHgrow	CKHgrow²	NCENTra..	NCENTr..	NCENTra..		
	2.430087	11.98645	-9.29038	86.31115	-1.164243	1.355462	-8.91082	79.40271	-8.12573		
	NCENTr..	CENTspr~m	CENTsp~²	Cjunjul~P	Cjunj~P²	RABCGROW	RABCGR~²	CENTHDI	LNpop_~m	LOGele~n	
	66.02748	-27.72373	768.6052	-27.35175	748.1182	-5.908034	34.90487	-.0915	2.484907	7.560081	
	CENTlat~e	SQRTru~f	SQRTrdi	SQRTFl~a	runoff~p	SQRTtr~v	roaddens	cl	fl	rk	rg
	-2.51709	0	4.326662	0	0	.7609727	.147541	.00608	0	0	0
110.	r	ln_lv_~t	CENTMAY	CENTMAY²	CKHgrow	CKHgrow²	NCENTra..	NCENTr..	NCENTra..		
	2.690709	12.57705	5.799621	33.6356	4.635758	21.49025	2.962514	8.77649	3.500937		
	NCENTr..	CENTspr~m	CENTsp~²	Cjunjul~P	Cjunj~P²	RABCGROW	RABCGR~²	CENTHDI	LNpop_~m	LOGele~n	
	12.25656	-52.72373	2779.792	-31.10175	967.3188	-18.5747	345.0195	-.1185	3.135494	6.09357	
	CENTlat~e	SQRTru~f	SQRTrdi	SQRTFl~a	runoff~p	SQRTtr~v	roaddens	cl	fl	rk	rg
	-.7970908	1	4.554119	.2244388	27.35833	1.797932	.353798	.121226	0	0	0

B.) Leverage – Analysis:

```
Stem-and-leaf plot for lev (Leverage)

lev rounded to nearest multiple of .001
plot in units of .001

  0** | 70,79,83,91,95
  1** | 00,04,04,06,10,14,17,20,22,23,25,26,27,28,29,33,37,37,38,39, ... (27)
  1** | 53,55,60,64,64,64,72,72,73,77,77,79,80,82,82,83,85,90,92
  2** | 00,03,07,07,08,10,11,12,20,21,22,28,35,39,39,43,47,49
  2** | 50,51,56,57,64,75,76,83,91
  3** | 00,01,11,13,19,31,32,37,38
  3** | 54,60,73,74,78,81,87
  4** | 06,34,49
  4** | 63,72
  5** | 38
  5** | 78,91
  6** | 15,33
  6** |
  7** | 11,16,48
  7** |
  8** | 44
  8** | 53,86

. hilo lev district, show(5) high
5 highest observations on lev
```

lev	district
.7162943	Karak
.7484037	Kohistan
.8437145	Abbottabad
.8532006	Islamabad
.8862448	Lower Dir

Leverage versus Residual plot:

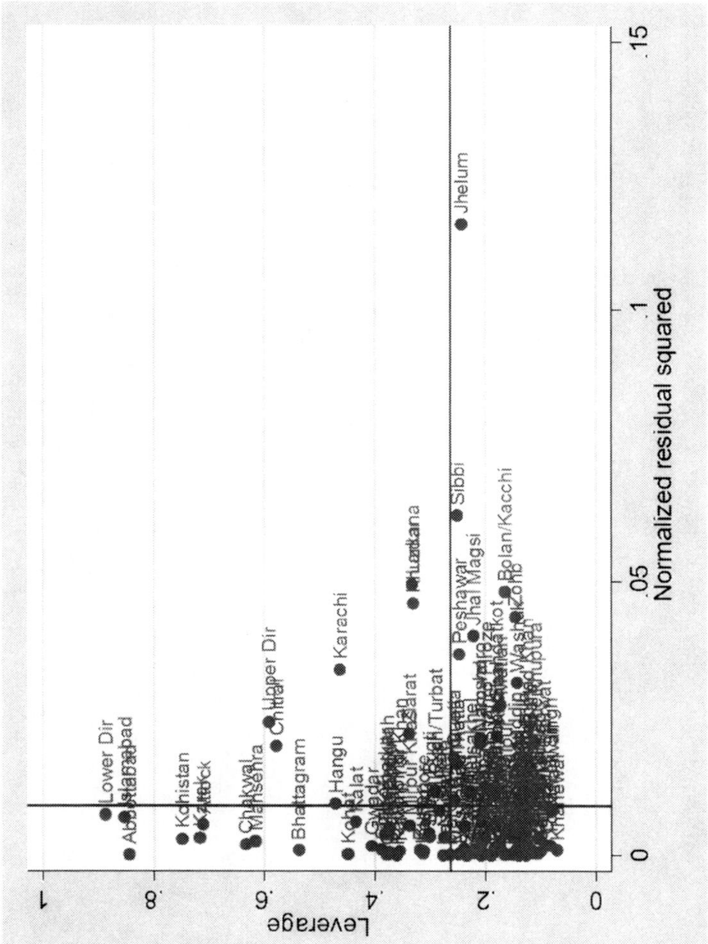

C.) Measures of overall influence: COOK's D and DFITS:

Cook's D:

104.

ln_lv_~t	CENTMAY	CENTMAY^2	CKHgrow	CKHgrow^2	NCENTra..	NCENTr..	NCENTra..	NCENTr..
13.72114	-7.71038	59.44995	-4.897577	23.98626	-6.854154	46.97942	-7.562397	57.18984

CENTspr~m	CENTsp~^{2}	Cjunjul~P	Cjunj~P^2	RABCGROW	RABCGR~^{2}	CENTHDI	LNpop_~m	LOGele~n	CENTlat~e
-6.353727	40.36985	-13.81925	190.9716	6.348632	40.30513	.0825	3.091043	7.644441	.0329092

SQRTru~f	SQRTrdi	SQRTFl~a	runoff~p	SQRTtr~v	roaddens	cl	fl	rk	rg
1.732051	3.773592	0	49.4	8.062258	.990909	0	0	0	0

district	d
Ziarat	.0476505

106.

ln_lv_~t	CENTMAY	CENTMAY^2	CKHgrow	CKHgrow^2	NCENTra..	NCENTr..	NCENTra..	NCENTr..
14.7181	-1.600381	2.561219	-1.63091	2.659869	-2.004153	4.01663	-1.832397	3.357677

CENTspr~m	CENTsp~^{2}	Cjunjul~P	Cjunj~P^2	RABCGROW	RABCGR~^{2}	CENTHDI	LNpop_~m	LOGele~n	CENTlat~e
75.27627	5666.517	114.3982	13086.96	36.75864	1351.197	.1855	6.781058	6.246107	3.352911

SQRTru~f	SQRTrdi	SQRTFl~a	runoff~p	SQRTtr~v	roaddens	cl	fl	rk	rg
15	0	0	4805.625	2.964848	38.5173	0	0	0	.697593

district	d
Islamabad	.7697053

107.

ln_lv_~t	CENTMAY	CENTMAY^2	CKHgrow	CKHgrow^2	NCENTra..	NCENTr..	NCENTra..	NCENTr..
14.11065	-.8003796	.6406075	.5690904	.3238638	-.2708196	.0733433	-.6657296	.443196

CENTspr~m	CENTsp~^{2}	Cjunjul~P	Cjunj~P^2	RABCGROW	RABCGR~^{2}	CENTHDI	LNpop_~m	LOGele~n	CENTlat~e
83.27627	6934.937	-16.85175	283.9814	21.4253	459.0435	.0225	7.381502	5.817111	3.65291

SQRTru~f	SQRTrdi	SQRTFl~a	runoff~p	SQRTtr~v	roaddens	cl	fl	rk	rg
3.741657	7.186098	.1434133	319.55	2.297816	3.10155	.929238	0	0	0

district	d
Peshawar	.0447612

108.

ln_lv_~t	CENTMAY	CENTMAY^2	CKHgrow	CKHgrow^2	NCENTra..	NCENTr..	NCENTra..	NCENTr..
13.90985	-9.600381	92.16731	-6.53091	42.65279	-7.070821	49.99651	-8.36573	69.98544

CENTspr~m	CENTsp~^{2}	Cjunjul~P	Cjunj~P^2	RABCGROW	RABCGR~^{2}	CENTHDI	LNpop_~m	LOGele~n	CENTlat~e
429.3763	184364	65.57325	4299.851	120.7586	14582.65	-.0335	6.118097	6.628041	3.65291

SQRTru~f	SQRTrdi	SQRTFl~a	runoff~p	SQRTtr~v	roaddens	cl	fl	rk	rg
4.358899	3.065942	.296314	288.6417	2.515185	3.06188	0	0	0	0

district	d
Lower Dir	1.423945

109.

ln_lv_~t	CENTMAY	CENTMAY^2	CKHgrow	CKHgrow^2	NCENTra..	NCENTr..	NCENTra..	NCENTr..
11.98645	-9.29038	86.31115	-1.164243	1.355462	-8.91082	79.40271	-8.12573	66.02748

CENTspr~m	CENTsp~^{2}	Cjunjul~P	Cjunj~P^2	RABCGROW	RABCGR~^{2}	CENTHDI	LNpop_~m	LOGele~n	CENTlat~e
-27.72373	768.6052	-27.35175	748.1182	-5.908034	34.90487	-.0915	2.484907	7.560081	-2.51709

SQRTru~f	SQRTrdi	SQRTFl~a	runoff~p	SQRTtr~v	roaddens	cl	fl	rk	rg
0	4.326662	0	0	.7609727	.147541	.00608	0	0	0

110.

ln_lv_~t	CENTMAY	CENTMAY^2	CKHgrow	CKHgrow^2	NCENTra..	NCENTr..	NCENTra..	NCENTr..
12.57705	5.799621	33.6356	4.635758	21.49025	2.962514	8.77649	3.500937	12.25656

CENTspr~m	CENTsp~^{2}	Cjunjul~P	Cjunj~P^2	RABCGROW	RABCGR~^{2}	CENTHDI	LNpop_~m	LOGele~n	CENTlat~e
-52.72373	2779.792	-31.10175	967.3188	-18.5747	345.0195	-.1185	3.135494	6.09357	-.7970908

SQRTru~f	SQRTrdi	SQRTFl~a	runoff~p	SQRTtr~v	roaddens	cl	fl	rk	rg
1	4.554119	.2244388	27.35833	1.797932	.353798	.121226	0	0	0

district	d
Sibbi	.0776444

DFITS:

1.	ln_lv~t	CENTMAY	CENTMAY²	CKHgrow	CKHgrow²	NCENTra..
	NCENTr..	NCENTra..	NCENTr..			
	11.32558	.59962	.3595441	-.0975776	.0095214	.6958476
	.4842038	.7009372	.491313			

THDI	CENTspr~m	CENTsp~²	Cjunjul~P	Cjunj~P²	RABCGROW	RABCGR~²	CEN
	LNpop_~m	LOGele~n	CENTlat~e				
1535	43.27627	1872.836	93.89825	8816.881	15.75863	248.3345	.
	5.56452	5.811141	2.58291				

cl	SQRTru~f	SQRTrdi	SQRTFl~a	runoff~p	SQRTtr~v	roaddens	
	fl	rk	rg				
5247	11.74734	6.900724	.5641942	3240.7	3.598795	3.53549	.48
	0	0	.01607				

dfit	district
-2.155324	Jhelum

2.	ln_lv~t	CENTMAY	CENTMAY²	CKHgrow	CKHgrow²	NCENTra..
	NCENTr..	NCENTra..	NCENTr..			
	12.16405	4.59962	21.1565	2.802426	7.85359	4.162515
	17.32653	4.200937	17.64787			

THDI	CENTspr~m	CENTsp~²	Cjunjul~P	Cjunj~P²	RABCGROW	RABCGR~²	CEN
	LNpop_~m	LOGele~n	CENTlat~e				
0675	-65.72373	4319.609	-34.60175	1197.281	-24.5747	603.9158	-.
	5.560682	3.871201	-2.797091				

cl	SQRTru~f	SQRTrdi	SQRTFl~a	runoff~p	SQRTtr~v	roaddens	
	fl	rk	rg				
4752	1	13.81448	.7988586	26.84167	8.474857	4.45771	.69
	.305248	0	0				

dfit	district
-1.781471	Larkana

3.	ln_lv~t	CENTMAY	CENTMAY²	CKHgrow	CKHgrow²	NCENTra..
	NCENTr..	NCENTra..	NCENTr..			
	11.27621	1.19962	1.439089	-.6642432	.4412191	5.329179
	28.40015	6.13427	37.62926			

THDI	CENTspr~m	CENTsp~²	Cjunjul~P	Cjunj~P²	RABCGROW	RABCGR~²	CEN
	LNpop_~m	LOGele~n	CENTlat~e				
1725	-60.72373	3687.371	-7.85175	61.64997	-20.24137	409.713	.
	7.935587	4.49981	-5.44709				

cl	SQRTru~f	SQRTrdi	SQRTFl~a	runoff~p	SQRTtr~v	roaddens	
	fl	rk	rg				
3349	1	4.679744	0	26.00833	2.66125	.485876	.27
	0	0	0				

dfit	district
-2.159476	Karachi

4.	ln_lv~t	CENTMAY	CENTMAY²	CKHgrow	CKHgrow²	NCENTra..
	NCENTr..	NCENTra..	NCENTr..			
	13.36068	-7.80038	60.84592	-6.53091	42.65279	-5.304152
	28.13403	-7.632397	58.25349			

	CENTspr~m	CENTsp~²	Cjunjul~P	Cjunj~P²	RABCGROW	RABCGR~²	CEN
THDI	LNpop_~m	LOGele~n	CENTlat~e				
	369.2763	136365	54.89825	3013.818	124.092	15398.82	-.
0575	5.049856	7.453562	4.762911				

	SQRTru~f	SQRTrdi	SQRTFl~a	runoff~p	SQRTtr~v	roaddens	
cl	fl	rk	rg				
	18.3303	0	0	5910.8	0	4.21486	
0	0	0	0				

dfit	district
-2.708978	Upper Dir

6.	ln_lv~t	CENTMAY	CENTMAY²	CKHgrow	CKHgrow²	NCENTra..
	NCENTr..	NCENTra..	NCENTr..			
	9.690597	.3896209	.1518045	3.135758	9.832977	-.70082
	.4911486	-.142397	.0202769			

	CENTspr~m	CENTsp~²	Cjunjul~P	Cjunj~P²	RABCGROW	RABCGR~²	CEN
THDI	LNpop_~m	LOGele~n	CENTlat~e				
	-62.72373	3934.266	-34.10175	1162.929	-21.90803	479.9619	-.
1815	3.401197	4.043051	-1.84709				

	SQRTru~f	SQRTrdi	SQRTFl~a	runoff~p	SQRTtr~v	roaddens	
cl	fl	rk	rg				
	0	6.807349	.4906485	0	1.234035	1.19337	.95
7559	0	0	0				

dfit	district
-1.11469	Jhal Magsi

8.	ln_lv~t	CENTMAY	CENTMAY²	CKHgrow	CKHgrow²	NCENTra..
	NCENTr..	NCENTra..	NCENTr..			
	13.15064	-7.700379	59.29584	-4.497577	20.2282	-6.804153
	46.2965	-8.332397	69.42883			

	CENTspr~m	CENTsp~²	Cjunjul~P	Cjunj~P²	RABCGROW	RABCGR~²	CEN
THDI	LNpop_~m	LOGele~n	CENTlat~e				
	135.2763	18299.67	-44.35175	1967.078	19.09197	364.5033	-.
0105	3.044523	7.87967	5.49291				

	SQRTru~f	SQRTrdi	SQRTFl~a	runoff~p	SQRTtr~v	roaddens	
cl	fl	rk	rg				
	3.464102	4.489989	0	201.8	1.841889	14.6096	.01
8075	0	.039864	0				

dfit	district
-2.331436	Chitral

	ln_lv_~t	CENTMAY	CENTMAY²	CKHgrow	CKHgrow²	NCENTra..
12.	12.65735	-8.80038	77.44669	-1.63091	2.659869	-6.07082

NCENTr..	NCENTra..	NCENTr..
36.85485	-6.899063	47.59707

THDI	CENTspr~m	CENTsp~²	Cjunjul~P	Cjunj~P²	RABCGROW	RABCGR~²	CEN
0355	182.2763	33224.64	111.6482	12465.33	72.09196	5197.251	.

LNpop_~m	LOGele~n	CENTlat~e
5.225747	5.97381	3.802912

cl	SQRTru~f	fl	SQRTrdi	rk	SQRTFl~a	rg	runoff~p	SQRTtr~v	roaddens	
0316	1	0	2.880972	.604915	.0863373	.320691	15.81667	3.913523	2.86322	.07

dfit	district
-1.985393	Attock

	ln_lv_~t	CENTMAY	CENTMAY²	CKHgrow	CKHgrow²	NCENTra..
93.	12.74285	1.299621	1.689014	.5690904	.3238638	.62918

NCENTr..	NCENTra..	NCENTr..
.3958675	.2676038	.0716118

THDI	CENTspr~m	CENTsp~²	Cjunjul~P	Cjunj~P²	RABCGROW	RABCGR~²	CEN
0005	41.27627	1703.731	27.89825	778.3124	8.4253	70.98567	.

LNpop_~m	LOGele~n	CENTlat~e
4.85203	6.522093	4.092909

cl	SQRTru~f	fl	SQRTrdi	rk	SQRTFl~a	rg	runoff~p	SQRTtr~v	roaddens	
2678	19.15724	0	0	.907322	0	0	8777.417	2.147318	.908207	.09

dfit	district
1.529845	Karak

	ln_lv_~t	CENTMAY	CENTMAY²	CKHgrow	CKHgrow²	NCENTra..
94.	13.00511	-11.80038	139.249	-4.464244	19.92948	-10.37082

NCENTr..	NCENTra..	NCENTr..
107.5539	-11.9324	142.3821

THDI	CENTspr~m	CENTsp~²	Cjunjul~P	Cjunj~P²	RABCGROW	RABCGR~²	CEN
1455	312.2763	97516.47	178.1483	31736.8	121.4253	14744.1	-.

LNpop_~m	LOGele~n	CENTlat~e
4.143135	7.5807	4.852911

cl	SQRTru~f	fl	SQRTrdi	rk	SQRTFl~a	rg	runoff~p	SQRTtr~v	roaddens
0	19.41649	0	3.832753	0	.0750808	0	4944.983	.1469046	.538677

dfit	district
1.703587	Kohistan

```
99. | ln_lv_~t  |    CENTMAY | CENTMAY² |   CKHgrow  | CKHgrow² | NCENTra..
  | NCENTr..  | NCENTra..  | NCENTr.. |
  | 13.25053  | -.7003793  | .4905311 | -4.330911  | 18.75679 |  .2625132
  | .0689132  | -.5323964  |  .283446 |

     | CENTspr~m | CENTsp~² | Cjunjul~P | Cjunj~P² |  RABCGROW | RABCGR~² | CEN
THDI | LNpop_~m  | LOGele~n | CENTlat~e |
     | 82.27627  | 6769.384 |  39.39825 | 1552.222 |  18.09197 | 327.3193 | -.
1225 | 5.659482  | 6.921658 |  3.182909 |

     | SQRTru~f  | SQRTrdi  | SQRTFl~a | runoff~p | SQRTtr~v | roaddens |
  cl |       fl  |      rk  |       rg |
     | 8.062258  | 1.292285 |        0 | 1496.625 | .5508493 |  6.25773 | .16
7071 |        0  | .832929  |        0 |

  |                         district |
  |        dfit                      |
  |                            Hangu |
  |     1.141048                     |

04. | ln_lv_~t  |    CENTMAY | CENTMAY² |   CKHgrow  | CKHgrow² | NCENTra..
  | NCENTr..  | NCENTra..  | NCENTr.. |
  | 13.72114  |  -7.71038  | 59.44995 | -4.897577  | 23.98626 | -6.854154
  | 46.97942  | -7.562397  | 57.18984 |

     | CENTspr~m | CENTsp~² | Cjunjul~P | Cjunj~P² |  RABCGROW | RABCGR~² | CEN
THDI | LNpop_~m  | LOGele~n | CENTlat~e |
     | -6.353727 | 40.36985 | -13.81925 | 190.9716 |  6.348632 | 40.30513 |   .
0825 | 3.091043  | 7.644441 |  .0329092 |

     | SQRTru~f  | SQRTrdi  | SQRTFl~a | runoff~p | SQRTtr~v | roaddens |
  cl |       fl  |      rk  |       rg |
     | 1.732051  | 3.773592 |        0 |     49.4 | 8.062258 |  .990909 |
   0 |        0  |        0 |        0 |

  |                         district |
  |        dfit                      |
  |                           Ziarat |
  |     1.188296                     |

06. | ln_lv_~t  |    CENTMAY | CENTMAY² |   CKHgrow  | CKHgrow² | NCENTra..
  | NCENTr..  | NCENTra..  | NCENTr.. |
  |  14.7181  | -1.600381  | 2.561219 |  -1.63091  | 2.659869 | -2.004153
  |  4.01663  | -1.832397  | 3.357677 |

     | CENTspr~m | CENTsp~² | Cjunjul~P | Cjunj~P² |  RABCGROW | RABCGR~² | CEN
THDI | LNpop_~m  | LOGele~n | CENTlat~e |
     | 75.27627  | 5666.517 |  114.3982 | 13086.96 |  36.75864 | 1351.197 |   .
1855 | 6.781058  | 6.246107 |  3.352911 |

     | SQRTru~f  | SQRTrdi  | SQRTFl~a | runoff~p | SQRTtr~v | roaddens |
  cl |       fl  |      rk  |       rg |
     |       15  |        0 |        0 | 4805.625 | 2.964848 |  38.5173 |
   0 |        0  |        0 |  .697593 |

  |                         district |
  |        dfit                      |
  |                        Islamabad |
  |     4.810738                     |
```

07.

ln_lv_~t	CENTMAY	CENTMAY²	CKHgrow	CKHgrow²	NCENTra..
14.11065	-.8003796	.6406075	.5690904	.3238638	-.2708196

NCENTr..	NCENTra..	NCENTr..
.0733433	-.6657296	.443196

THDI	CENTspr~m	CENTsp~²	Cjunjul~P	Cjunj~P²	RABCGROW	RABCGR~²	CEN
0225	83.27627	6934.937	-16.85175	283.9814	21.4253	459.0435	.

LNpop_~m	LOGele~n	CENTlat~e
7.381502	5.817111	3.65291

cl	SQRTru~f	SQRTrdi	SQRTFl~a	runoff~p	SQRTtr~v	roaddens	
9238	3.741657	7.186098	.1434133	319.55	2.297816	3.10155	.92

fl	rk	rg
0	0	0

dfit	district
1.160962	Peshawar

08.

ln_lv_~t	CENTMAY	CENTMAY²	CKHgrow	CKHgrow²	NCENTra..
13.90985	-9.600381	92.16731	-6.53091	42.65279	-7.070821

NCENTr..	NCENTra..	NCENTr..
49.99651	-8.36573	69.98544

THDI	CENTspr~m	CENTsp~²	Cjunjul~P	Cjunj~P²	RABCGROW	RABCGR~²	CEN
0335	429.3763	184364	65.57325	4299.851	120.7586	14582.65	-.

LNpop_~m	LOGele~n	CENTlat~e
6.118097	6.628041	3.65291

cl	SQRTru~f	SQRTrdi	SQRTFl~a	runoff~p	SQRTtr~v	roaddens
0	4.358899	3.065942	.296314	288.6417	2.515185	3.06188

fl	rk	rg
0	0	0

dfit	district
6.60608	Lower Dir

09.

ln_lv_~t	CENTMAY	CENTMAY²	CKHgrow	CKHgrow²	NCENTra..
11.98645	-9.29038	86.31115	-1.164243	1.355462	-8.91082

NCENTr..	NCENTra..	NCENTr..
79.40271	-8.12573	66.02748

THDI	CENTspr~m	CENTsp~²	Cjunjul~P	Cjunj~P²	RABCGROW	RABCGR~²	CEN
0915	-27.72373	768.6052	-27.35175	748.1182	-5.908034	34.90487	-.

LNpop_~m	LOGele~n	CENTlat~e
2.484907	7.560081	-2.51709

cl	SQRTru~f	SQRTrdi	SQRTFl~a	runoff~p	SQRTtr~v	roaddens	
0608	0	4.326662	0	0	.7609727	.147541	.0

fl	rk	rg
0	0	0

dfit	district
1.70833	Khuzdar

10.	ln_lv_~t	CENTMAY	CENTMAY²	CKHgrow	CKHgrow²	NCENTra..
	12.57705	5.799621	33.6356	4.635758	21.49025	2.962514

	NCENTr..	NCENTra..	NCENTr..
	8.77649	3.500937	12.25656

	CENTspr~m	CENTsp~²	Cjunjul~P	Cjunj~P²	RABCGROW	RABCGR~²	CEN
	-52.72373	2779.792	-31.10175	967.3188	-18.5747	345.0195	-.

THDI	LNpop_~m	LOGele~n	CENTlat~e
1185	3.135494	6.09357	-.7970908

	SQRTru~f	SQRTrdi	SQRTFl~a	runoff~p	SQRTtr~v	roaddens	
	1	4.554119	.2244388	27.35833	1.797932	.353798	.12

cl	fl	rk	rg
1226	0	0	0

dfit	district
1.557289	Sibbi

D.) DFBETA – Analysis –Effect of Variable exclusion:

	district	_dfbeta_1	_dfbeta_2	_dfbeta_3	_dfbeta_4	_dfbeta_5	_dfbeta_6	_dfbeta_7	_dfbeta_8
1.	Jhelum	-.1248929	.0974256	.0933987	.0799813	.0890354	-.13174	-.088872	.1219255

_dfbeta_9	_dfbet~10	_dfbet~11	_dfbet~12	_dfbet~13	_dfbet~14	_dfbet~15	_dfbet~16	_dfbet~17
.0820176	-.4567365	-.7670341	-.1414247	-.139884	.6509132	-.8209531	.6499426	-.3282271

_dfbe~18	_dfbet~19	_dfbet~20	_dfbet~21	_dfbet~22	_dfbet~23	_dfbet~24	_dfbet~25	_dfbet~26
-.041513	-.0173267	-.0932798	-.6916241	.0233196	-.118942	.0504304	-.2640803	.8292215

_dfbet~27	_dfbet~28
.3760262	.6692916

	district	_dfbeta_1	_dfbeta_2	_dfbeta_3	_dfbeta_4	_dfbeta_5	_dfbeta_6	_dfbeta_7	_dfbeta_8
2.	Larkana	.2379603	-.2393747	.0123958	-.2141492	-.2930196	-.2405661	.136868	.4220059

_dfbeta_9	_dfbet~10	_dfbet~11	_dfbet~12	_dfbet~13	_dfbet~14	_dfbet~15	_dfbet~16	_dfbet~17
-.0469241	.0289624	.4866699	-.1488523	.0520339	-.0957873	.594766	.3837582	.1628247

_dfbe~18	_dfbet~19	_dfbet~20	_dfbet~21	_dfbet~22	_dfbet~23	_dfbet~24	_dfbet~25	_dfbet~26
.1711381	-.3532196	-.6779674	-.2646624	.2453532	-.9904184	-.2844285	.0756316	-.1394406

_dfbet~27	_dfbet~28
-.3622537	-.326476

	district	_dfbeta_1	_dfbeta_2	_dfbeta_3	_dfbeta_4	_dfbeta_5	_dfbeta_6	_dfbeta_7	_dfbeta_8
3.	Karachi	.4970506	.4042648	-.2791764	.030771	.3024497	.4130647	-.4447099	-.599086

_dfbeta_9	_dfbet~10	_dfbet~11	_dfbet~12	_dfbet~13	_dfbet~14	_dfbet~15	_dfbet~16	_dfbet~17
-.0679201	-.06332	.2237186	-.3643946	.1830492	.0107978	-.4988519	-1.478154	-.8016412

_dfbe~18	_dfbet~19	_dfbet~20	_dfbet~21	_dfbet~22	_dfbet~23	_dfbet~24	_dfbet~25	_dfbet~26
.3252877	.2455789	.3825981	-.0681613	-.1936637	.5216751	.4309504	.5545934	-.0593354

_dfbet~27	_dfbet~28
-.0162758	.0167611

4.	district	_dfbeta_1	_dfbeta_2	_dfbeta_3	_dfbeta_4	_dfbeta_5	_dfbeta_6	_dfbeta_7	_dfbeta_8
	Upper Dir	-.2199857	.1317735	.1199194	.1981194	-.4218527	.2939001	.5336418	-.3646912

_dfbeta_9	_dfbet~10	_dfbet~11	_dfbet~12	_dfbet~13	_dfbet~14	_dfbet~15	_dfbet~16	_dfbet~17
.2056471	.2803599	-.006657	1.200627	-.2951931	-.7736226	-.1944778	.3065146	-.2100986

_dfbe~18	_dfbet~19	_dfbet~20	_dfbet~21	_dfbet~22	_dfbet~23	_dfbet~24	_dfbet~25	_dfbet~26
.3528185	-.0179024	.4885339	-.1216204	-.2174788	.1176366	-.3026688	-.4027236	-.485671

_dfbet~27	_dfbet~28
.0739368	-.1208212

5.	district	_dfbeta_1	_dfbeta_2	_dfbeta_3	_dfbeta_4	_dfbeta_5	_dfbeta_6	_dfbeta_7	_dfbeta_8
	Bolan/Kacchi	.0658042	.241096	-.5617582	-.6119926	.1652088	-.0401302	-.099759	-.0540541

_dfbeta_9	_dfbet~10	_dfbet~11	_dfbet~12	_dfbet~13	_dfbet~14	_dfbet~15	_dfbet~16	_dfbet~17
-.0354566	-.0065094	-.1476892	.0267359	-.0592316	.0793905	.3523833	.0085043	.0994976

_dfbe~18	_dfbet~19	_dfbet~20	_dfbet~21	_dfbet~22	_dfbet~23	_dfbet~24	_dfbet~25	_dfbet~26
.0219189	.1273342	.0473815	-.1529618	-.1427946	-.0077378	-.0139442	-.0560355	.0507627

_dfbet~27	_dfbet~28
.0625617	-.0297478

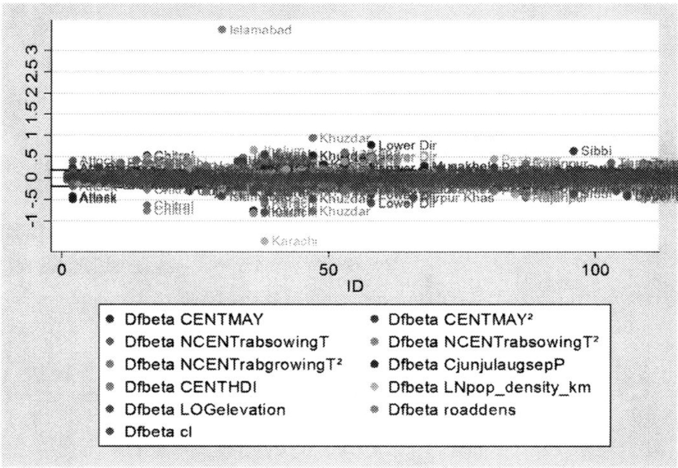

E.) Tests for Normality of Residuals :

Kernel-Density plot:

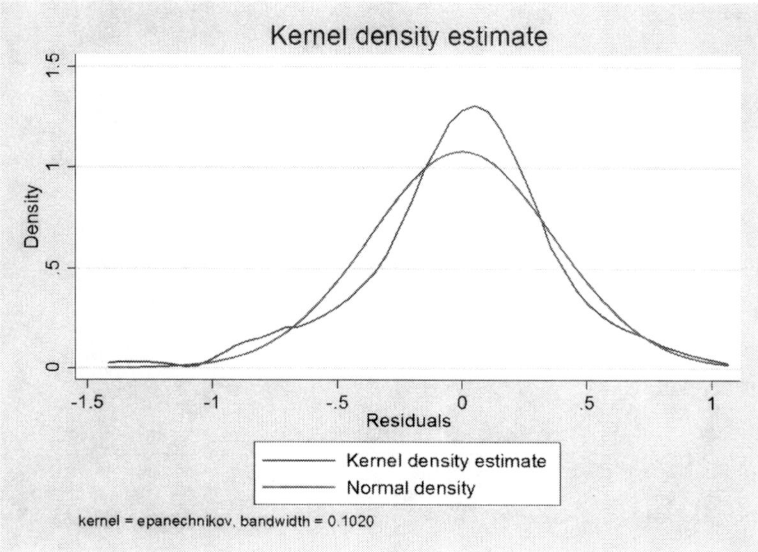

Standardized normal probability plot (P-P):

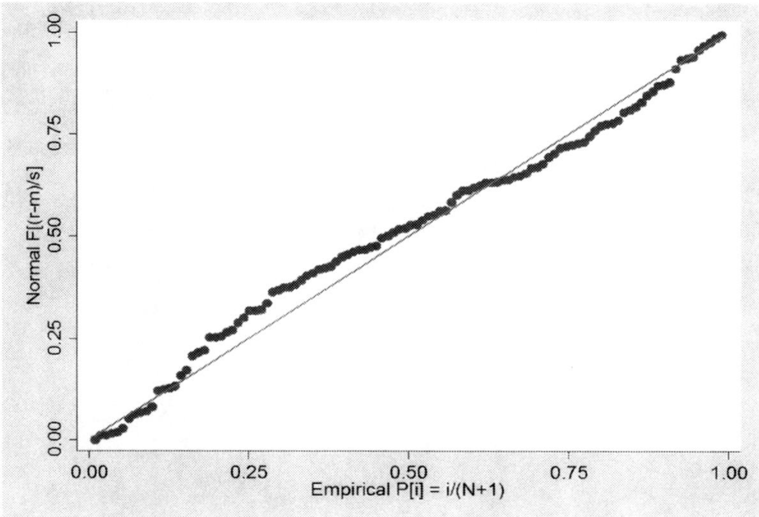

Quantile-Quantile plot (Q-Q):

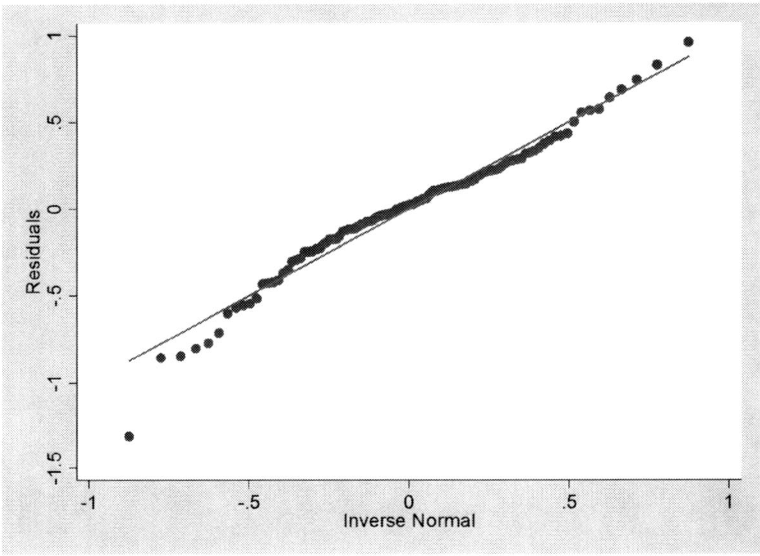

Hamilton's Inter-Quartile-Range Test:

```
     mean= -4.3e-10          std.dev.=  .3697           (n= 110)
   median=  .0216     pseudo std.dev.=  .2901         (IQR=  .3913)
  10 trim=  .0135
                                                   low          high
                                                  -------------------
                                 inner fences     -.7618        .8035
                            # mild outliers       5             2
                            % mild outliers       4.55%         1.82%

                                 outer fences     -1.349          1.39
                            # severe outliers     0             0
                            % severe outliers     0.00%         0.00%
```

F.) Added Variable Plots

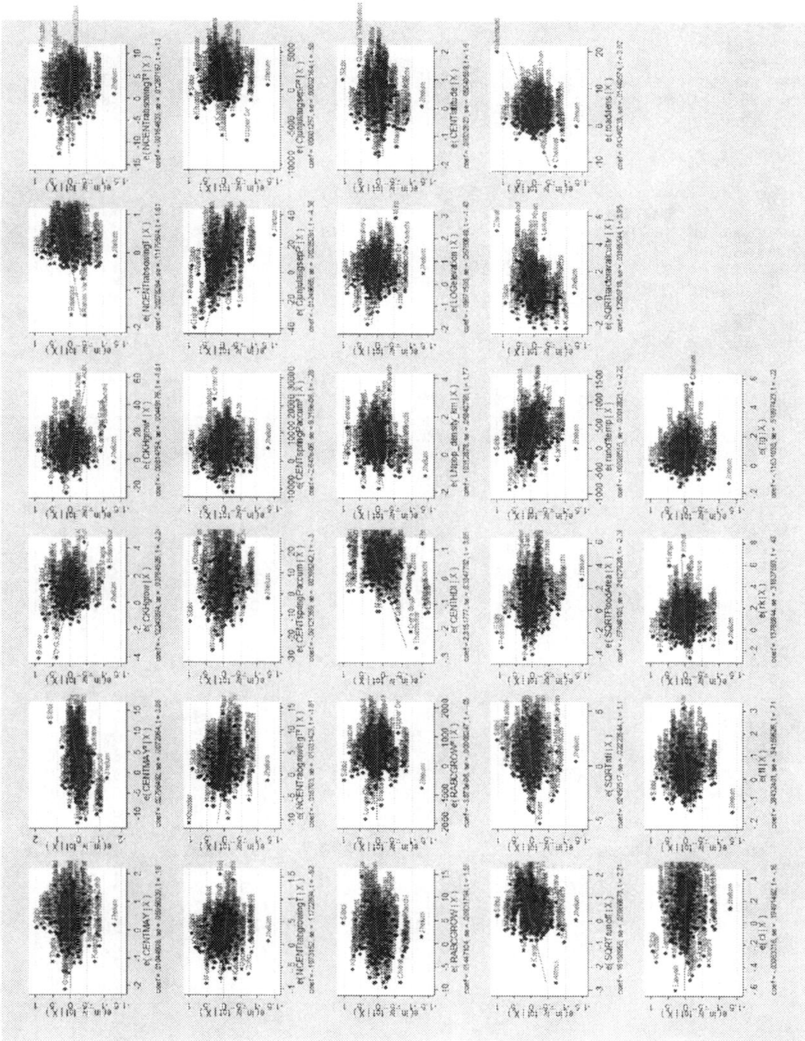

G.) Spatial Dependence Regressions and diagnostic tests:

```
Regression
SUMMARY OF OUTPUT: ORDINARY LEAST SQUARES ESTIMATION
Data set            :  Study_area_final
Dependent Variable  :    LN_LV_TOT  Number of Observations:  110
Mean dependent var  :      11,9273  Number of Variables   :   30
S.D. dependent var  :      1,00643  Degrees of Freedom    :   80


R-squared           :     0,798325  F-statistic           :     10,9199
Adjusted R-squared  :     0,725218  Prob(F-statistic)     :1,33017e-017
Sum squared residual:      22,4703  Log likelihood        :    -68,7275
Sigma-square        :     0,280879  Akaike info criterion :     197,455
S.E. of regression  :      0,52998  Schwarz criterion     :     278,469
Sigma-square ML     :     0,204275
S.E of regression ML:     0,451968
```

Variable	Coefficient	Std.Error	t-Statistic	Probability
CONSTANT	10,4603	0,6021267	17,37226	0,0000000
CENTMAY	0,02681058	0,07297052	0,3674166	0,7142776
V16	0,02339394	0,009147065	2,557536	0,0124305
CKHGROW	-0,2044474	0,05462166	-3,742973	0,0003410
V18	-0,01149169	0,005843617	-1,966538	0,0527044
NCENTRABSO	0,1487512	0,1223565	1,21572	0,2276665
NCENTRAB_A	-0,0008509784	0,01533852	-0,05547984	0,9559109
NCENTRABGR	-0,05868325	0,115666	-0,507351	0,6133031
NCENTRAB_B	-0,01510243	0,01150636	-1,31253	0,1930944
CENTSPRING	-0,0003606957	0,005027613	-0,07174293	0,9429947
CENTSPRI_A	-1,183717e-005	1,206924e-005	-0,9807719	0,3296623
CJUNJULAUG	-0,006459933	0,003257595	-1,983038	0,0507950
CJUNJULA_A	-8,934976e-006	2,632663e-005	-0,3393893	0,7352076
RABCGROW	0,01310354	0,01197959	1,093822	0,2773154
V28	6,144973e-005	0,0001077217	0,5704488	0,5699731
CENTHDI	0,03859833	0,01727316	2,234584	0,0282344
LNPOP_DENS	0,1743556	0,06611868	2,63701	0,0100456
LOGELEVATI	-0,02095171	0,07799945	-0,2686136	0,7889186
CENTLATITU	0,1150004	0,08405602	1,36814	0,1750988
SQRTRUNOFF	0,01969743	0,02805587	0,7020789	0,4846670
SQRTRDI	0,09002087	0,05852012	1,538289	0,1279239
SQRTFLOODA	0,04059541	0,04541814	0,8938149	0,3741020
RDITEMP	-0,0001661845	0,0001752228	-0,9484181	0,3457726
IRRITEMP	1,848189e-006	4,26495e-006	0,4333436	0,6659334
SQRTTRACTO	0,08580538	0,03727222	2,302127	0,0239282
ROADDENS	0,05446811	0,01532122	3,555076	0,0006379
CL	-0,1611424	0,1120009	-1,43876	0,1541196
FL	0,02766827	0,04166581	0,664052	0,5085667
RK	0,424925	0,597379	0,7113155	0,4789570
RG	-0,09434681	0,09870848	-0,9558126	0,3420463

```
REGRESSION DIAGNOSTICS
MULTICOLLINEARITY CONDITION NUMBER    47,808898
TEST ON NORMALITY OF ERRORS
TEST                  DF          VALUE              PROB
Jarque-Bera            2            3,48656          0,1749456
```

RVF-Plot:

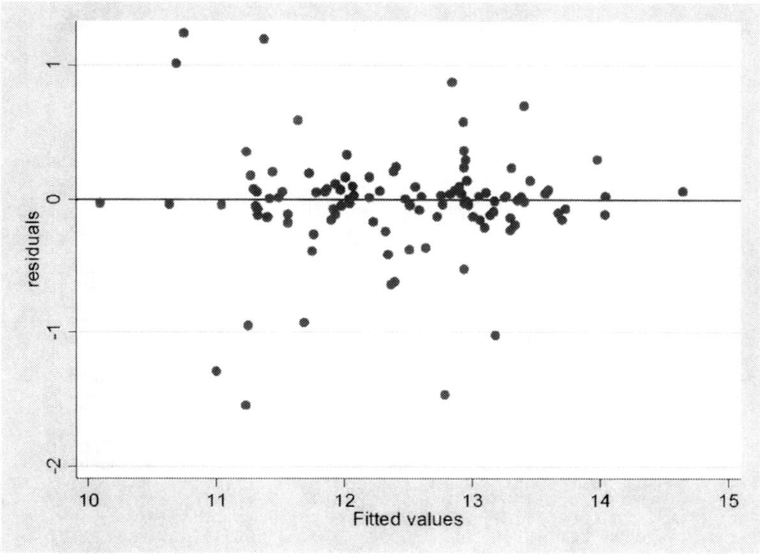

Tests for Heteroskedasticity:

```
DIAGNOSTICS FOR HETEROSKEDASTICITY
RANDOM COEFFICIENTS
TEST                  DF          VALUE          PROB
Breusch-Pagan test    29          47,26436       0,0174976
Koenker-Bassett test  29           35,8909       0,1767643
```

Spatial Lag Model:

```
Regression
SUMMARY OF OUTPUT: SPATIAL LAG MODEL - MAXIMUM LIKELIHOOD ESTIMATION
Data set              : Study_area_final
Spatial Weight        : Study_area_final.gal
Dependent Variable    :    LN_LV_TOT  Number of Observations:  110
Mean dependent var    :      11,9273  Number of Variables   :   31
S.D. dependent var    :      1,00643  Degrees of Freedom    :   79
Lag coeff.   (Rho)    :   0,00485354


R-squared             :     0,798633  Log likelihood        :     -68,6438
Sq. Correlation       : -             Akaike info criterion :      199,288
Sigma-square          :     0,203963  Schwarz criterion     :      283,002
S.E of regression     :     0,451623
```

Variable	Coefficient	Std.Error	z-value	Probability
W_LN_LV_TOT	0,004853539	0,01190416	0,4077179	0,6834808
CONSTANT	10,41516	0,5258501	19,80632	0,0000000
CENTMAY	0,02923775	0,06240862	0,468489	0,6394349
V16	0,02317433	0,007806753	2,968498	0,0029927
CKHGROW	-0,2043333	0,0465463	-4,389893	0,0000113
V18	-0,01114132	0,005046484	-2,207739	0,0272624
NCENTRABSO	0,1478084	0,1042728	1,417516	0,1563322
NCENTRAB_A	-0,0003486735	0,01312044	-0,02657482	0,9787988
NCENTRABGR	-0,06075782	0,0986986	-0,6155895	0,5381654
NCENTRAB_B	-0,01540989	0,009837356	-1,566467	0,1172394
CENTSPRING	3,747363e-005	0,00438736	0,008541271	0,9931851
CENTSPRI_A	-1,262341e-005	1,046139e-005	-1,206667	0,2275606
CJUNJULAUG	-0,006406675	0,002778146	-2,306097	0,0211051
CJUNJULA_A	-7,448427e-006	2,275885e-005	-0,327276	0,7434593
RABCGROW	0,01271725	0,01024405	1,241428	0,2144479
V28	6,127411e-005	9,179787e-005	0,6674894	0,5044594
CENTHDI	0,03781779	0,01483567	2,549112	0,0107998
LNPOP_DENS	0,1785029	0,05733013	3,113596	0,0018484
LOGELEVATI	-0,02033977	0,06650862	-0,3058215	0,7597407
CENTLATITU	0,1094297	0,07296948	1,499664	0,1337015
SQRTRUNOFF	0,01770191	0,02443257	0,724521	0,4687458
SQRTRDI	0,08604782	0,0508861	1,690989	0,0908389
SQRTFLOODA	0,03910969	0,03884595	1,006789	0,3140360
RDITEMP	-0,0001610398	0,0001499409	-1,074022	0,2828130
IRRITEMP	1,88144e-006	3,635259e-006	0,5175533	0,6047700
SQRTTRACTO	0,08521539	0,03179421	2,680218	0,0073575
ROADDENS	0,05486004	0,01309287	4,190071	0,0000279
CL	-0,1650764	0,09603998	-1,71883	0,0856452
FL	0,02945096	0,03575533	0,8236803	0,4101211
RK	0,4086784	0,5102698	0,8009066	0,4231856
RG	-0,09592931	0,08421269	-1,139131	0,2546485

```
DIAGNOSTICS FOR SPATIAL DEPENDENCE
SPATIAL LAG DEPENDENCE FOR WEIGHT MATRIX : Study_area_final.gal
TEST                                        DF       VALUE        PROB
Likelihood Ratio Test                       1       0,1673493     0,6824784
======================== END OF REPORT ==============================


Regression
SUMMARY OF OUTPUT: SPATIAL ERROR MODEL - MAXIMUM LIKELIHOOD ESTIMATION
Data set              : Study_area_final
Spatial Weight        : Study_area_final.gal
Dependent Variable    :    LN_LV_TOT   Number of Observations:  110
Mean dependent var    :   11,927273  Number of Variables    :   30
S.D. dependent var    :    1,006426  Degrees of Freedom     :   80
Lag coeff. (Lambda)   :    0,041721
```

H.) Summary Statistics Standard Ricardian Model

		Total Farms				Rainfed Farms				Irrigated Farms			
Variable	**Unit**	**Mean**	**Std. Dev.**	**Min**	**Max**	**Mean**	**Std. Dev.**	**Min**	**Max**	**Mean**	**Std. Dev.**	**Min**	**Max**
Land Value	$/ha	1882.563	1864.853	81.80612	12479.68	1461.097	2024.16	30.73118	13391.06	2306.156	3352.872	30.07491	30506.6
Summer Sowing Temp	(°C)	29.40038	4.48073	17.6	35.2	29.01413	4.611361	17.6	35.2	29.39764	4.501334	17.6	35.2
Summer Growing Temp	(°C)	30.03091	2.883603	18.23333	34.66667	29.88633	3.094001	22.56667	35.36667	30.17031	2.980283	22.56667	35.36667
Winter Sowing Temp	(°C)	18.5	3.879542	9.633333	24.56667	17.99936	4.049795	8	24.53333	18.31948	4.025999	8	24.6
Winter Sowing Temp²	(°C)	14.91242	4.175187	5.233334	21.73333	14.34573	4.346796	5.233334	21.73333	14.7122	4.408709	2.833333	21.6
Summer Sowing Precip	(mm)	76.72373	92.73935	5	506.1	89.82051	104.875	5	506.1	77.29	92.97647	5	506.1
Summer Growing Precip	(mm)	51.10175	47.22913	1.25	229.25	53.08651	52.07329	1.25	229.25	51.54764	47.21411	1.25	229.25
Winter Precipitation	(mm)	29.5747	32.61	2	153.6667	25.83013	27.46765	1.333333	115	29.63807	32.75382	2	153.6667
HDI	(HDI)	.6165	.0857195	.31	.802	.6037179	.0913116	.31	.802	.6180275	.084598	.31	.802
Population Density	(No./km²)	337.6909	481.1303	4	3566	249.0641	276.8187	4	1606	340.6514	482.345	4	3566
Elevation	(m)	550.5182	692.1619	5	3314	620.4359	700.5658	5	3314	554.8807	693.838	5	3314
Latitude	(Decimal°)	30.34709	2.974395	24.67	35.84	30.51359	3.159575	24.67	35.84	30.39505	2.945104	24.67	35.84
Runoff	(mm/year)	42.18182	106.7898	0	627	53.07692	123.3121	0	627	42.56881	107.2055	0	627
River Density	(km)	32.876	32.30021	0	190.84	30.86013	32.93246	0	190.84	33.11679	32.35007	0	190.84
Flood	(% affected area)	.140413	.2102281	0	.9734642	.1258673	.1782819	0	.8287163	.1417012	.2107625	0	.973464
Tractor Density	(No./ha)	.0982185	.1262116	0	.6467662	1009.753	2177.007	0	9917.05	.0991071	.1264484	0	.6467662
Road Density	(km/ha)	2.908689	4.097185	.004	38.51734	2.705572	4.648455	.004	38.51734	2.933349	4.1079	.004	38.51734
Soil Calcisols	(%/area)	.4357614	.3542353	0	1	.3574277	.3365988	0	1	.4353572	.3558461	0	1
Soil Fluvisols	(%/area)	.0694739	.1185034	0	.5094377	.044863	.095333	0	.4134663	.0701113	.1188612	0	.5094377
Soil Rock outcrop	(%/area)	.0443401	.1669379	0	1	.034207	.1358952	0	.738439	.0447469	.1676542	0	1
Soil Regosols	(%/area)	.0242558	.1152798	0	.738439	.0625309	.1956978	0	1	.0244784	.1157886	0	.738439

Schriften zur Internationalen Entwicklungs- und Umweltforschung

Herausgegeben vom

Zentrum für internationale
Entwicklungs- und
Umweltforschung
der Justus-Liebig-Universität Gießen

Band 1 Hans-Rimbert Hemmer / Rainer Wilhelm: Fighting Poverty in Developing Countries. Principles for Economic Policy. 2000.

Band 2 Lorenz King / Martin Metzler / Tong Jiang (eds.): Flood Risks and Land Use Conflicts in the Yangtze Catchment, China and at the Rhine River, Germany. 2001.

Band 3 Ingrid-Ute Leonhäuser (ed.): Women in the Context of International Development and Cooperation. Review and Perspectives. Selected Papers and Abstracts presented at the Justus-Liebig-University Gießen 26.-28. October 2000. 2002

Band 4 Margit Schratzenstaller: Internationale Mobilität von und internationaler fiskalischer Wettbewerb um Direktinvestitionen. 2002.

Band 5 Armin Bohnet u.a.: Theoretische Grundlagen und praktische Gestaltungsmöglichkeiten eines Finanzausgleichssystems für die VR China. Unter Mitwirkung von Chen Biyan, Chen Shixin, Ge Licheng, Ge Naixu, Ge Zhuying, Ma Shuanyou, Markus Peplau, Yang Zhigang, Zhu Qiuxia. 2003.

Band 6 Armin Bohnet / Matthias Höher (eds.): The Role of Minorities in the Development Process. 2004.

Band 7 Thi Phuong Hoa Nguyen: Foreign Direct Investment and its Contributions to Economic Growth and Poverty Reduction in Vietnam (1986–2001). 2004.

Band 8 Andreas Böcker / Roland Herrmann / Michael Gast / Jana Seidemann: Qualität von Nahrungsmitteln. Grundkonzepte, Kriterien, Handlungsmöglichkeiten. 2004.

Band 9 Christina Mönnich: Tariff Rate Quotas and Their Administration. Theory, Practice and an Econometric Model for the EU. 2004.

Band 10 Reimund Seidelmann / Ernst Giese (eds.): Cooperation and Conflict Management in Central Asia. 2004.

Band 11 Claudia Ohly: Das Steuersystem im ungarischen Transformationsprozess. Ein Beitrag zur Transformationstheorie. 2004.

Band 12 Nicole Mau: Umweltzertifikate. Der Einsatz von Umweltzertifikaten in der Landwirtschaft am Beispiel klimarelevanter Gase. 2005.

Band 13 P. Michael Schmitz (Hrsg.): Water and Sustainable Development. 2005.

Band 14 Ira Pawlowski: Die Wettbewerbsfähigkeit der ukrainischen Milchwirtschaft. Messung von Marktverzerrung und Politikeinfluß im Transformationsprozeß. 2005.

Band 15 Kirsten Westphal (ed.): A Focus on EU-Russian Relations. Towards a close partnership on defined road maps? 2005.

Band 16 Andreas Langenohl / Kirsten Westphal (eds.): Conflicts in a Transnational World. Lessons from Nations and States in Transformation. 2006.

Band 17 Rosemarie von Schweitzer: Home Economics Science and Arts. Managing Sustainable Everyday Life. 2006.

Band 18 Dörthe List: Regionale Kooperation in Zentralasien. Hindernisse und Möglichkeiten. 2006.

Band 19 Michael Gast: Determinanten ausländischer Direktinvestitionen. OECD-Länder als Investoren und besonderer Aspekte der Ernährungswissenschaft. 2007.

Band 20 Kim Schmitz: Die Bewertung von Multifunktionalität der Landschaft mit diskreten Choice Experimenten. 2008.

Band 21 Kerstin Kötschau / Thilo Marauhn (eds.): Good Governance and Developing Countries. Interdisciplinary Perspectives. 2008.

Band 22 Armin Bohnet (ed.): Poland on its Way to a Federal State? 2008.

Band 23 Johannes Harsche: Regionale Inzidenz und ökonomische Bestimmungsgrößen der Gemeinsamen Europäischen Agrarpolitik. 2009.

Band 24 Aikaterini Kavallari: Agricultural Trade Policy Reforms and Trade Liberalisation in the Mediterranean Basin. A Partial Equilibrium Analysis of Regional Effects on the EU-27 and on the Mediterranean Partner Countries. 2009.

Band 25 Marc Christopher Kramb: Sanitäre und phytosanitäre Handelsbeschränkungen unter dem Einfluss des WTO-Abkommens. Ein Gravitationsansatz unter besonderer Berücksichtigung des EU-Rindfleischsektors. 2009.

Band 26 Lorenz King / Giorgi Khubua (eds.): Georgia in Transition. Experiences and Perspectives. 2009.

Band 27 Frank Schüssler (Hrsg.): Geographische Energieforschung. Strukturen und Prozesse im Spannungsfeld zwischen Ökonomie, Ökologie und sozialer Sicherheit. 2010.

Band 28 Kerstin Kötschau: Impact of Land Reform Strategies on Rural Poverty in the Commonwealth of Independent States. Comparison between Georgia and Moldova. 2012.

Band 29 Eva Greschek: Die evolutive Auslegung völkerrechtlicher Verträge am Beispiel des GATT. 2012.

Band 30 Manuchehr Goibov: An Analysis of Farmers Preferences for Different Land Use Options in Northern Tajikistan. A Choice Experiment Approach. 2013.

Band 31 Mirza Nomman Ahmed: A Structural Ricardian Valuation of Climate Change Impacts on Agriculture in Pakistan. 2013.

www.peterlang.com